ENCYCLOPAEDIA OF HELMINTHES

ENCYCLOPAEDIA OF HELMINTHES

Vol. 1

Platy Helminthes

By

Arvind N. Shukla

Rajiv Tyagi

ANMOL PUBLICATIONS PVT. LTD.

NEW DELHI - 110 002 (INDIA)

ANMOL PUBLICATIONS PVT. LTD.
4374/4B, Ansari Road, Daryaganj
New Delhi - 110 002
www.anmolbooks.com

Encyclopaedia of Helminthes

First Edition, 2002

ISBN 81-261-1043-0 (Set)

PRINTED IN INDIA

Published by J.L. Kumar for Anmol Publications Pvt. Ltd., New Delhi - 110 002 and Printed at Mehra Offset Press, Delhi.

Contents

Preface

Helminthes constitute a large assemblage of *worms* of comparatively simple and varied organization. Helminthes is derived from the Greek word *helmins* means *worms*. The term worms is not correct because it is applied to elongated invertebrates without appendages and with bilateral symmetary. In 18th Century all the worm-like animals were included in an old phylum- *Vermes* (*Linnaneus*, 1735). But this was a very heterogenous group which included a highly diverse assemblage of forms. In the modern classification this group is divided into several independent phyla such as Platyhelminthes, Aschelminthes, Annelida etc.

The term helminthes is restricted to a few phyla of invertebrate animals, all of which are superficially worm like but they differ markedly in their morphology, life-history and bionomics. The helminthes are restricted to two phyla of Animal Kingdom. These are:

1. Platyhelminthes or flat-worms.
2. Aschelminthes or round-worms.

The present work is designed to present all vital information on Helminthes. The entire information is gathered in three volumes namely:

- Platy Helminthes
- Aschelminthes
- Physiology of Helminthes

Profusely illustrated in the light of latest researches, this work will prove a dependable reference work for students, teachers and researchers in the discipline.

—**Authors**

Chapter—1

Platyhelminthes

Phylum - Platyelminthes

Term *Platyhelminthes* was proposed by Gegenbaur in 1859, means flatworms which refers to their characteristic contour of flattened body. The term is derived from two Greek words-*platys* = flat; *helminthes* = worms. The animals show low organisation as they are without anus, skeletal, respiratory and circulatory system. The body is filled with mesenchymal cells which are mesodermal in origin.

Phylum: Plathelminthes (Flatworms)

This phylum groups dorso-ventrally flattened worms; their body is covered with a dermomuscular sac which consists of a single epithelial layer overlaying a layer of muscles (longitudinal, circular, and oblique).

The plathelminths have no special organs of locomotion. Their mobility is effected by contractions of the dermomuscular sac. The body is oval or ribbonlike, strongly elongated.

The loose parenchyma inside the dermomuscular sac surrounds internal organs. The digestive system is primitive; it is totally absent in cestodes. The excretory system consists of protonephridia. The nervous system consists of paired cerebral ganglia and longitudinal nerve cords. The plathelminths are hermaphroditic. Their reproductive system is very well-developed and fills almost the entire body of a worm.

Numerous fish parasites can be found in the following classes: the *Monogenea, Cestodaria. Cestoda,* and *Trematoda.*

Class: Monogenea (Monogenetic trematodes)

Monogeneans are mostly small parasites, ranging from about 1 to several mm in length. A few marine species are longer than 20 mm.

The body is flattened dorso-ventrally (seldom cylindrical), elongated, oval, or resembling a circular disc, slightly concave ventrally and convex dorsally.

The monogeneans have characteristic adhesive organs used to attach the parasite to the host's body and to aid in translocation on the host.

The attachment organs on the head occur as suckers and processes. Additionally, the anterior part of the body contains glands producing a viscous secretion ensuring adhesion to the host when it is feeding or moving.

The posterior end of the parasite's body is provided with a complex adhesive organ called the opisthaptor, differing widely from species to species: it may contain suckers, hooks, clamps, all of them forming an attachment disc fastening the parasite to its host.

The monogenean body is very contractile: it may double in length or shorten by half compared to a resting animal. When moving from place to place, the parasite fastens its anterior end to the fish body, tears the opisthaptor off and transfers it forwards, then releases the anterior end and attaches it to a different spot on the fish body. The movements resemble those of a leech or a measuring worm.

The digestive tract begins with the mouth, placed ventrally near the anterior end of the worm body. The mouth leads to the muscular pharynx acting as a sucking pump for food. The pharynx passes into a short oesophagus and further into the intestine, usually in the form of two bluntly ended caeca. Occasionally, the intestine is single, sac-like, or the two caeca fuse in the posterior part of the body. In some species, the two intestinal caecas have many branches.

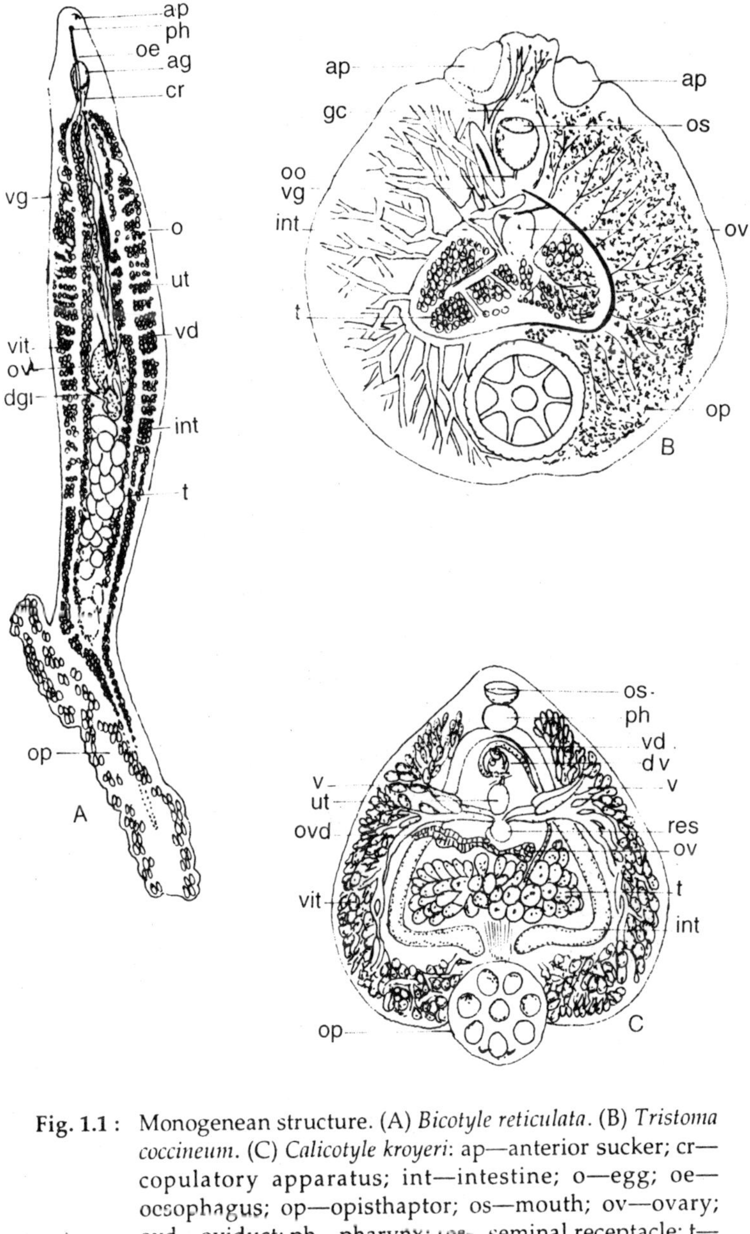

Fig. 1.1 : Monogenean structure. (A) *Bicotyle reticulata*. (B) *Tristoma coccineum*. (C) *Calicotyle kroyeri*: ap—anterior sucker; cr—copulatory apparatus; int—intestine; o—egg; oe—oesophagus; op—opisthaptor; os—mouth; ov—ovary; ovd—oviduct; ph—pharynx; res—seminal receptacle; t—testes; ut—uterus; v, vg—vagina; vit—vitellary.

Sense organs occur in many species as two or more pairs of pigmented eye spots provided with lenses. On the other hand, numerous species lack eye spots altogether. Additionally, the surface of the body may be covered with sensory papillae.

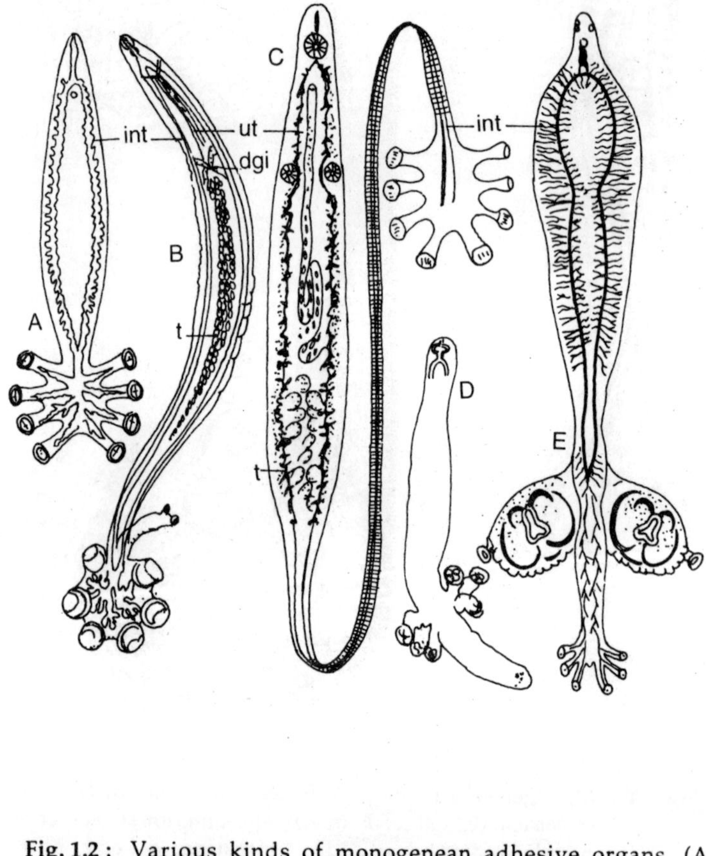

Fig. 1.2 : Various kinds of monogenean adhesive organs. (A) *Diclidophoropsis taschenbergi*. (B) *Rajonchocotyle* sp. (C) *Chimaericola leptogaster*. (D) *Pedocotyle morone*. (E) *Anthocotyle merluccii:* int—intestine; ut—uterus; t—testes.

The monogeneans are hermaphroditic. The male reproductive system consists of a single testis (or of numerous testes) and the vas deferens ended with a copulatory apparatus of a very complex structure. The latter, along with the opisthaptor, is an important key character of these parasites.

The female system consists of a single ovary (ootype) where sperm contacts and fertilizes eggs, and the uterus through which eggs are expelled outside. Eggs are formed inside the ootype into which open numerous vitellary glands, providing the eggs with yolk.

Only one egg is produced at a time. Oval or triangular eggs usually have lids and threads, sometimes very long, on one or both poles. The thread facilitates egg's adhesion to the fish mucus and aids the egg in floating in the water.

The *Monogenea* are mostly oviparous; some gyrodactylids only (e.g. *Gyrodactylus elegans, Macrogyrodactylus polypteri*) are viviparous. Development is direct, without metamorphosis.

On hatching, the egg releases a ciliated larva, the oncomiracidium, its cilia arranged in zones. The larva possesses eye spots, pharynx, head, glands, and a developed, but inactive, opisthaptor with emerging hooks. Generally, the larval structure resembles that of an adult. The larva swims freely for some time, viable for 6—8 hours; on encounter with an appropriate fish host it adheres to the fish skin, loses its cilia and, crawling on the skin, gets to certain organs (the gills, mouth, cloaca) to settle there and develop, without metamorphosing, into an adult.

In spite of producing a single egg at time, the monogenetic trematodes multiply very rapidly. *Dactylogyrus vastator*, a common carp parasite, was observed to live for 10–12 days, from egg to death. During this time a single specimen produced 16 new ones, which in turn gave rise, before they died, to 310 offsprings.

In the viviparous species, the egg in the maternal uterus develops into an embryo; another one grows within it, the

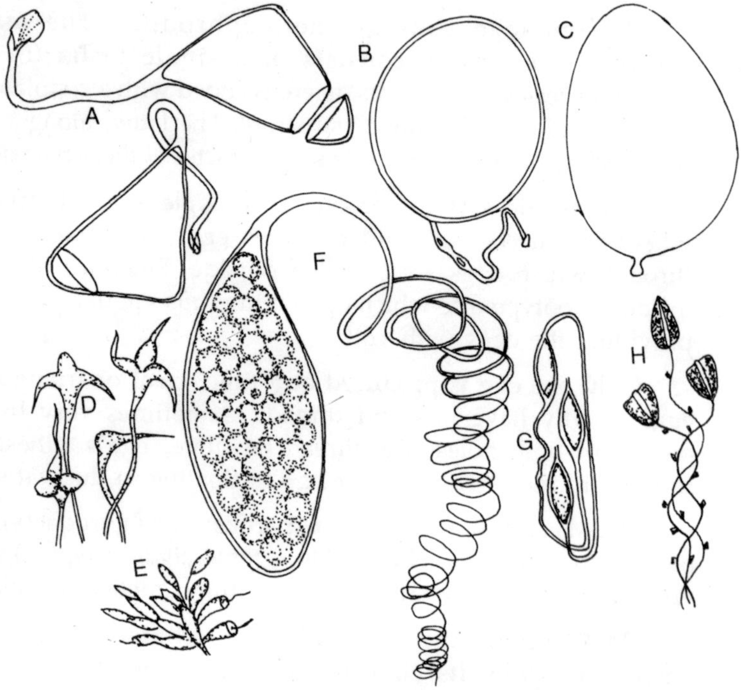

Fig. 1.3 : Monogenean eggs (after various authors). (A) *Amphibdella torpedinis*. (B) *Ancylodiscoides vistulensis*. (C) *Dactylogyrus anchoratus*. (D) *Encotyllabe pagelli*. (E) *Udonella* sp. (F) *Diplozoon paradoxum*. (G) *Erpocotyle catenulata*. (H) *Entobdella soleae*.

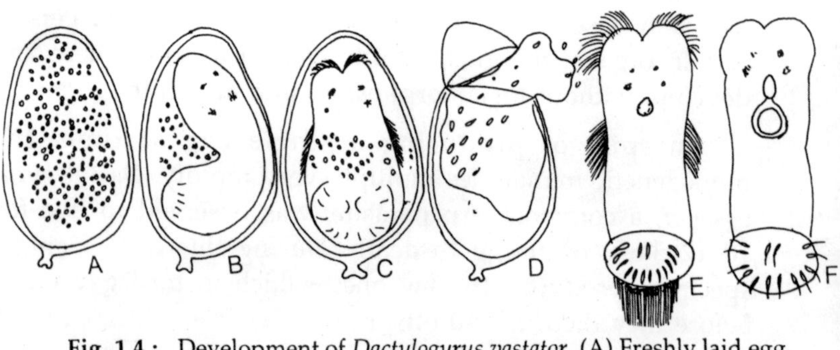

Fig. 1.4 : Development of *Dactylogyrus vastator*. (A) Freshly laid egg. (B) Same, after 2 days. (C) Same, 2.5 days later. (D) Hatching egg. (E) Oncomiracidium. (F) Sessile larva.

third embryo inside that, and sometimes there are four embryos one within another. Having finished growing in the uterus, a juvenile monogenean advanced in its overall development (subadult) gets out and lives. On its own.

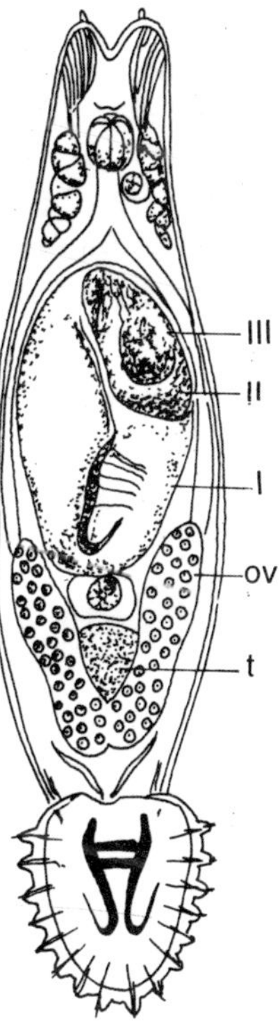

Fig. 1.5 : *Gyrodactylus elegans*: ov—ovary; t—testis; I—first generation embryo; II—second generation embryo; III—third generation embryo.

The *Gyrodactylidae* mulptiply very rapidly. According to Byhkovskii (1957), a single specimen of *Gyrodactylus elegans* produces, after 30 days, 2453 offsprings.

The monogenetic trematodes attack fish, amphibians, and reptiles. Only exceptionally are they found in other animals, for instance *Oculotrema hippopotami,* fund under the eyelid of a hippopotamus, and *Isancistrum loliginis* observed to live in the gills of the squid *Loligo media.*

Most species parasitise fish, both freshwater and marine. Yamaguti (1963) lists more than 1300 monogenean species known from fish.

The *Monogenea* are external parasites dwelling on the fish body surface, fins, in the mouth and nasal cavities, and most often on the gills. Some species, however, tend to settle in their host's internal organs, as for instance *Amphibdella torpedinis* (Chatin, 1874) (Fig. 1.6), a worm usually dwelling in the gill cavity of the electric rays, *Torpedo ocellata* and *T. marmorata*. Ruszkowski (1931) found the parasite, both adult individuals and eggs, in the *Torpedo* heart, *T. ocellata* invasion incidence reaching 35% and the eggs being found in more than 50% of the electric ray's specimens examined.

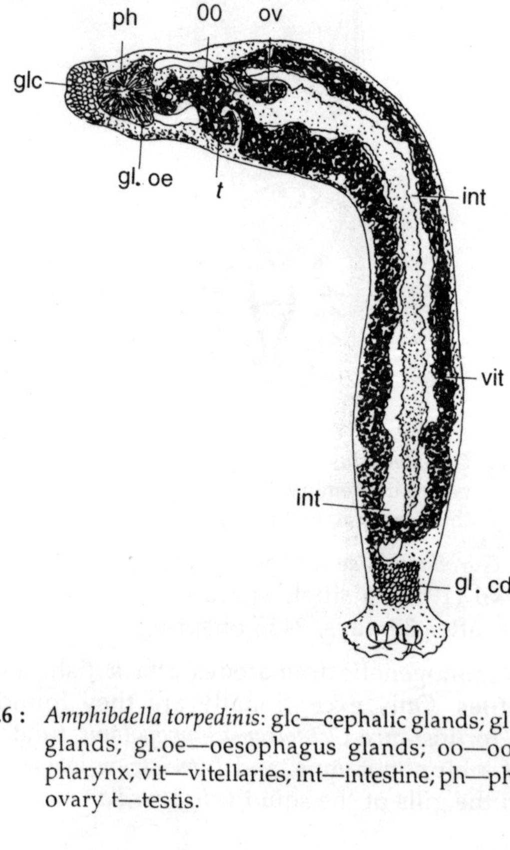

Fig. 1.6 : *Amphibdella torpedinis*: glc—cephalic glands; gl.cd—caudal glands; gl.oe—oesophagus glands; oo—ootype; ph—pharynx; vit—vitellaries; int—intestine; ph—pharynx; ov—ovary t—testis.

Members of the genus *Acolpenteron* live in the urinary duct, while numerous species of *Calicotyle* settle in the cloaca of rays. Paperna (1963) described a new monogenean species, *Enterogyrus cichlidarum*, dwelling in the anterior part of the intestine of *Tilapia zilli* and *T. nilotica* from Israeli rivers.

These monogeneans are actually internal parasites and even show some adaptations to this mode of life. the body of *Enterogyrus cichlidarum* is covered with a thick cuticle; the cephalic adhesive organ is vestigial and the opisthaptor poorly developed. Such typically monogenean characters as the eye spots and posterior hooks are retained. On the other hand, no morphological adaptations to an altered environment were found in *Amphibdella torpedinis*.

The examples described demonstrate one of the possible pathways to endoparasitism.

The monogeneans feed on fish muchus, peeled epithelium, and blood. The group contains numerous pathogenic species.

The parasites attached to the host's body injure it with their hooks and produce ecchymoses; they damage gill and skin epithelium, the latter often peeling off in lobes as a strong invasion. Juvenile fish are particularly vulnerable. Frequently the irritation caused by the parasites results in epithelial outgrowths and tissue swelling, as was often observed in fish gills affected by *Dactylogyrus vastator*. Serious epizootics and mortality are known to occur in pond cultures. The monogeneans can be dangerous for aquaria-kept fish and also in maricultures.

It is exceptionally rare, however, for a monogenean invasion to cause a disease outbreak and mortality in fish under natural conditions. On the other hand, a high fish population density, undernourishment, water deoxygenation, altered pH, and pollution act in favour of parasites' proliferation: the parasites attack those fishes weakened by detrimental conditions of the environment and then produce substantial losses.

Pond cultures in Poland often fall victim to invasions of gill-attacking monogeneans *Dactylogyrus vastator* producing such pathological conditions as dactylogyriasis, particularly dangerous for juvenile carp (the so-called July fry), and gyrodactyliasis, frequent in juvenile trout. Those interested in these diseases and in other species parasitising freshwater fish should consult fish diseases textbooks.

The largest monogenean specres and the most dangerous ones for fish belong to the family *Capsalidae*. The capsalids possess a large opisthaptor, frequently divided by septa into a number of radically arranged grooves which increase the parasite's force of adhesion. Moreover, the opisthaptor contains 14 small marginal hooks and 2—6 central ones. The opisthaptor clings to the host's body so strongly that it is sometimes easier to tear the parasite into pieces than to remove it from the fish body.

Capsala martinieri (Bosc., 1811) belongs to the largest capsalid species. It is a flat oval monogenean, up to 30 mm long and 25 mm wide. It occurs on the gills of *Mola mola* and *Diodon* sp.

Tristoma coccineum (Cuvier, 1817) frequently found on the gills of *Xiphias gladius*, is a large parasite with a flat, disc-like circular body more than 1 cm in diameter. The animal has three suckers near its mouth (hence the name *Tristoma*) and a large sucker-like opisthaptor divided by septa into additional adhesive areas. Frequently, an imprint of the parasite's opisthaptor, or even of its whole body, is seen on fish gills. The species of *Benedenia* show two large bow-like suckers on the head. *B. sciaenae* (Van Beneden, 1852) often occurs on the skin of meagre, *Sciaena aquila*. (Jahn and Kuhn 1932) described a strong invasion of *B. melleni* (MacCallum, 1927) in the serranids and lutjanids kept in New York marine aquaria, up to 2000 monogeneans being found on one fish specimen. The parasites attacked fish eyes, nasal cavity, and gills. Numerous fishes died as a result of the invasion. The parasites were introduced to the aquaria with newly delivered Atlantic fishes.

B. derzavini (Layman, 1930) a small (about 5 mm long) parasite is common in the gill cavity of *Sebastodes schlegeli* in the Sea of Japan.

Japanese maricultures have often been affected by a mass occurrence of *B. seriolae* (Yamaguti, 1934) on the species of *Seriola* serious pathological changes and fish mortality being the outcome.

The family *Capsalidae* also includes *Nitzschia sturionis* (Abildgaard, 1794) a parasite of sturgeons in the Ponto-Caspian basin (the Caspian, Black, and Aral Seas). The parasite occurs on the gills, lips, palate and tongue, and even in the anterior part of the intestine. This large (1-2.5 cm long) monogenean has two attachment cylinders on the head and a sucker-like opisthaptor; the opisthaptor, in contrast to other capsalids, lacks septa. It contains 14 marginal hooks and 3 pairs of central ones.

An extensive invasion of *N. sturionis* brings about emaciation and even death of the host. The parasite mechanically damages gill lobes, which results in hypertrophy of gill epithelium and causes necrosis in the damaged spot.

N. sturionis is common in Atlantic, Mediterranean, Caspian, and Black Sea sturgeons and usually produces no disease symptoms. However, a strong epizootic caused by the species was recorded in 1936; the disease was brought about by transplanting *Acipenser stellatus*, together with the parasite, from the Caspian to Aral Sea. The Aral Sea had been inhabited by an indigenous sturgeon species, *A. nudiventris*, *Nitzschia*-free and therefore lacking immunity to the parasite. *Nitzschia* attacked the defenseless Aral sturgeon and developed aboundantly on the gills, 100-600 and even 1000 parasites being recorded on a single specimen. The fishes were dying with symptoms of suffocation. According to Dogiel and Lutta (1937), a single parasite sucks out about 0.5 cm^3 of fish blood a day; thus a fish inhabited by 300–400 parasites loses 150–200 cm^3 of blood during 24 hours. According to Packov (1954), blood makes up, at the mos,

5.2% of sturgeon weight, so those sturgeons weighing more than 5 kg are able to survive such a large loss of blood. The epizootic described reduced the Aral sturgeon population to the extent that sturgeon fishing had to be banned in the area and the species protected by law for many years.

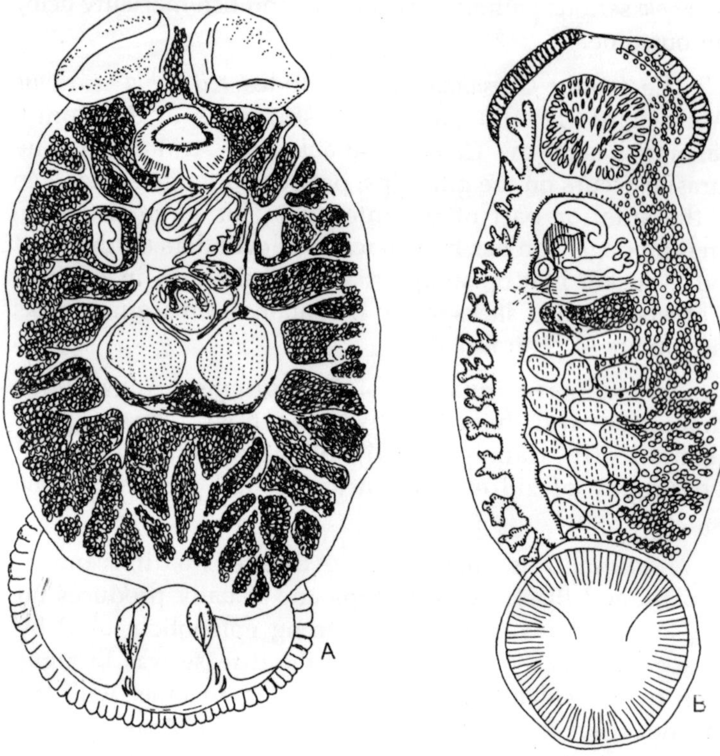

Fig. 1.7 : (A) *Benedenia derzavini.* (B) *Nitzschia sturionis.*

Class: Cestodaria

The *Cestodaria* are primitive unsegmented plathelminths, having a flattened body and a contractile proboscis in place of the scolex. The newly hatched lycophoratype larva is ciliated and has 100 hooklets on its posterior end. Inside the body there is a large aggregation of gland cells opening on the anterior end.

Earlier, the *Cestodaria* had been placed within the class *Cestoda*. Owing to a specific character of their structure and the originally non-strobilated body, they were later separated as a class, subdivided into two orders: the *Amphilinoidea* and *Gyrocotyloidea*. Bykhovskii (1957) elevated the *Gyrocotyloidea* to the level of a separate class and demonstrated that they were more closely related to monogeneans than to cestodes. In 1974, Dubinina rejected the name *Cestodaria* as redundant and formed the class *Amphilinoidea*, in her opinion phylogenetically related to monogeneans as well.

The classification system adopted here is based on that developed by Baer and Joyeux (1961); the class *Cestodaria* is subdivided into the orders *Amphilinoidea* and *Gyrocotyloidea*. The attempts to alter the plathelminth classification system have only been briefly mentioned here.

Order: Amphilinoidea

This small parasitic taxon contains 7 genera with 10 species. The three species of the genus *Amphilina* are coelomic parasites of Holarctic acipenserids: *Amphilina foliacea* and *A. japonica* Goto and Ishii, 1936 parasitise European and Asian sturgeons, while the North American sturgeons are attacked by *A. bipunctata* Riser, 1948.

The remaining species live in freshwater and marine tropical fish *e.g. Gephyrolina paragonopora* in catfish from India, *Gigantolina magna* in the Indopacific marine perciforms.

Amphilina foliacea is the best-known member of the order. The parasite is common in sturgeons in the basins of the Black and Caspian Seas, Lake Bailkal, and Siberian rivers (the Ob, yenisey, Irtish, Angara). The principal host is *Acipenser ruthenus*. Juvenile sturgeons show the strongest invasion: according to Dubinina (1974), invasion incidence in fish not older than 6 years may reach 70-89.9%.

Other acipenserids: *Huso huso, A. stellatus, A. gueldenstaedti, A. nudiventris, A. baeri, A. schrencki, A. sturio,* and *A. medirostris* are less affected.

Amphilina foliacea is oval, dorsoventrally flattened, 60-65 mm long and 25-30 mm wide. The body is slightly convex dorsally and flat ventrally, milky white or yellowish in colour.

The anterior end of the body is provided with a contractile proboscis, on the top of which open numerous unicellular glands located centrally in the body, between the gonads. The uterus opens next to the proboscis and releases eggs outside. The rear end of an adult body has 10 hooklets partly burrowed in the dermomuscular sac. The parasite is hermaphroditic. The male gonadal opening and the vaginal pore are placed close to each other on the posterior end of the body. Numerous testes are scattered

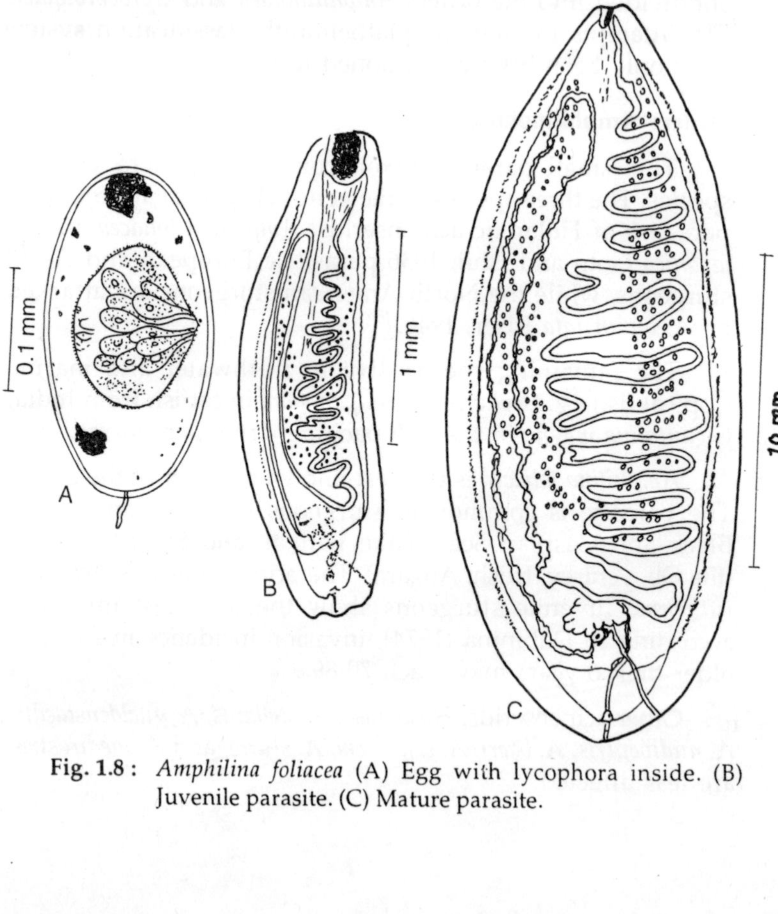

Fig. 1.8 : *Amphilina foliacea* (A) Egg with lycophora inside. (B) Juvenile parasite. (C) Mature parasite.

between the uterine loops. The single ovary with bubble-like vitellaries on its sides lies posteriorly in the body. The intestine is absent.

Eggs are oval, lacking lids, each provided with a short process on one pole. Embryonic development proceeds in the uterus so that each egg, when ejected, contains a well-developed larva, the so-called lycophora. It is ciliated, with 10 hooks on its rear end. Once in the intestine of the intermediate host, an amphipod crustacean, the lycophora leaves the egg.

Amphipods such as *Dikerogammarus haemobaphes, Pandorites platycheir, Corophium curvispinum, Carcinogammarus roeseli,* and *Rivulogammarus pulex* can serve as intermediate hosts.

The development of *A. foliacea* was studied by the Polish biologist Konstanty Janicki during his stay in Saratov on the river Volga in 1927-1928.

Janicki put forward an interesting hypothesis that mesozoic reptiles were the original definitive hosts for *A. foliacea* and housed its adults, while the second larval stage, the plerocercoid, developed in sturgeons. The adult forms were extinct along with the reptiles and the plerocercoid acquired the faculty of neothenic reproduction, becoming thus the sexually mature *Amphilina* of today.

Recently, Dubinina (1974) supplied a number of new pieces of information on development and structure of *A. foliacea*.

The parasite's eggs get into the water via the fish abdominal pores and sink down to the bottom where they are consumed by amphipods. The egg shell breaks under the pressure of gammarid maxillae and the motile lycophora, measuring 165–230 um × 70–190 μm, is liberated. The lycophora pushes through the gammarid intestine into its coelom and grows there into a 2–4 mm long larva, able to invade a sturgeon, which happens after 30 days at 27–30°C. This stage is a morphological equivalent of the adult *Amphilina*.

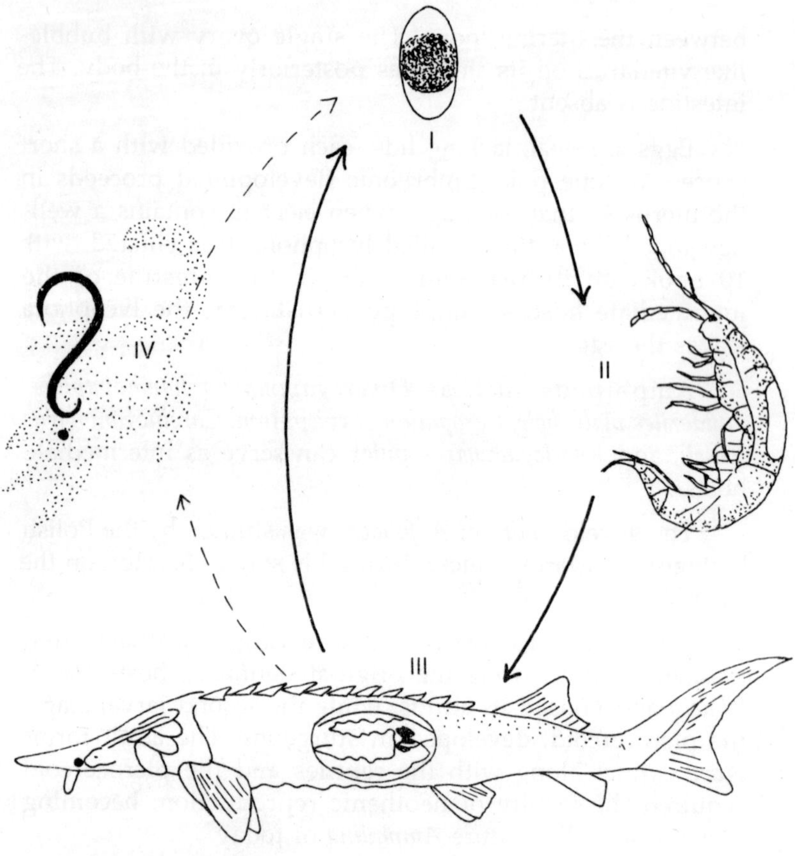

Fig. 1.9 : Hypothetical life cycle of *Amphilina foliacea*: I—egg; II—plerocercoid in a gammarid; III—plerocercoid in sturgeon; IV—hypothetical definitive host, a piscivorous mesozoic reptile.

The fish become infected by consuming crustaceans, the basic item in their diet. Once in the fish intestine, the parasite makes its way to the body cavity where it undergoes further develpment, matures, and produces eggs passed to the water through the fish abdominal pores. The eggs develop in water.

By mechanically injuring fish gonads, *A. foliacea* can bring about a partial or even complete castration, which has undoubtedly a bearing on the sturgeon population biology.

The *Amphilinoidea* are phylogenetically very old animals, as are their hosts, both the definitive (acipenserids) and the intermediate (relict gammarids) ones.

Order: Gyrocotyloidea

This is a very small taxon consisting of 3 genera with 10 species only: the *Gyrocotyle, Amphiptyches,* and *Gyrocotyloides.*

The *Gyrocotyloidea* parasitise exclusively chimaerids (*Holocephali*). The body consists of a single flattened segment, its sides strongly crenate. The parasites are rather large, up to a few cm long. The anterior end of the body is provided with the anterior sucker of an unknown function, while the posterior end has a muscular funnel-like attachment organ of more or less wrinkled margins. The funnel fastens the parasite to the chimaerid intestinal spiral valve. Some parasites (*Gyrocotyle*) show transverse striation on the surface, which then resembles segmentation. The intestine is absent, the parasite absorbing nourishment with the whole surface of the body.

The *Gyrocotyloidea* are hermaphroditic. Their development is known only incompletely; observations made by the Polish parasitologist Ruszkowski (1932) have greatly contributed to the present knowledge. Ruszkowski described the development of *Amphiptyches urna* (Fig. 1.10), a common parasite of *Chimaera monstrosa* in the Norwegian Sea. The invasion incidence in that area was very high, up to 100%.

A. urna is a rather large parasite. The living animal is translucent, pinkish in colour, very active and contractile.

The *Amphiptyches* eggs, measuring 78–93 μm × 60–67 μm, have lids. Embryos develop within 25–30 days (under experimental conditions) after the eggs have been placed in water. The egg hatches into a lycophora-type larva, completely ciliated and with 10 hooks on the rear end, similar to the *Amphilina foliacea* larva. However, the 120 μm long and 30 um wide *Amphiptyches* larva, as opposed to that of

Amphilina, leaves the egg on its own by opening the lids with strong movements. The newly released larva swims rapidly, its movements resembling those of trematode miracidia.

The presence of the lycophora in the *Gyrocotyloidea* life cycle is the only character they have in common with the *Amphilinoidea*.

The further larval development is unknown. The smallest larvae Ruszkowski found in the chimaerid intestines were 3-5 mm long and showed gonad formation in progress.

Information on the *Gyrocotyloidea* pathogenicity is lacking. The parasites are of no importance for the fishing industry

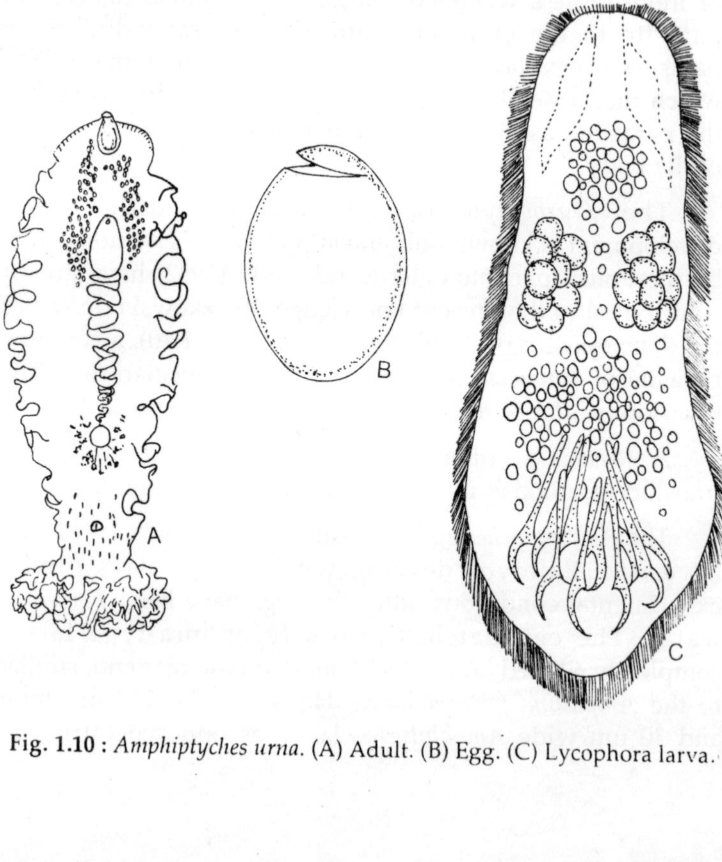

Fig. 1.10 : *Amphiptyches urna.* (A) Adult. (B) Egg. (C) Lycophora larva.

as the chimaderids are not caught for consumption, their flesh being unpalatable.

The group has been discussed here as an interesting example of marine fish parasites, and for the sake of a complete presentation of parasitic plathelminths. The present state of knowledge of their biology is largely a result of contributions of Polish parasitologists.

Class: Trematoda (Flukes)

The trematodes are endoparasitic unsegmented flatworms provided with adhesive organs in the form of suckers or attachment discs, frequently divided with septa. The oral sucker surrounds the mouth; the other sucker is located ventrally and has no connection with the digestive system. Both suckers are bowl-like, muscular, strongly contractile organs. The *Aspidogastrea* flukes have their ventral sucker developed as a large attachment disc, divided with septa into secondary suckers (alveolae).

The mouth leads to a strongly muscular pharynx opening to the oesophagus, of length dependent on the species; from the oesophagus the digestive tract gives rise to two blind intestinal caeca which may be branched.

The trematodes are, as a rule, hermaphroditic, a tendency toward separation of sexes being observed in the subclass *Didymozoida*.

The trematode development is very complex and involves metamorphosis and change of hosts.

The class is divided into three subclasses: the *Aspidogastrea, Digenea,* and *Didymozoida*.

Subclass: Aspidogastrea

The key character is the presence of a sucker-like attachment disc, divided with thick septa into the alveolae. Details of the disc structure are of taxonomic importance.

The parasites are small (less than 10 mm long); only exceptionally do some species (*Macraspis, Stichocotyle*) grow to attain a length of several centimetres.

Most *Aspidogastrea* are parasites of molluscs, a few species only being found in marine fish: *Macraspis* in the intestine of *Chimaera monstrosa: Stichocotyle* in bile ducts of *Raja;* and *Cotylogaster* in the intestine of certain marine teleosts. Many cyprinids in the Black, Caspian, and Aral Seas are parasitized by *Aspidogaster limacoides* Diesing, 1835.

Development of these trematodes is not known in detail. Presumably the development is direct, but juvenile stages of *Stichocotyle nephropis* (Cunningham, 1887) were found encysted externally on the intestine of lobsters *Nephrops norvegicus* and *Homarus americanus*. The adult trematodes dwell in bile ducts of *Raja clavata* and *R. laevis*. It is not, however, known how the fish becomes infected.

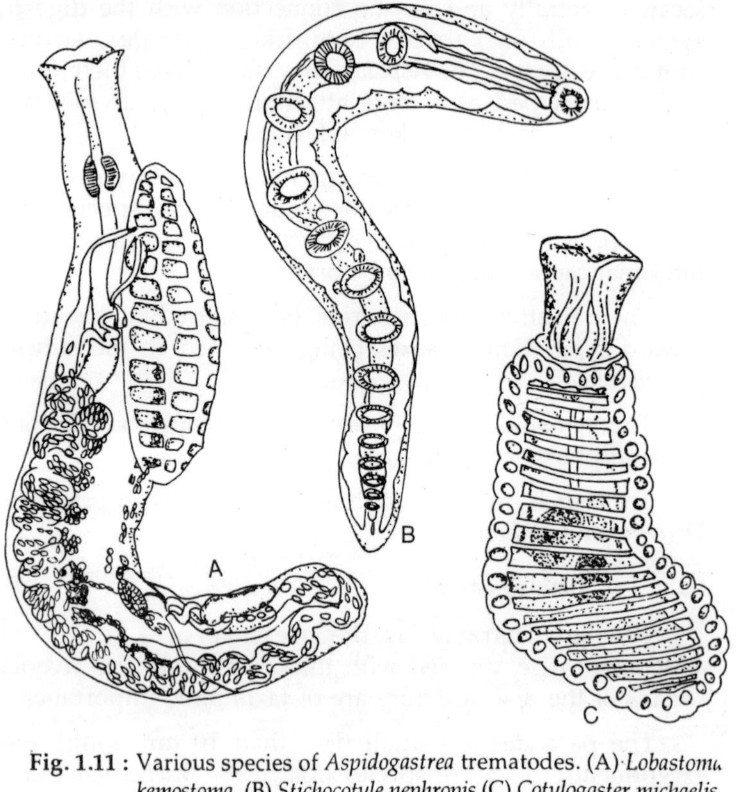

Fig. 1.11 : Various species of *Aspidogastrea* trematodes. (A) *Lobastoma kemostoma*. (B) *Stichocotyle nephropis* (C) *Cotylogaster michaelis*.

The subclass is of negligible importance for the fishing industry.

Subclass: Digenea (Digenetic trematodes)

The subclass groups plathelminths which have two circular suckers with strong muscular walls: the oral sucker, placed on the anterior end of the body close to the mouth, and the ventral sucker. More than 2500 species dwelling in fishes have been described so far.

The body is covered with the dermomuscular sac having a thin, elastic cuticle on the surface. Besides the suckers, the cuticle of some species shows additionally spines (*Acanthocolpidae* and *Stephanochasmidae*) and pores of unicellular glands formed by deeply burrowed epithelial cells. The cuticle is lined with three layers of muscles: circular, diagonal, and longitudinal their contractions producing worm-like movements of the animals. The body is filled with a loose parenchyma surrounding all the viscera. The digestive system begins with the mouth leading to the muscular pharynx. The pharynx opens into the oesophagus of a length depending on the species. The oesophagus passes into two intestinal caeca in the form of straight or branching blind canals.

Two symmetrical ganglias interconnected with transverse commisures are placed between the oral sucker and the pharynx. Each ganglion gives rise to three nerve cords which run toward the anterior and posterior ends of the body and branch off to all the organs. Digenean sensory organs occur as a few sensory bodies, located mostly near the oral sucker. The eye spots are present at the larval stage only, in the miracidia and cercariae, the adults lacking them altogether.

The excretory system consists of protonephridia and canaliculi fusing into two symmetrical lateral excretory vessels; those in turn fuse into a common vessel ending with the excretory pore on the rear end of the body. The digenean excretory system is of key importance in species identification.

The digeneans are hermaphroditic, *i.e.*, each individual contains both male and female gonads.

The female system consists, as a rule, of a single ovary; the oviduct leads from the ovary to the ootype, where the fertilisation takes place. The ootype is also connected with vitellaries, their secretion providing the embryo with yolk and participating in the egg shell formation. The eggs, fertilised in the ootype, are passed to the uterus which expells them outside via the genital pore located in the anterior section of the body.

The male system consists of testes, most often paired. Sometimes a single testis (*Asymphylodora*) or multiple testes (*Aporocotyle*) are present. The testes are connected, via the vasa efferentia, with the vas deferens opening to the ejaculatory duct which frequently ends with the penis. The ejaculatory duct opens near the uterine pore. Both gonad systems may contain such additional elements as the seminal vesicle and bursa in the male system and/or the seminal receptacle in the female one.

The digenean life cycle is very complex; it involves both metamorphosis (heterogenesis) and change of hosts.

The adult parasites live in the intestine, bile and urinary ducts, and blood vessels of vertebrates (fishes, amphibians, reptiles, birds, and mammals including man). The digenean *Transversotrema patialense* is known to occur in the scale capsules of a fish *Brachydanio rerio*. It is thus an ectoparasite whose larvae are tissue parasites.

The digenean life cycle usually involves two intermediate hosts and a single definitive one.

The following developmental stages occur: egg, miracidium, sporocyst, redia cercaria, metacercaria, and adult.

The miracidium larva hatching from an egg is oval and elongated, its cilia enabling it to swim freely in water. Eggs of some species, sinking down to the bottom, are eaten by snails. In the intestine of a snail species, typical of a given digenean, the eggs hatch into miracidia which wander to the snail digestive gland to develop further. In this case, the miracidium lacks cilia. An X-shaped pigmented eye spot

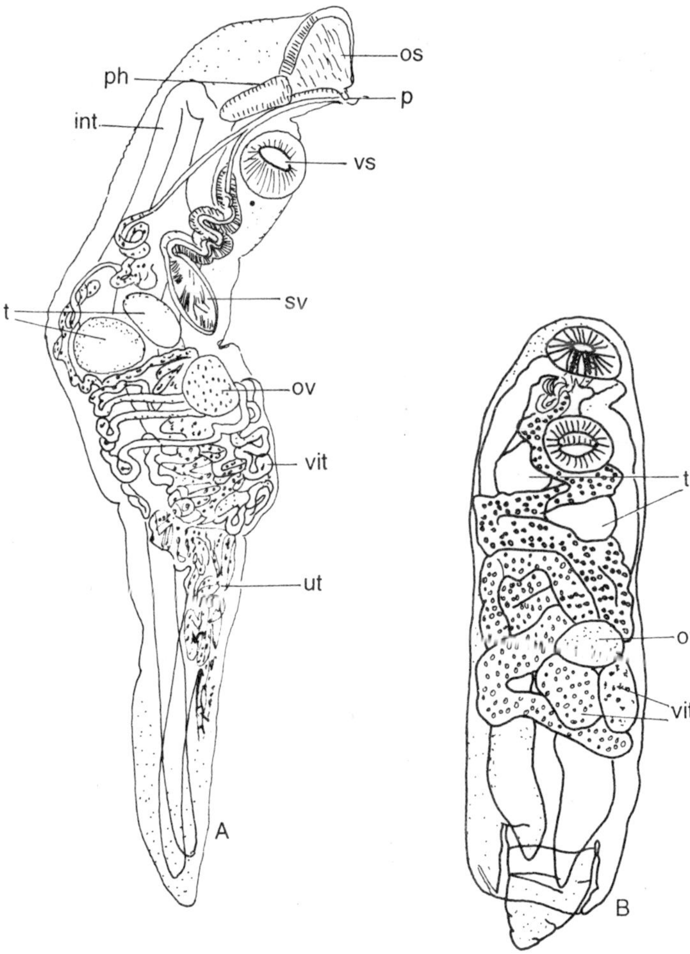

Fig. 1.12 : Digeneans of the family *Hemiuridae* (A) *Lecithocladium excisum* from stomach of mackerel. (B) *Brachyphallus crenatus* from stomach of salmon: os—oral sucker; vs—ventral sucker; int—intestine, ov—ovary; p—penis; ph—pharynx; t—testes; ut—uterus; vit—vitellaries; sv—seminal vesicle.

is visible on its head. The larva contains glands opening anteriorly. Their secretion has lytic properties enabling the larva to penetrate the intermediate host's body wall. Miracidia may show beginings of digestive, excretory, and nervous systems.

The miracidium lives freely for some time (usually not longer than 24 hours) and dies unless the first intermediate host, typically a mollusc, is found. In the latter's digestive gland the miracidium loses its cilia and metamorphoses into the next stage, the sporocyst.

The sporocyst is a sac-like, poorly differentiated structure containing in its rear part a group of germinal cells developing into daughter sporocysts, rediae, or cercariae. The redia, also sac-like but more advanced in development, has a distinct pharynx and the fore section of the intestine. The redia also has an opening to release the next developmental stages. The redia contains germinal cells, too; give rise to either the daughter rediae or next juvenile stages, the cercariae.

Thus the cercariae may be formed in the sporocysts or in the rediae, depending on the digenean species. The cercaria has a well-developed intestine and beginnings of all other organs of the adult. Additionally, this stage contains some larval organs lacking in the adults, namely the X-shaped eye spot and two groups of penetration glands on both sides of the mouth, the glands secreting lytic enzymes which dissolve host's tissues. Cercariae of many species have additionally a stiff rod near the mouth, the so-called stylet, used to pierce the host's skin. The cercaria, as a rule, is provided with a locomotory organ, the tail, differing in shape and size from species to species. It enables the cercaria to swim freely for some time after leaving the mollusc host's body before the second intermediate host is encountered. For many species it is a fish. The cercaria penetrates its skin to get to the muscles, loses its tail and encysts to metamorphose into the next stage, the metacercaria. Some cercariae enter the fish body swallowed with food.

The metacercariae, encysted or not, settle in various organs of the second intermediate host and wait there for their definitive (primary) host—another fish species, an amphibian, reptile, bird, or mammal (including man). In the definitive host's intestine the metacercaria loses its encystment, develops gonads to their final form and metamorphoses into a mature fluke capable of producing eggs.

Not every developmental cycle involves all the stages described above. Some species lack a sporocyst or redia; even a cercaria may metamorphose directly into adult, the metacercaria stage being skipped.

The ability to reproduce as a sporocyst and redia is a characteristic feature in the digenean life cycle. As a consequence, a single egg may eventually give rise to very numerous offspring, thus increasing the parasite's chance of finding a proper host and perpetuate the species.

There is a diversity of opinions as to the sporocyst and redia reproduction. Many authors regard it as larval parthenogenesis, while others consider it as polyembryony, the first opinion, however, prevailing.

More than 15000 digenean species are known from fish. It is difficult to give a precise number, as new species are being continually described.

The pathogenicity of adult digeneans living in a fish intestine is usually weaker than that of the tissue-dwelling larvae. This is particularly evident in the case of the cercariae actively penetrating the fish skin; the damage caused by the skin-piercing metacercariae can be fatal to newly hatched and juvenile fish. The damage in tissues is both mechanical and toxic, the latter being caused by the lytic secretion of the penetration glands. Occasionally, swimmers happening to come across a cercarian "cloud" emitted by molluscs can be affected by the cercariae whose attack produces an itching and burning sensation on the skin accompanied by a rash.

The settlement of metacercariae in fish organs is species-specific; they encyst or not, depending on the species.

Diplostomum spathaceum is a very common species in freshwater fish of Europe; Asia, and North America; it is common also in brackish lagoons and bays of the Baltic.

The adult digeneans live in the fish-feeding birds (*Larus, Sterna, Ardea, Phalacrocorax, Sula, Alca,* and others). On the coast it is mainly seagulls that are the definitive hosts. The first intermediate hosts are such lymnaeid gastropods as *Lymnaea stagnalis, Radix auricularia,* and *Galba palustris.* Within the snails, sporocysts develop with cercariae in them. The cercariae liberate themselves and attack fishes. They pierce the fish skin and then move within the fish body to settle finally in the eye lens where they remain unencysted, awaiting further development in the intestine of a fish-feeding bird. They often occur in masses, several or few hundred individuals in a single lens.

They inflict a disease called diplostomulosis. The eye lenses, affected by the metacercariae invasion, are destroyed, become opaque and brittle. The invasion is often accompanied by exophthalmus, the cornea breaking down and the lens falling out of the eye ball. A secondary infection, bacterial or fungal, may ensue, the mold covering then the entire eyeball. The blinded fish moves haphazardly, keeps to the water surface and thus becomes an easy prey of piscivrous birds.

D. spathaceum dwell in many freshwater fishes and also in those living in marine coastal waters, *e.g.*, flounder (*Platichthys flesus*), various gobiids (*Gobius niger, Pomatoschistus minutus, P. microps*), broad-nosed pipefish (*Syngnathus typhle*) and straight-nose pipefish (*Nerophis ophidion*). Particularly vulnerable to the invasion are salmonid pond cutures, frequently experiencing high losses.

It is very difficult to control the parasite. The most efficient method is to eliminate snails, the intermediate hosts for *Diplostomum,* and thus to break the parasite's life cycle.

Parasite-borne melanosis is a disease inflicted by metacercariae of various digeneans settling on or just underneath the fish skin. The metacercariae are encysted and surrounded by accumulating black pigment, melanin, forming black patches on the skin. At the same time, the rest of the skin is often depigmented, which makes the black spots still more conspicuous. The condition is induced by several species. Some of them are discussed below.

Posthodiplostomum cuticola (Nordmann, 1822) (= *Neascus cuticola*), a freshwater species common in Poland, brings about melancsis in cyprinids. The disease is particularly common in the vicinity of heron rookeries as the grey heron is the definitive host of the species.

Apophallus muehlingi occurs in coastal waters of the Baltic and is particularly common in the Szczecin Lagoon and Lake Dabie. Its metacercariae attack such cyprinids as bream, rudd, and vimba and settle in the subcutaneous tissue and in the fins. Melanin accumulates around the metacercariae and forms fine black spots. The species' life cycle was studied by Odening (1970). The first intermediate host for *A. muehlingi* is a prosobranch snail *Lithoglyphus naticoides* of the family *Hydrobiidae,* originally a Black Sea and East Baltic species introduced to western Europe. As a consequence, the parasite's range is widening westward, too. The second intermediate hosts are cyprinids, the seagull *Larus ridibundus* being the definitive host. Mammals may be infected, too.

A. donicus (= *Rossicotrema donicum*), a species related to *R. muehlingi* and occurring in the same areas, more often affects the percids. Particularly heavy invasions are recorded in perch; very frequently, the whole body and fins are covered with fine black dots.

The parasite's life cycle, described also by Odening (1970), proceeds similarly to that of *A. muehlingi.* The first intermediate host is the same snail, *Lithoglyphus naticoides,* while the role of the second intermediate host may be played by piscivorosu birds (mostly seagulls) and mammals (dogs and cats). Malczewski (1962) found the presence of *A. donicus* in farm-bred fur-bearing animals (minks and foxes) fed with fish.

Cryptocotyle lingua occurs in marine fish, causing melanosis symptoms similar to those caused by previously discussed species. As in the previous cases, the affected fish's skin shows black spots with metacercariae inside.

The adult parasite, 0.5–2.0 mm × 0.2–0.9 mm in size, lives in the intestine of seagulls (*Larus argentatus, L. fuscus, L. ridibundus* and others) and also in the intestine of mammals such as seals, dogs, and foxes.

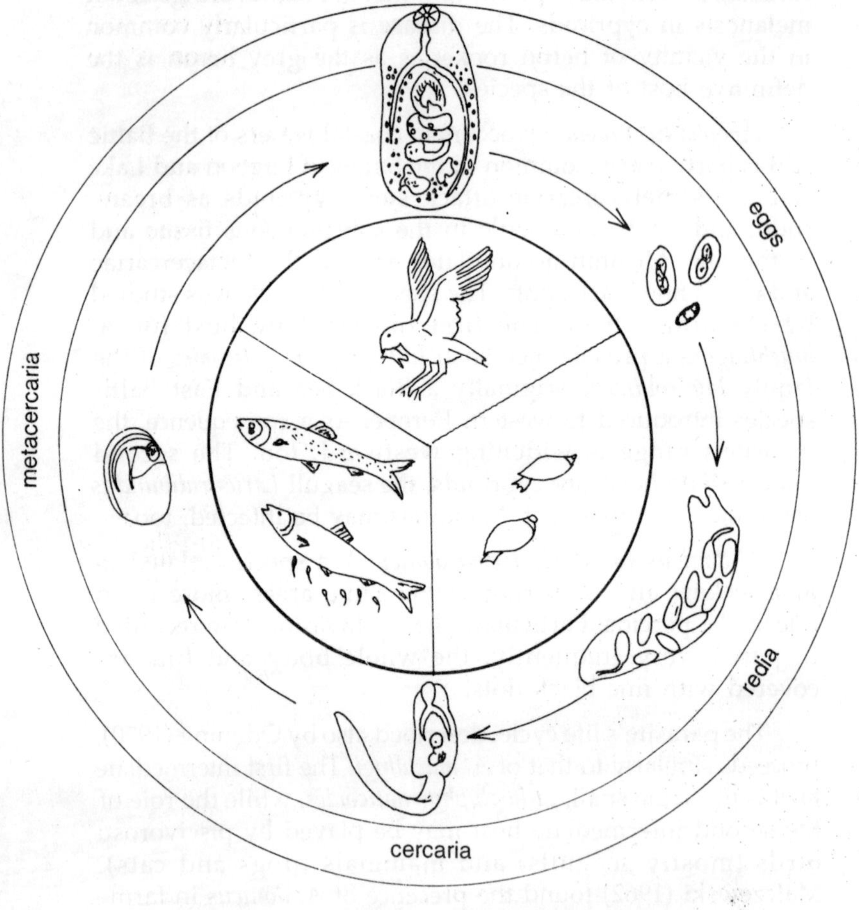

Fig. 1.13 : Life cycle of *Cryptocotyle lingua*.

The parasite's eggs get to the water with feces, sink down to the bottom, and are subsequently swallowed by snails, mainly *Littorina littorea*. The hatched larvae migrate from the snail intestine to the digestive gland where rediae develop with cercariae inside. The cercariae leaves the snail and attacks fishes. They settle and encyst underneath the fish skin and metamorphose into metacercariae. These are surroundded by melanophores.

A number of different fish (herring, mackerel, and others), mainly in the NW Atlantic and off the Atlantic coasts of Europe, can be infected. The parasite was also recorded off Japan. When abundant, the metacercariae cover the entire fish body, fins, and even the eye cornea, thus blinding the affected fish.

The parasite is most dangerous, particularly for juvenile fishes, when the cercariae migrate within the fish body. The herring fry is particularly vulnerable and frequently dies as a result of the invasion. Thus the parasite has a detrimental affect on the fish population.

The fishes infected by *C. lingua* should be barred from consumption for aesthetic reasons and, more important, because cases of human infestation have been recorded.

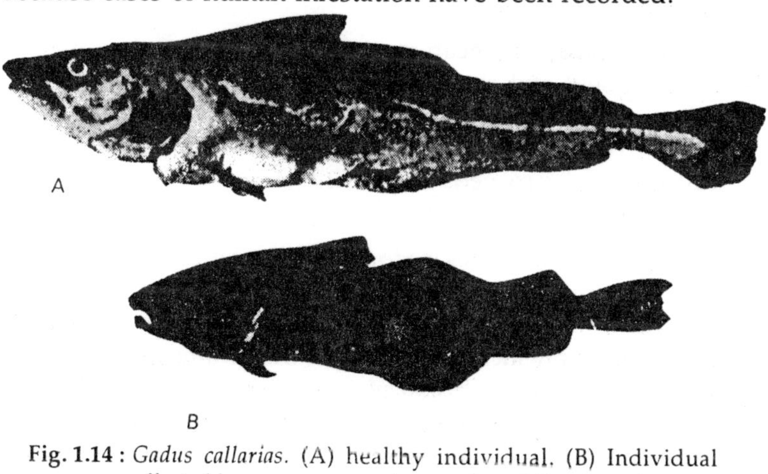

Fig. 1.14 : *Gadus callarias*. (A) healthy individual. (B) Individual affected by numerous metacercariae of *Cryptocotyle* sp.

The two *Apophallus* species and *Cryptocotyle* belong to the family *Heterophyidae,* containing numerous species known to be human parasites (see: Anthropozoonoses).

Hsiao (1940) described an interesting case of digenean infection in cod (*Gadus callarias*) caught off Race Point (Provincetown). The fish was completely black, its body surface being very rough, hardened, and devoid of mucus. A closer examination revealed a complete melanism caused by the mass occurrence of digenean metacercariae on the skin. The number of cysts was enormous and the melanocytes occurred in great numbers; they were particularly abundant around the cysts. The cysts conained metacercariae at various stages. On the other hand, neither the muscles nor the body contained the parasites. The cysts occurred within the epidermis; the connective tissue was at least three times thicker than normal.

The cysts and melanophores were abundant also in the eyes, thus causing blindness.

The author referred to was unable to identify the metacercariae; presumably they were a species of *Cryptocotyle.* Difficult to explain is also the occurrence of an infection on such an immense scale.

Numerous digenean metacercariae are not surrounded by a pigment and as a consequence become invisible to the naked eye. They settle in the fish skin, in muscles or in viscera (hert, intestinal serous membrane). Their further development proceeds in the intestine of predatory fish, birds, or piscivorous mammals.

One of such species is *Stephanostomum baccatum* whose life cycle is well-known as a result of the studies of Wolfgang (1955). The digenean, 3.34 × 0.75 mm in size, has two rows of spines around the oral sucker. Adults live in the intestine of marine predatory fishes, *e.g., Hemitripterrus americanus, Hippoglossoides platessoides, Pleurogrammus azonus, Myoxocephalus scorpius.* The parasite's range extends from the Atlantic coasts of North America to the shores of Europe, including the Baltic Sea.

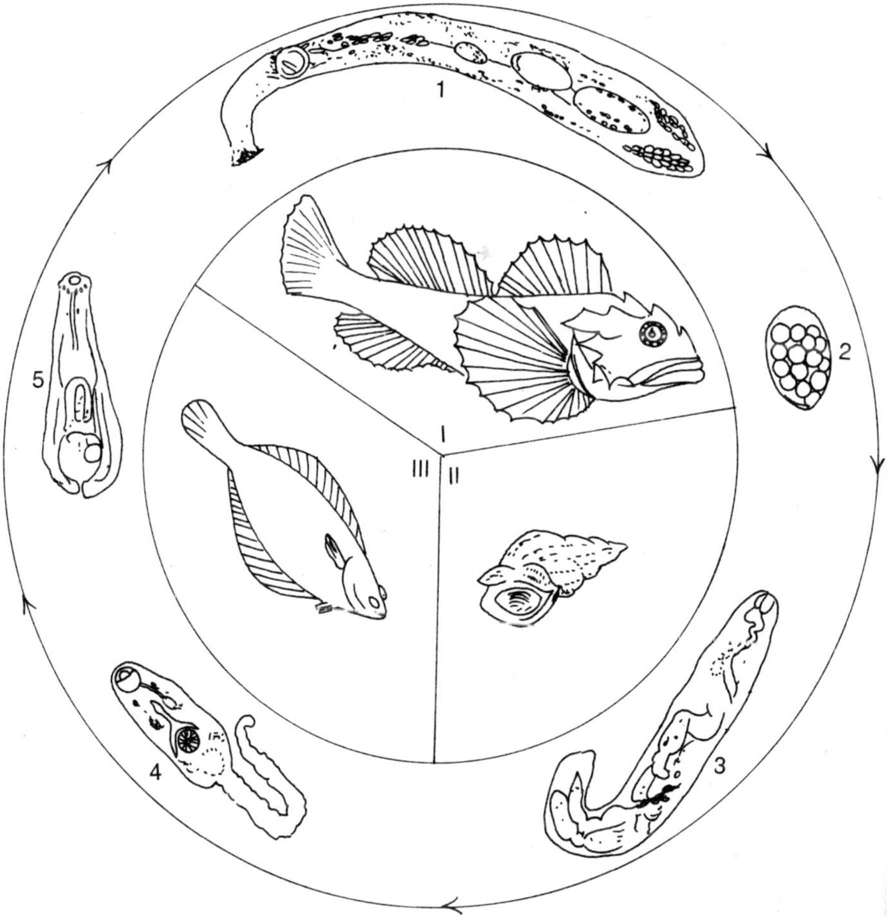

Fig. 1.15 : Life cycle of *Stephanostomum baccatum* 1—adult digenean; 2—egg; 3—redia; 4—cercaria; 5—metacercaria; I—*Myoxocephalus scorpius,* definitive host; II—*Neptunea despecta,* first intermediate host; III—fish, second intermediate host.

The cercariae develop in buccinind snails (*Buccinum undatum* and *Neptunea decemcostatum*), and then actively attack fish of the family *Pleuronectidae (Limandla limanda, Platessa platessa,* and others), piercing through their skin and

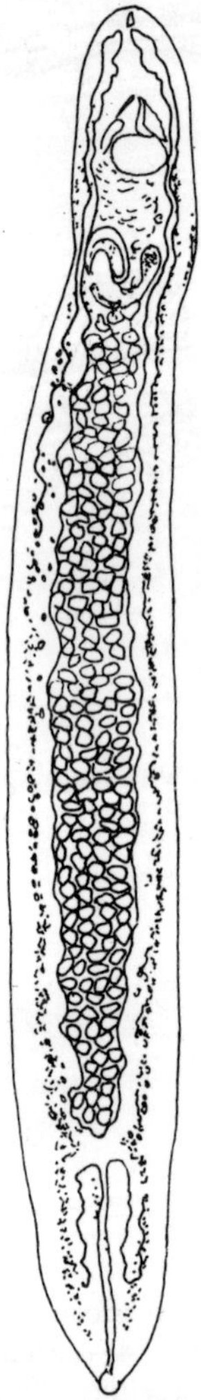

Fig. 1.16 : *Aporocotyle simplex* from blood of *Hippoglossoides platessoides.*

mostly settling under it, in muscles of the blind side, and in the fins. They occur in masses, particularly in large fishes, encyst and metamorphose into metacercariae infective for other fishes, their definitive hosts.

The fish blood system may contain digeneans of the family *Sanguinicolidae,* consisting of 7 genera with 21 species known to date. They are elongated, fusiform, and lack suckers and pharynx. Their development involves two hosts. Molluscs are intermediate hosts providing a site for the development of cercariae, while marine and freshwater fish are the definitive hosts. The life cycle of the freshwater *Sanguinicola* species was described by the Polish parasitologist Ejsmont (1926).

Most often, the eggs are released into the water through the broken gill epithelium. The eggs hatch into miracidia which actively search for their first intermediate hosts, usually pulmonate snails, and metamorphose in them into sporocysts. The cercariae with forked tails (furcocercariae) develop in the sporocysts and thus release themselves from the snails. Then, using their penetration glands, they make their way

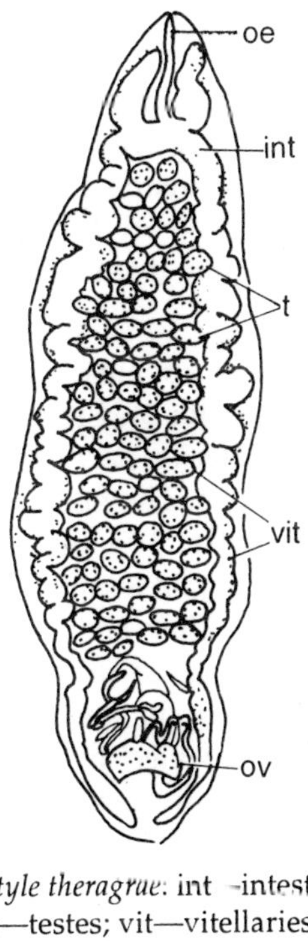

Fig. 1.17 : *Aporocotyle theragrae*: int—intestine; oe—oesophagus; ov—ovary; t—testes; vit—vitellaries.

through the fish skin and reach blood vessels to metamorphose into adults.

The family *Sanguinicolidae* contains species of the genus *Sanguinicola,* typical of freshwater fish. Marine fish parasites include *Aporocotyle simplex* (Odhner, 1900) and *A. theragrae* (Ichihara, 1970). *A. simplex* was found for the first time and described by Odhner in the gill of dab (*Limanda limanda*) off Sweden. *A. theragrae* is frequently recorded in the Alaska pollock (*Theragra chalcogramma* of the North Pacific. This is a small (up to 7 mm in length and 1—2 mm in width) parasite having a characteristic, H-shaped intestine. Numerous (more than 100) testes fill the space between the intestinal caeca. The single, large ovary is located posteriorly in the body.

These digeneans are often encountered extravasated on the gills, intestine, kidney or between the viscera during dissection or processing of fish, hence occasionally the fillets may be accidentally contaminated.

Very serious infections are observed at times in *Theragra chalcogramma.* Soviet authors reported on about 50% invasion incidence and about 54 digeneans per host. There are no data on the pathogenicity of *A. theragrae,* but a strong infection of the blood system doubtless has a bearing on fish condition.

Various *Phyllodistomum* species dwell in the urinary baldder and ducts of freshwater and marine fish. Their body is tapered anteriorly, the posterior part often being broadened leaf-like, with wrinkled edges. The species belonging to this genus are highly host-specific. The first intermediate hosts in their life cycle are bivalves (*Dreissena, Anodonta* and others) harbouring cercariae which later encyst either in the same bivalves in daughter sporocysts or in dragonfly larvae. Thus the second intermediate host is not always present. The definitive host—a fish—is infected by cating metacercariae-containing bivalves or dragonfly larvae.

Most *Phyllodistomum* species parasitize freshwater fish; marines species, however, are known as well, for instance *Ph. marinum* (Layman, 1930) from the urinary duct of

Spheroides borealis, and *Ph. pacificum* (Yamaguti, 1951) from *Caranx equula*. Both species were found in the Sea of Japan.

In the same sea, a related species, *Urinatrema hispidum* (Yamaguti, 1934) was described from the urinary bladder of *Hexagrammos otakii.*

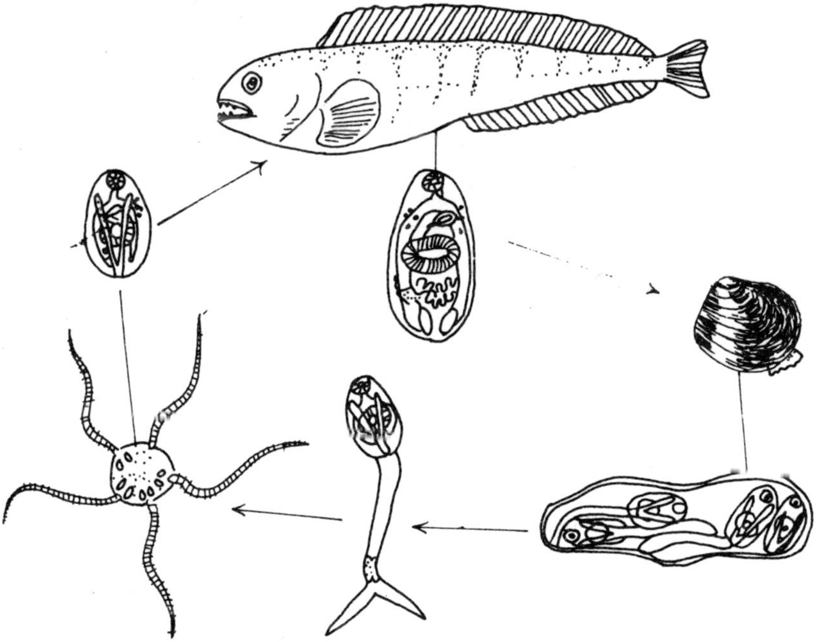

Fig. 1.18 : Life cycle of *Fellodistomum fellis*.

The gall bladder of *Anarrhichas lupus* often contains the already mentioned digenean *Fellodistomum fellis*.

Most digenetic trematodes dwell in the intestine of marine fish.

Digeneans of the family *Hemiuridae* are the commonest species. They usually inhabit the anterior part of the digestive system (stomach). They are small (a few mm in length) parasites whose posterior part is developd as a retractable tail.

Their life cycle requires two intermediate hosts. Molluscs are the first hosts and contain cercariae which have characteristic adaptations to the planktonic mode of life in the form of long filamentous processes arranged fan-or umbrella-like. The processes enable the cercariae to float in the water for 3-4 days, which is exceptionally long for this larval stage. At this time, the cercariae may be eaten by such plankters as crustaceans, medusae, and chaetogenaths, which in turn are consumed by plankton-feeding fish.

Hemiurus luehei (Odhner, 1905) is often reported from European waters, occurring in stomachs of herring and sprat and encountered also, although much less commonly, in salmon and trout. *H. levinseni* (Odhner, 1905) is a frequent parasite of *Theragra chalcogramma* in the Pacific. *Lecithocladium excisum* dwells in the stomach of mackerel (*Scomber scombrus*), the parasite being recorded both from the Atlantic and Pacific. Another widely distributed parasite is *Derogenes varicus* (Muller, 1784), occurring in the intestine of salmon (*Salmo salar*) and many other fish species of the genera *Gadus, Hippoglossus, Cottus,* and *Clupea* and others in the Arctic seas. The intestine of marine *Clupeiformes, Gadiformes, Gobiiformes, Perciformes,* and others is often inhabited by *Lecithaster gibbosus* a widely distributed species found off the coasts of Europe North America, and Asia.

All the species listed above belong to the family *Hemiuridae*. The *Acanthocolpidae* groups trematodes with elongated, almost cylindrical, spine—coverd body, such as that of *Deropristis inflata* frequently found in the intestine of European (*Anguilla anguilla*) and American (*A. rostrata*) eels.

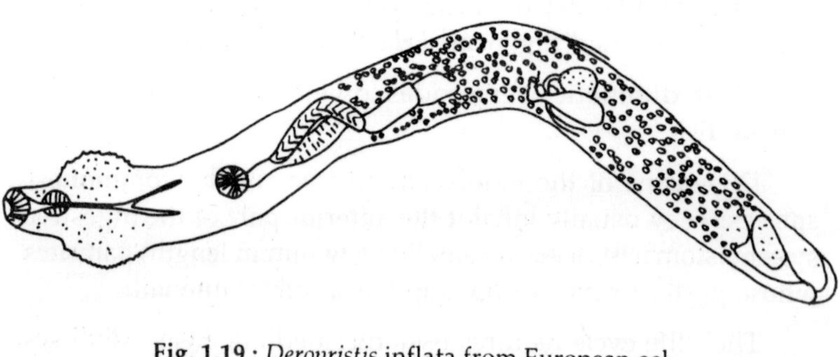

Fig. 1.19 : *Deropristis* inflata from European eel.

The life cycle of *Deropristis inflata* is known from North America. The cercariae, resembling trichocercariae with tails covered with short spines, form in rediae, developing in snails *Bittinium alternatum* and penetrate polychaetes *Nereis virens* and *N. diversicolor*. They encyst there and metamorphose into metacercariae. Eels are infected after consuming polychaetes with metacercariae.

Adult digeneans dwelling in fishes are harmless to man. They are usually removed together with viscera on gutting; however, fillets and carcasses may be accidentally contaminated at a massive invasion, which is undesirable for sanitary reasons.

Subclass: Didymozoida

Members of this subclass, except for the three freshwater fish parasites known to date (*Nematobothrium labeonis*[1], *Philopinna higai*[2], and *N. texomensis*[3]), parasitize marine fishes. They are widely distributed in pelagic subtropical and tropical waters of the Atlantic, Pacific, and Indian Oceans and adjacent seas. According to Nikolaeva (1978), they make up about 7% of all the trematodes.

Most of them occur in cysts produced by affected fish as a defensive reaction; more seldom they remain unencysted. The cysts may be located in gills, mouth cavity, in eye sockets, under the skin, and in muscles. The muscle-dwelling species are usually found unattached in the muscle tissue. When encysted, they always occur in pairs and exhibit various types of reproduction, from hermaphroditic to dioecious. When hermaphroditic, some individuals show domination of the male system, while the female system prevails in others (gonochorism, *i.e.*, incomplete hermaphroditism).

The cysts are variously shaped, the shape being typical of the genus or even of the species.

The *Didymozoida* are elongated, filamentous (*Nematobothrium*) or ribbon-like (*Atalastrophion*). In some species, the body is divided into two sections: the anterior one is usually thin and filamentous or ribbon-like and flattened, while the posterior one is large, swollen, completely filled with gonads and eggs (*Koellikeria, Didymozoon, Didymocystis*).

The didymozoid life cycle is known only in fragments, hence their taxonomic status is not fully established.

Their bean-shaped eggs are very small, only about 10 µm long, and have lids. The eggs hatch into the miracidium, having an anterior ring of about 30 spines surrounding the apical sucker. The whole miracidium is covered by setae. The miracidium development is unknown. It is not known, either, how the eggs and larvae get out of the fish muscles. Presumably, this happens after the fish has died. According

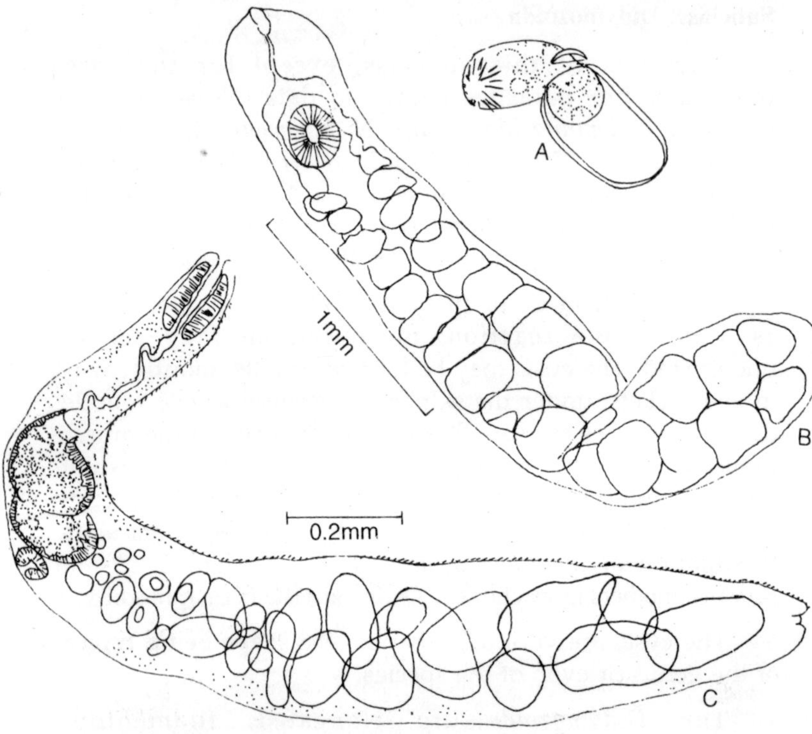

Fig. 1.20 : *Didymozoidae* larvae. (A) Miracidium of *Didymozoon fasciale* hatching from egg (B) *Torticaecum* sp., larva from stomach of *Raja miraletus*. (C) *Monilicaecum* sp., larva from intestine of *Cypselurus bachyensis*.

to Nikolaeva (1965), the miracidium develops within a gastropod, the first intermediate host. In her opinion, rediae or sporocysts and cystophora-type cercariae develop in snails, similarly to the related family *Hemiuridae.* Crustaceans are the suspected second intermediate hosts providing a site for the development of metacercariae. The metacercariae were being found in cirripeds (*Lepas* sp.) and in copepods *Paracalanus aculeatus.*

The metacercariae belong to either of the two following types: the *Monilicaecum* and the *Torticaecum.* Those are collective names for the didymozoid metacercariae. Yamaguti (1971) differentiates between three stages among them:

1. *Premonilicaecum* and *pretorticaecum,* developing in crustaceans, the second intermediate hosts.
2. *Monilicaecum* and *Torticaecum,* juvenile stages dwelling in various fishes, some of which serving as proper hosts and others playing a role of parathenic hosts.
3. *postmonilicaecum* and *Posttorticaecum,* stages developing into the final form within their proper definitive hosts (fishes).

The subclass groups have about 200 species. They are usually of medium size, but some very large parasites are encountered as well. *Nematobothrium filarina* (Van Beneden, 1858) occurs, incysts the size of a fist, in the gill cavity of meagre (*Sciaena aquila*) and grows to more than 1 m. *Gonapodasmius okushimai* (Ishii, 1935), described from the muscles of *Pagrosomus major* off the Japanese coasts reaches up to 590 cm (the male is much smaller). *Nematobibothrioides histoidii* Noble, 1974 is narrow, ribbon-like, and exceeds 12 m in length.

Unitubulotestis sardae (G.A. and W.G. MacCallum, 1916) (= *Nematobothrium sardae*) up to 3 cm in length, is one of the noteworthy and relatively well-known species. Pairs of individuals live in 7 mm long and 3–6 mm wide, orange cysts located among the gill lobes of the Atlantic bonito (*Sarda sarda*) in the North Sea. The parasite was also recorded in the NW Atlantic. Studies of E. and J. Grabda (1939) and E. Grabda (1947) showed the infection to be quite substantial, up to 64.8%. Single gill arches container, apart from juvenile parasites invisible to the naked eye, up to 10 cysts of various sizes. The cyst walls are very delicate and thin, frequently provided with fine capillary vessels. The cyst colour is imparted by eggs filling the parasites.

Each cyst contains two coiled filamentous individuals of *U. sardae,* their maximum length and width reaching 29.7 mm and 1.5 mm, respectively. Both individuals are hermaphroditic, but with a clear domination of male gonads in one and female gonads in the other.

The earliest stage of development is unknown. On the other hand, juveniles 7-14 mm in size and cyst formation were observed. The parasites, first one and then the other, get into gill lobe blood vessels and produce the arteriectasis in the lobe artery, which—in the course of further development of the parasite—leads to the formation of a typical sacculated aneurism (*aneurisna verminosa saccata*). Some cysts may contain blood extravasated from arteries.

Undoubtedly, the parasites settling in the blood vessel lumen bring about disorders ingill circulation. However, no anatomo-pathological changes were observed in the gills except for a single case of a strong invasion (10 cysts in a gill arch) in a juvenile 20 cm long fish were clearly pale and anemic.

Koellikeria orientalis (= *Wedlia orientalis*) is a typically dioecious trematode. Its paired individuals occurring in cysts located in various parts of the digestive system of *Germo macropterus, Euthynnus pelamys,* and *Thunnus thynnus* in the Pacific. The trematode body is divided into two parts. The anterior section is thin and elongated, slightly widened and flattened around the oral sucker. The posterior section is large, swollen, and almost spherical. The female posterior part is particularly enlarged due to the enormous amount of eggs filling it.

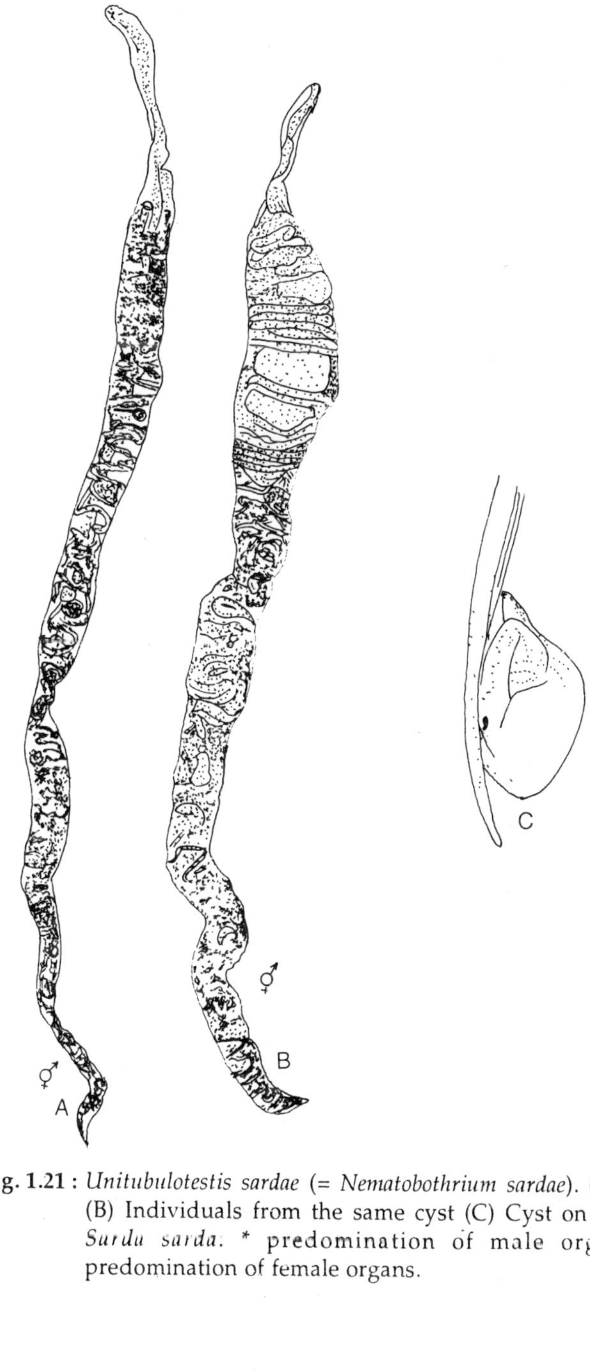

Fig. 1.21 : *Unitubulotestis sardae* (= *Nematobothrium sardae*). (A) and (B) Individuals from the same cyst (C) Cyst on gills of *Sarda sarda*. * predomination of male organs; * predomination of female organs.

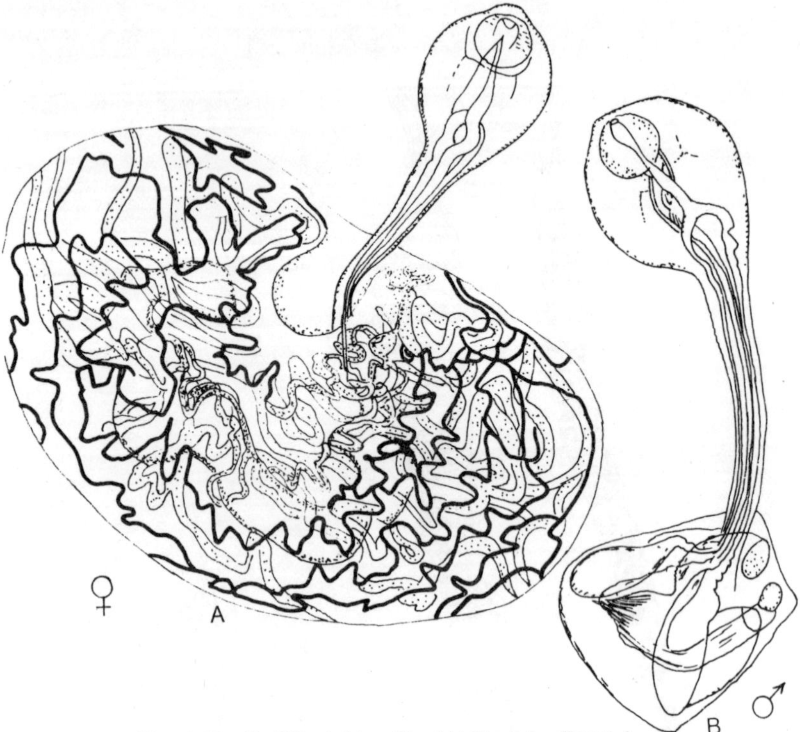

Fig. 1.22 : *Koellikeria bipartita*. (A) Female. (B) Male.

Neodiplotrema pelamydis is an interesting case of two individuals being permanently accreted in a single cyst, similar to the freshwater monogenean *Diplozoon paradoxum*. *N. pelamydis* dwells in the gill cavity of *Euthynnus pelamys* from the Pacific. Both individuals in the cyst are hermaphroditic, their posterior parts being fused.

Without doubt, none of the trematode species is indifferent to the fish health and condition. However, those parasites settling in muscles are the main concerns for fish processing. One of those is the largest known trematode, *Nematobibothrioides histoidii* mentioned earlier, occurring unattached in muscles of *Mola mola* off California. These trematodes interlace fish muscels with numerous loops. Up to 12 individuals per host were found in fish specimens weighing, on average, 16 kg. The fish flesh was literally

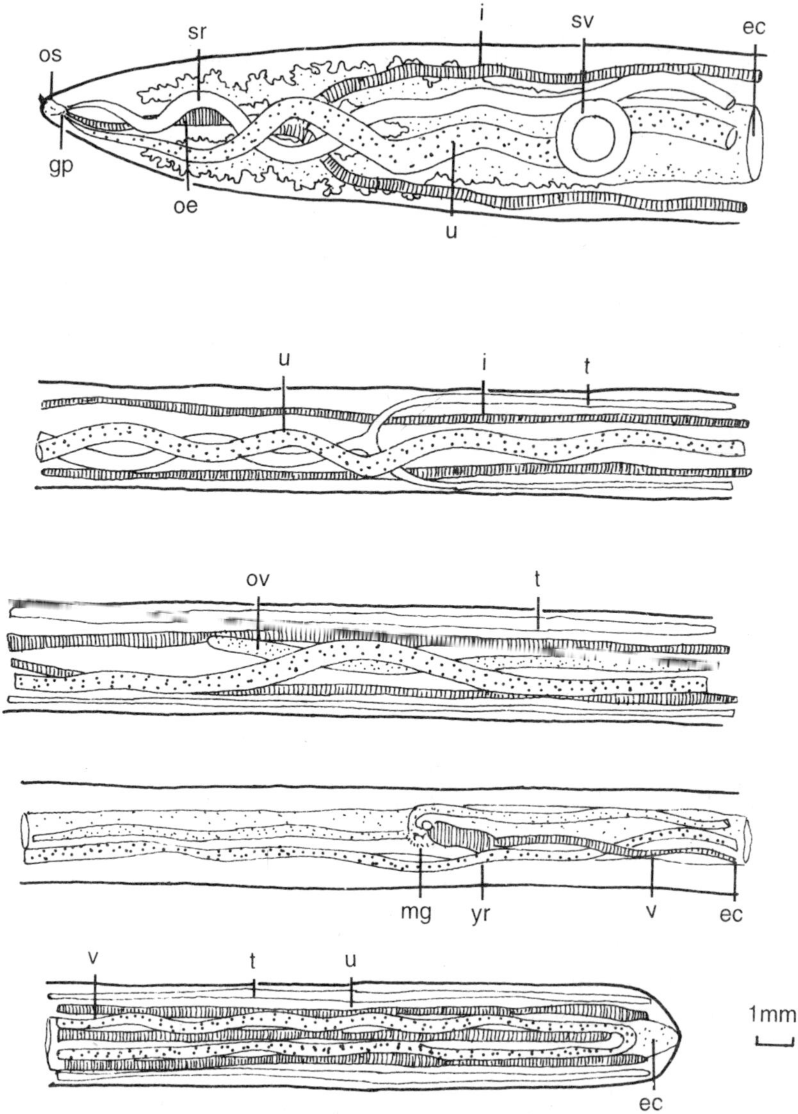

Fig. 1.23 : *Nematobibothrioides histoidii* Mg—Mehlis gland; i—intestine; ec—excretory canal; u—uterus; gp—genital pore; ov—ovary; oe—oesophagus; os—oral sucker; sv—ventral sucker; sr—seminal receptacle; yr—yolk receptacle; t—testis; v—vitellary.

packed with thin (1.5-4 mm), very long (up to 1200 cm in length) parasites difficult to separate from fish without damaging the latter.

Besides the adults, the affected fish contained aggregations of eggs (17 × 15 μm in size, on the average) in its connective tissue, which additionally contaminated the flesh.

The *M. mola* behaviour was abnormal: the lazily moving fish kept to the water surface, returning there when pushed deeper with an oar. It should be added that, besides *N. histoidii* described, the *M. mola* individuals examined also showed the presence of numerous parasites of various taxa: protozoans, digeneans, cestodes, and copepods. The abnormal behaviour was thus a net result of the combined effects of all the parasites.

Muscles of *Pagrosomus major* in the Pacific may contain *Gonadopodasmius okushimai* (Ishii, 1935). The species, about 0,5 mm wide, has a long (up to about 6 m) body. Muscles of the Pacific tuna *Thunnus alalonga* sometimes contain *Metanematobothrium guernei* 177 mm long and 1 mm wide. *Didymozoon tenuicolle* was encountered in muscles of *Lampris guttatus* from the Europeen waters. The parasite's "neck" is clearly separated from its "trunk." The total length is 63.8 mm, the "trunk" measuring 47 mm.

Wierzbicka (1980) described a new didymozoid, *Neolampididymozoon tenuicolle* from *Lampris guttatus* in the SE Atlantic. The parasites occur encysted in pairs on the gills and gill membrane as well as in the dorsal fin base muscles. It is interesting to note that the parasite's size depends on its location in host.

For the first time in the Atlantic, the same author described the occurrence in *Lampris guttatus* of another species, *Metadidymocystis cymbi*. The parasite occurs unattached among the subcutaneous muscles down to 5 mm and frequently forms large aggregations of hermaphroditic individuals 10-26 mm long.

In recent years, muscles of the hairtail (*Lepidopus caudatus*) from off South Africa have often been found to contain trematodes of the family *Didymozoonidae,* not identified any closer. They frequently occur only as aggregations of eggs, 16–18 µm x 30–32 µm in size, typical of the family. The hairtail invasion incidence in 1977 reached 50%. Mackerel (*Scomber scombrus*) caught in the Georges Bank was found to be infected by *Atalastrophion* trematodes, the parasites intertwining with the fish muscle fibres and forming orange-coloured centres. The colour is imparted by bean-shaped tiny eggs, enormous amounts of them filling the parasite's body. Those fillets abundantly infected with the parasites were soft, watery and greyish in colour.

Nikolaeva (1972), too, drew attention to a frequent presence of didymozoid trematodes in muscles of marine fish. She found unencysted and unattached trematodes in many fish species from the Aegean, Red, and Mediterranean Seas and from the Atlantic and Indian Ocean. "Balls" of parasites interlaced the fish skeletal muscles. It was not possible to dissect out all of the trematodes and identify them to species; some could be identified to the generic level only. For example, muscles of the Red Sea *Lutianus bohar,* the Indian Ocean *L. lineatus, Letrinus miniatus,* and *Epinnula orientalis* as well as those of *Cypselurus* sp. from the Aegean Sea and *C. furcatus* from the Atlantic were found to contain *Gonapodasmius* trematodes. Muscles of the Indian Ocean *Platycephalus pristis* were infested by *Didymozoon* sp.

The fish quality was unquestionably lowered by the presence of these parasites. Although the parasites are harmless to man, attention should be paid to them; badly infected fishes should be rejected. On weak invasion, and when the muscles contain encysted species, the cysts should be removed from fillets.

Class: Cestodes (Tapeworms)

Cestodes arc plathelminths whose body is ribbon-like, flattend dorsoventrally, and consists of the scolex, neck, and strobila, the latter made up of a number of sections (proglottids) arranged one behind another. Unsegmented cestodes are known as well (*Caryophyllaeus, Khavia, Caryophyllaeides*).

The class contains exclusively parasites. Adult forms dwell in the intestine of vertebrates.

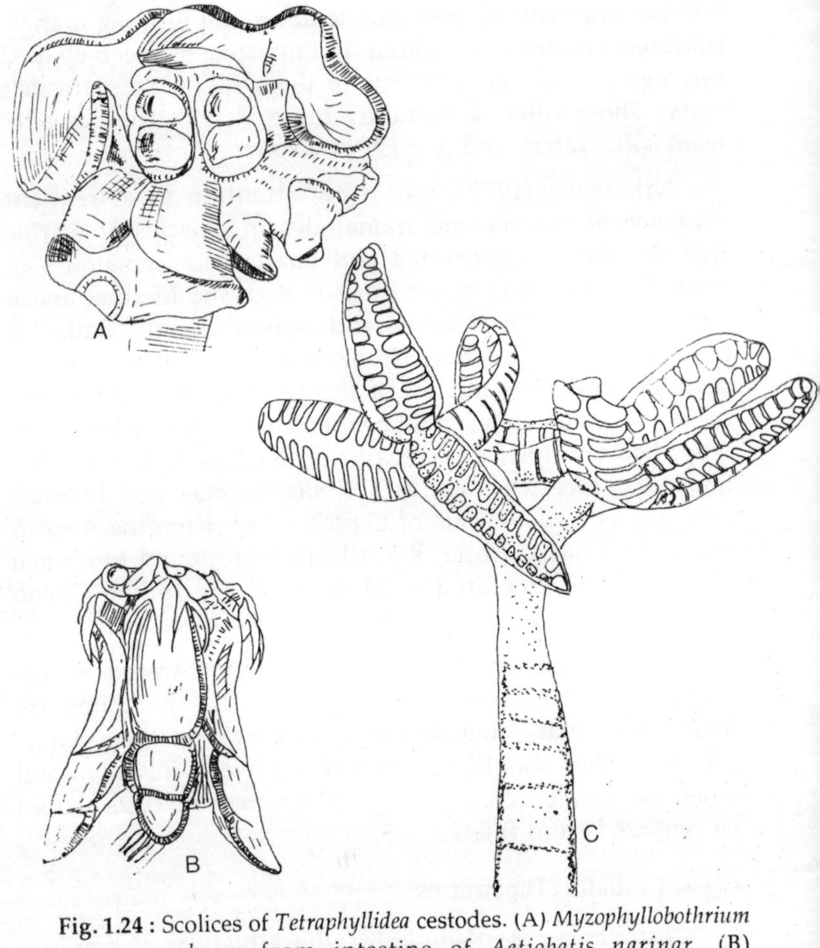

Fig. 1.24 : Scolices of *Tetraphyllidea* cestodes. (A) *Myzophyllobothrium rubrum* from intestine of *Aetiobatis narinar*. (B) *Calliobothrium verticillatum* from *Mustela laevis* (C) *Rhinebothrium corymbum* from *Dasyatis americana* 1975.

The scolex is an attachment organ used to fasten a cestode to the host's intestinal mucosa; in this way the parasite overcomes the peristaltic movements and the danger of being removed from the gut. Consequently, the scolex is provided with various additional holdfast structures such as suckers (acetabulae) and bothria in the form of 2 (*Pseudophyllidea*) or 4 (*Tetraphyllidea*) grooves. More complex are bothridia; they are grooves, having their own muscles and divided by septa into secondary grooves, frequently supported by stalks (*Tetraphyllidea*)

Additionally, the scolex may possess various rings of hooks and retractable proboscids covered with numerous rows of fine hooks (*Tetraryhynchidea*). The scolex structure is an important identification character of cestodes.

The cestode neck is a growth zone. New proglottids are produced on the borderline between the neck and the strobila.

The body is filled with a loose parenchymal tissue, its cells containing numerous calcareous bodies.

The central nervous system is located in the scolex. The system consists of two ganglia interconnected with transverse commisures. The ganglia give rise to anteriorly-directed nerves of the scolex attachment structures and to 10 nerve cords running parallel backwards along the strobila. Two of them are the main nerves, much thicker than the rest, located symmetrically along the sides of the body. All the nerve cords are interconnected with transverse commissures; their branches extend to the organs within the strobila.

The excretory system consists of protonephridia and a network of capillary canaliculi opening to the main excretory canals (most often two) located laterally and connected with transverse canals in each proglottid. The excretory ducts open in the terminal proglottid.

Digestive system. Cestodes have no intestine, the nourishment being absorbed by the whole surface of the

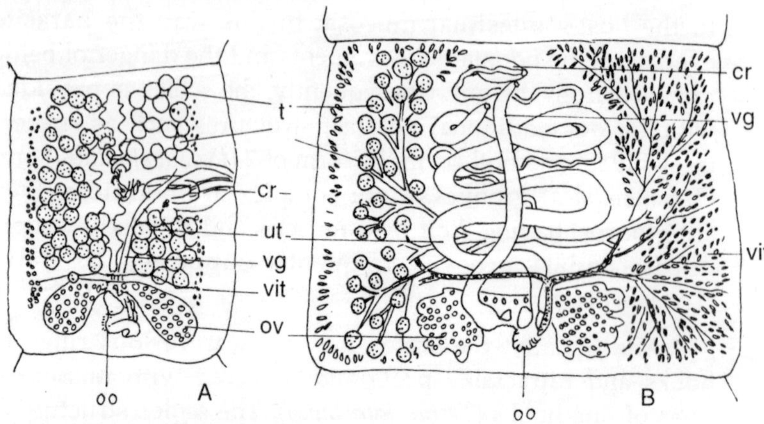

Fig. 1.25 : Cestode gonads: a diagram. (A) *Proteocephala*. (B) *Pseudophyllidea*: cr—bursa cirri; oo—ootype; ov—ovary; t—testes; ut—utcrus; vg—vagina; vit—vitellaries.

body. Food absorption is facilitated by the structure of tegument, its surface being thrown into numerous folds (the microtriches) and pierced by numerous pores.

With a few exceptions, as for examples *Dioicocestus* sp., cestodes are hermaphroditic, each proglottid having its own set of male and female gonads. In some species the segmentation is obliterated externally, but sets of gonads are replicated several times. As a rule, male gonads develop first, followed by developing female gonads.

The male system consists of numerous testes, spherical or oval, scattered in the parenchyma. The vasa efferentia leave the testes and fuse into a single vas deferens, its terminal part functioning as a copulatory organ (cirrus), opening into the cirrus bursa (*bursa cirri*). The vas deferens often widens in its terminal part and forms the seminal vesicle where spermatozoa are gathered.

The female system includes a single ovary, the Mehlis gland surrounding the ootype, vitellaries, and the seminal receptacle. Fertilization of egg cells and egg formation takes place in the ootype, to which open the oviduct, vitellary ducts, and the seminal receptacle. Eggs are collected in the uterus and are subsequently excreted via the gonopore. The vagina is a canal receiving spermatozoa introduced by the cirrus during copulation.

Both the male and female systems form the basis of cestode identification. The cestode development is complex; it involves metamorphosis and change of hosts, a vertebrate always being the definitive host. The larval phase proceeds in one (*Diphyllidea, Proteocephalidea*) or two (*Pseudophyllidea, Tetraphyllidea, Tetrarhynchidea*) intermediate hosts.

The first intermediate host in both lines of development is always an invertebrate; in the *Pseudophyllidea* cestodes affecting the fish, it is most often a copepod. The second intermediate host is a fish. The array of definitive hosts covers fishes, birds, and mammals (including man).

When in water, eggs of pseudophyllid cestodes hatch to release a spherical and completely ciliated larva provided with 6 hooklets. Due to the presence of cilia, this coracidium larva can swim for some time (from several hours to one day, depending on the conditions) and dies unless an appropriate host, a copepod, is found. When eaten by a crustacean, the coracidium loses its ciliary cover in the host's stomach, pierces the intestine with its hooks and gets to the host's body cavity as the so-called oncosphere.

The oncosphere grows longer, its hooklets grouped in the posterior end of the body set off as a cercomere; the latter soon tears off. Protonephridia and the excretory canaliculi develop, and within 2–3 weeks a procercoid larva emerges, its further development taking place in the second intermediate host.

A procercoid, swallowed together with its crustacean host by a fish, moves to the latter's abdominal cavity where it grows further. The scolex is formed, and nervous and excretory systems develop, before a plerocrercoid larva, invasive for the definitive host, appears. The cestode reaches maturity and produces eggs in the definitive host's intestine.

In different groups of cestodes the life cycle as described undergoes modifications with respect to the duration of various stages, advancement in development, larval structure, etc.

Cestode pathogenicity. Both the larval and adult cestodes are found in the fish as pathogens. Plerocercoids of many species, growing in the fish body cavity, exert mechanical pressure on the liver, kidney, and gonads, thus causing malfunctioning of the organs affected. Large plerocercoids of *Ligula* induce sterility in fish, their growth occasionally causing the fish abdominal cover to break down. Toxins excreted by those plerocercoids give rise to fish behavioural disorders, *e.g.*, the *Ligula* plerocercoid-infected bram do not migrate to spawn.

The presence of adult cestodes in the host's intestine is not unimportant to the host, since, especially inlarge species, cestodes may cause anaemia and emaciation.

Marine fish may contain cestodes belonging to five orders: the *Pseudophyllidea, Diphyllidea, Tetraphyllidea, Tetrarhynchidea,* and *Proteocephalidea*. The division is based mainly on the scolex structure. Of the greatest importance for the fishing industry are the *Pseudophyllidea* and *Tetrarhynchidea*.

Order: Pseudophyllidea

Cestodes grouped in this order have scolex with two bothria: dorsal and ventral; sometimes the scolex is funnel- (*Cyathocephalus*) or fan-shaped (*Caryophyllaeus*).

The strobila usually has numerous proglottids, although unsegmented cestodes are encountered, too, viz the family *Caryophyllaeidae*. The segmentation of strobila is usually pronounced and corresponds to internal segmentaion. In some species (*e.g.*, the *Ligulidae*), however, external segmentation does not coincide with the internal one.

Life cycle involves three hosts: two intermediate and one definitive.

The order contains numerous species dwelling in freshwater, marine, and migratory fish. Some species are pathogenic for man as well; these will be dealt with in the chapter on anthropozoonoses.

The commonest species include *Bothriocephalus scorpii* (Muller, 1776), a frequent parasite of flatfish, attacking other marine fish as well. It was found in the Southern Baltic in the flounder (*Platichthys flesus*), turbot (*Scophthalmus maximus*), and sculpin (*Myoxocephalus scorpius*). A massive invasion of the cestode is frequently observed; 100% of the North Sea *Rhombus laevis* can be affected. The cestode was also recorded in the Atlantic and Pacific as well as in the Mediterranean, White, and Black Seas.

B. scorpii is a rather large (up to 1 m long) cestode whose adult form lives, attached most often to the pyloric caeca mucosa, in the fish intestine. Its life cycle was followed in the laboratory by Markowski (1935).

The first intermediate host is a copepod, *Eurytemora hirundo;* the next host is a gobiid, *Pomatoschistus minutus* (more seldom the straightnose pipefish, *Nerophis ophidion*), the flatfish being definitive hosts.

The parasite affects fish condition considerably; a massive invasion may retard fish growth.

The eel in Puck Bay is frequently attacked by *Bothriocephalus claviceps* (Goetze, 1782), encountered also in European and American inland waters. Presumably, the eel becomes infected at its freshwater phase of life, which is inferred from the fact that the parasite's first intermediate host is, as observed experimentally by Jarecka (1959), a freshwater copepod *Macrocyclops albidus*. The second intermediate host, *i.e.*, the one passing the infection on to the eel, is not known to date.

The intestine of salmonids (i.a. salmon, trout) may be a suitable habitat for *Eubothrium crassum*. This rather large (up to 60 cm long) cestode with a well-developed scolex and two distinct bothria is very common in European, North Amercian, and Caspian basin waters.

The cestode settles most often in pyloric caeca and attaches itself by means of the scolex to the bottom of the caeca, the strobila falling down into the intestine. The infection is usually very heavy, up to several hundred cestodes per host.

The *E. crassum* life cycle was described by the Norwegian parasitologist Vik (1963). Typically of the *Pseudophyllidea,* it involves two intermediate hosts: the first is a copepod, either *Cyclops strenuus* or *Eucyclops serrulatus,* while the other is perch (*Perca fluviatilis*), eventually infecting salmon and trout.

There is no detailed data on the *E. crassum* pathogenicity. Even a heavily infected (up to 400 cestodes per host) fish retains good condition and has fat deposits lining the intestine. Mikhailov (1975) is of the opinion that *E. crassum* is pathogenic to the Caspian salmon. A very heavy invasion of up to 3000 cestodes per host was recorded in that species, resulting in the intestinal occlusion and death of the individuals affected.

The intestine of cod (*Gadus morhua*), haddock (*G. aeglefinus*), and other Atlantic gadids as well as *G. macrocephalus* and *Theragra chalcogramma* in the Pacific may be inhabited by *Abothrium gadi* van Beneden, 1871. The cestode has a deformed scolex with poorly developed bothria.

The order discussed includes also certain cestodes (*Diphyllobothrium latum, D. minutus, D. pacificum, D. ursi,* and others) whose larvae are intestinal parasites of vertebrates, while plerocercoids occur in fish. They are dangerous for man, infected via fish (see: Anthropozoonoses). The group includes also *Pyramicocephalus phocarum* whose adult forms live in the intestine of seals *Phoca barbata, Ph. hispida, Cystophora cristata, Eumetopias jubatus,* and *Enhydra lutris.* The plerocercoid of the species has a large (up to 4 mm in length and 3 mm in width as its base) triangular, pyramid-shaped scolex with large, strongly wrinkled bothria and a 5—6 cm long tail. The *P. phocarum* plerocercoids occur in numerous marine fish such as *Gadus morhua, Theragra chalcogramma,*

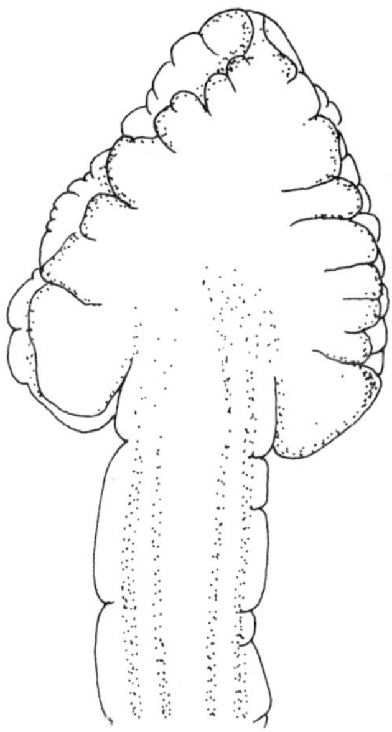

Fig. 1.26 : Plerocercoid of *Pyramicocephalus phocarum*.

Myoxocephalus verrucosus, and *Eleginus gracilis*. Particularly frequent are *P. phocarum* invasions in the Alaska pollock (*Theragra chalcogramma*).

The plerocercoids settle in the fish abdominal cavity, between the pyloric caeca, on the intestinal serous memberane, on gonads, and even in muscles.

P. phocarum tapeworms occur in Atlantic and Pacific fishes as well as in those in the White and Barents Seas. They are particularly abundant in the vicinity of seal colonies, seals being their definitive hosts. Those settling in the liver destroy it. Cases of human infections are common.

Order: Diphyllidea

This order groups cestodes which have a very specific scolex with two bothridia (dorsal and ventral) in the anterior part and two groups of large hooks set on the top of the rostellum.

The posterior part of the scolex resembles a handle or a stalk with vertical rows of fine hooks. The very short strobila consists of a few proglottids only.

The order contains only one genus, the *Echinobothrium*, with 11 species living in the intestinal spiral valves of rays (*Raidae*).

The *Echinobothrium* life cycle involves a single intermediate host, a crustacean or a gastropod, providing a site for the development of a cysticercoid larva with a fully formed scolex inside.

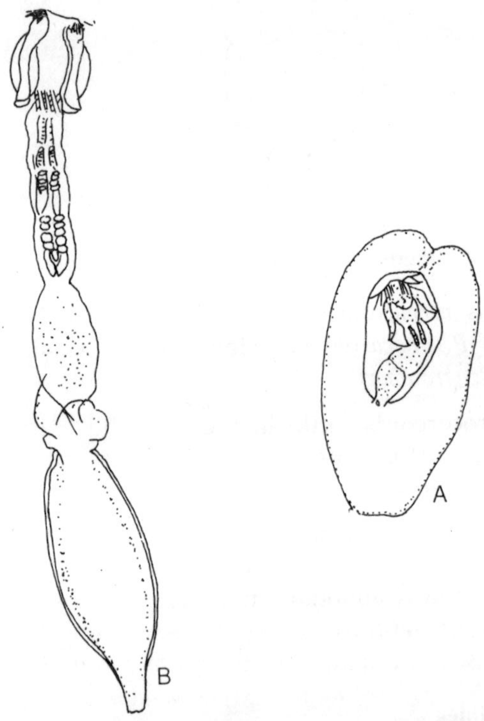

Fig. 1.27 : *Echinobothrium benedeni* larva from crustacean *Hippolyte varians* (A) Cysticercoid. (B) Everted larva.

Ruszkowski (1928) has greatly contributed to the knowledge on the *Echinobothrium* life cycle. He described *E. benedeni* from *Raja asterias* and *R. punctata* in the English Channel off Roscoff, and studied its development. When examining stomach contents of the infected rays, he found the stomachs to be filled with decapods *Hippolyte varians,* which contained *E. benedeni* larvae; in this way the intermediate host of the parasite was discovered.

The *Diphyllidea* cestodes are of no economic importance, but they are an interesting group with respect to their morphology and development.

Order: Tetraphyllidea

The *Tetraphyllidea* are cestodes having scolex with 4 bothridia of various shapes, frequently placed on a stalk. The attachment surface is large and often divided with septa. The scolex has additional attachment devices such as suckers and hooks.

The tetraphyllid life cycle is poorly known. Presumably, the cestodes have, similar to the *Pseudophyllidea* and *Tetrarhynchidea,* two intermediate hosts and a definitive one. A marine crustacean is the first intermediate host the procercoid grows in. The plerocercoid stage develops in teleost fishes, the mature cestodes living attached to the spiral valves of sharks.

Scolex pleuronectis Muller, 1780 (= *S. polymorphus*) is the commonest tetraphyllid larva. It occurs in the intestine, abdominal cavity, gall bladder, and muscles of salmonids, flatfish, and others. It is frequently encountered in the Barents, White, Black, and other European seas. Presumably, *S. polymorphus* is a collective name of various tetraphyllid larvae.

Other larva, *Pelichnibothrium caudatum* occurs in the intestine and pyloric caeca of marine fishes and anadromous salmonids of the river Amur basin. The larva was also found in Pacific cephalopods.

Adult tetraphyllids occur in the lancetfish (*Alepisaurus ferox*) in the Atlantic, as well as in the blue shark (*Prionace glauca*), sunfish (*Lampris regius*), and occasionally in tuna (*Thunnus thynnus*) off Japan.

Both the adult and larval *Tetraphyllidea* are of no importance to the fishing industry.

Order: Tetrarhynchidea

Synonym: Trypanorhyncha

The order groups relatively small tapeworms, living, as adults, in the stomach and spiral valves of sharks. The larvae are very common in marine teleosts.

The key character of these cestodes is the structure of the scolex. The scolex is relatively large, with 2 or 4 bothridia anteriorly and 4 or—less frequently—2 very contractile and retractable proboscids, each being retracted and everted by

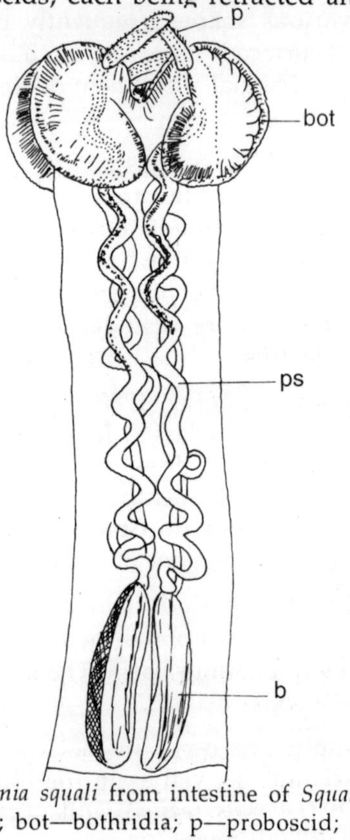

Fig. 1.28 : *Guilquinia squali* from intestine of *Squalus sucklii*: b—bulbus; bot—bothridia; p—proboscid; ps—proboscid sheath.

means of special muscles in the form of 4 bean-shaped structures (bulbus) connected with the retractor muscles drawing the proboscids into their canals in the scolex parenchyma. The proboscids are covered by numerous rows of hooks of a shape and size depending on the species and on the location on the proboscid.

The scolex, owing to its complex structure, can be thus divided into three parts: bothridial (*pars bothridialis*), vaginal (*pars vaginalis*), and muscular (*pars bulbosa*).

The tetrarhynchid life cycle is still poorly known. To date, it is only the Polish parasitologist J.S. Ruszkowski who, in his work on *Grillotia erinaceus* (Van Beneden, 1858) observed and described the first stage of the cycle, which made it possible to infer the full course of it. The *Tetrarhynchidea*, similarly to the *Pseudophyllidea*, require two intermediate hosts (a crustacean and a teleost) and a definitive one.

During his work at the Biological Station in Bergen, Ruszkowski (1934) cultured eggs collected from mature cestodes found in the intestine of *Raja oxyrhynchus*. Subsequently, he infected various planktonic organisms with typical coracidia hatched from those eggs. The following copepods succumbed to the infection: *Acartia longiremis*, *Pseudocalanus elongatus*, *Paracalanus parvus*, and *Temora longicornis*. While in copepods, the coracidia lost cilia and made their way to the body cavity to metamorphose there into procercoids provided with a cercomere with 6 hooklets.

The further development takes place in teleosts and involves plerocercoids having a fully developed scolex with retractable proboscids and bothridia. The scolex is located anteriorly in a special receptacle (*receptaculum scolecis*).

This larva does not fully correspond to the pseudophyllidean plerocercoid, hence various authors call it *plerocercus* or *blastocystis*.

In the intestine of a shark or a ray the scolex is everted, the strobila grows and the cestode takes on its final adult form.

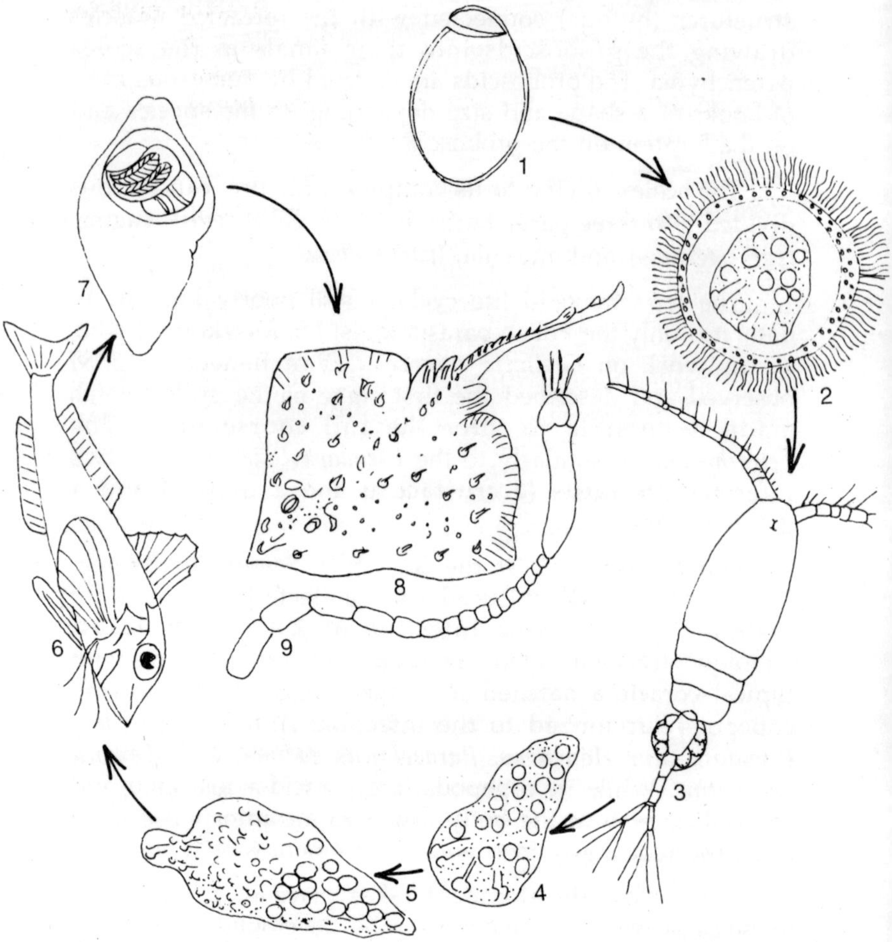

Fig. 1.29 : Life cycle of *Grillotia erinaceus* 1—egg; 2—coracidium; 3—first intermediate host, *Pseudocalanus* sp.; 4 to 5—procercoid; 6—second intermediate host, *Trigla* sp.; 7—cysticercoid; 8—definitive host, *Raja oxyrhynchus;* 9—adult cestode.

The larval *Tetrarhynchidea* are very common in marine fish and often occur in masses causing high losses; particularly harmful in this respect are the muscle-dwelling species.

They are not dangerous to man, but the relatively large, well-visible larvae make fish muscles look repulsive. For this reason, the infected fishes are processed into fish meal rather than marketed for consumption.

The already mentioned species *Grillotia erinaceus* is common in cod, hake, halibut, and other teleost fish in the North Atlantic as well as North, Irish, and Norwegian Seas, and in the North Pacific and adjacent waters. The larvae settle in abdominal cavity organs, on the peritoneum, and in muscles, particularly along the backbone. The larvae have 2 bothridia and 4 proboscids. The adult cestodes live in the spiral valve of *Raja oxyrhynchus*.

larvae of *Guilquinia squali* were found in the eyes of the North Sea whiting (*Odontogadus merlangus*). The larvae, frequently with their scolices everted and provided with 4 bothria and 4 proboscids, lie unattached in the hyaline body of the eye. A single eye may contain up to ten 3.5-8.0 mm long and 0.7-1.6 mm wide larvae.

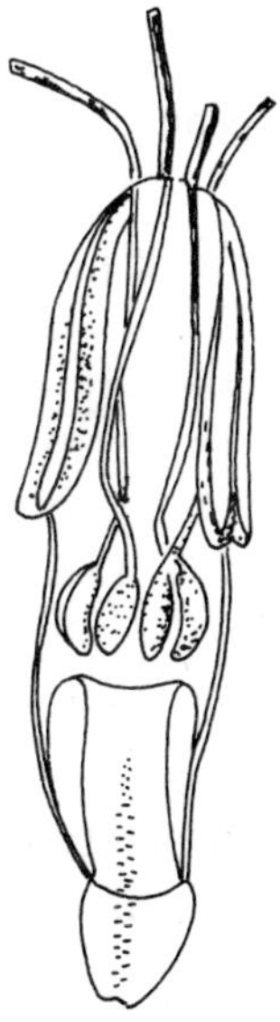

Fig. 1.30 : (A) Larval *Nybelinia surmenicola* (B) Isolated larva.

The whiting becomes infected with those larvae at a young age (before its third year of life), invasion incidence being much lower in older fish, which is probably related to a switch in their diet at that age.

The adult cestodes were recorded in the North Atlantic, Mediterranean Sea, and in the Pacific off American coasts, mainly in *Squalus acanthias* and *S. sucklii*, less frequently in other sharks.

The *Nybelinia surmenicola* larvae form small spherical or oval white cysts 2-3 mm in diameter. They are common in viscera and muscles of the Alaska pollock (*Theragra chalcogramma*), a Pacific gadid. They may also be encountered in other fish such as *Ophiodon elongatus, Merluccius productus, Pleurogrammus azonus, Sebastes flavidus. Cleisthenes herzensteini, Myozocephalus brandti*, and *Hippoglossoides elassodon*. Moreover, the larvae were found in squids *Ommatostrephes sloani pacificus*.

The *Nybelinia* plerocercoids show a characteristic fold (a kind of mantle), the so-called velum which surrounds the posterior end of the body. The scolex has 4 bothridia and 4 proboscids.

The *N. surmenicola* larvae are most commonly recorded in *Theragra chalcogramma,* the invasion incidence reaching occasionally 100% at the intensity of up to 205 larvae per host. The larvae occur in different organs: stomach wall, intestinal serous membrane, pyloric caeca, gonads, and muscles. They are particularly abundant, occasionally even as grape-like structures, in the abdominal cavity near the anus. At this site they are also most common in muscles.

The *N. surmenicola* plerocercoids are common in the North Pacific, in the Okhotsk and Bering Seas as well as in the Sea of Japan, and on the shelf off Asia and America.

Although the cestodes are harmless for man, any strong invasion is a reason to reject the Alaska pollock affected due to their repulsive appearance.

In the Soviet Union, where considerable amounts of the Alaska pollock are landed, a special processing technique is employed to get rid of the parasites and use the fish for consumption. The ventral part containing most parasites is cut off, the dorsal part—much less infected—being retained.

Plerocercoids of *Gymnorhynchus gigas* are large larvae, each having a very long (up to 50 cm) larval tail. The anterior vesicular section, 1.5 to 2 cm long, contains a retractable scolex which, when everted, reaches up to 5 cm in length (together with proboscids), the proboscids being upto 18 mm long.

The *G. gigas* larvae are found in the muscles of *Brama raii, Mola mola, Pagellus centrodntus,* and also in other Mediterranean and Atlantic fish species. They are particularly common in *B. raii.* In 1974, a heavy infection was recorded in the Canary Islands fishing grounds off NW Africa. More than 90% of the *B. raii* caught there were infected, the invasion intensity reaching 17 parasites per host. The

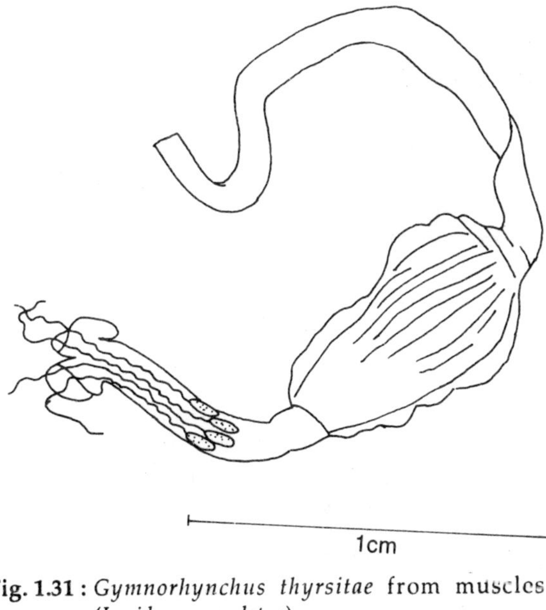

Fig. 1.31 : *Gymnorhynchus thyrsitae* from muscles of hairtail (*Lepidopus caudatus*).

plerocercoids gathered near the backbone, their scolex-containing vesicles being directed posteriorly and grouped in the ventral part of the fish body. The larval tails piereced posteriorly and grouped in the venral part of the fish body. The larval tails pierced all the fish muscles several times. Such arrangement of the tails suggested larval migrations within the fish body immediately after entering it. The affected muscles were loose in texture. Sanitary requirements did not allow them to be marketed.

The definitive host of the species is not known.

Larval forms of *Gymnorhynchus thyrsitae* (Robinson, 1959) are very common in muscles of snoek (*Thyrsites atun*) and hairtail (*Lepidopus caudatus*) off New zealand and on the South African shelf.

L. caudatus caught in 1974 off South Africa was infected in 95% 1-18 parasites being found per host, while up to 90% of *The. atun* contained up to 18 cestodes per fish. Off New Zealand, the invasion intensity in *The. atun* muscles reached 286 larvae per host.

The larvae have vesicular blastocysts measuring 5×3 mm and ribbon-like larval tails more than 10 cm in length. The blastocyst contains the scolex, about 5 mm long, provided with 4 large flat bothridia and 4 very long (more than 3 mm) proboscids protracting from the apical part of the bothridia. Hooks covering each proboscid are diverse in size and shape. The proboscid is naked at its base, the armed part beginning with an asymmetrical ring of large hooks.

Infected parts of the fish muscles are loose and moistened. The *G. thyrsitae* larvae occur mainly in the abdominal muscles, but can also be found in the dorsal ones in the entire fillet. Although not infective for man, the larvae seriously affect fish quality, a heavy invasion disqualifying the fish as a marketable product.

Molicola horridus Goodsir, 1841 is a close relative of *G. gigas;* some authors even regard *Molicola* as synonymous with *Gymnorhynchus*.

The anterior part of the *Molicola* plerocercoid is vesicular, too, the vesicle containing the scolex; the larval tail is ribbon-like and very long. The larvae are very common and abundant in the liver of *Mola mola;* very often the entire liver is tightly packed with packed with them.

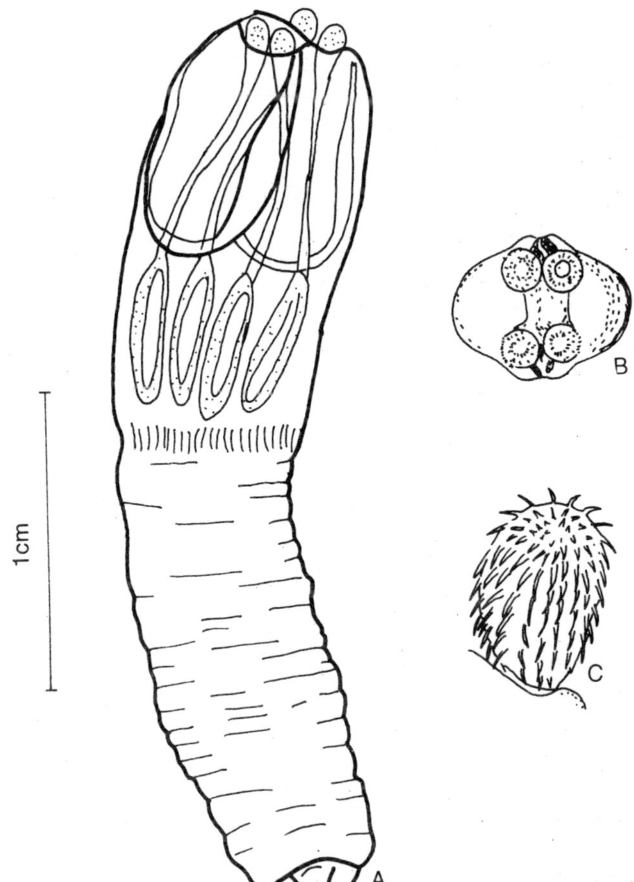

Fig. 1.32 : *Hepatoxylon trichiuri* from abdominal cavity of *Lepidopus caudatus.* (A) Whole specimen. (B) Anterior region of scolex with 4 proboscids in frontal view (C) Proboscid in lateral view.

Adults occur in the intestine of *Isurus glaucus.*

Large stout larvae of *Hepatoxylon trichiuri* (synonym: *H. squali*) occur commonly in the Atlantic, Pacific, and Indian Ocean fish. The tapeworms are particularly frequent in the West African shelf and off New Zealand. They occur among the abdominal cavity organs in snoek (*Thyrsites atun*), hairtail (*Lepidopus caudatus*), hakes (*Merluccius* spp.), *Genypterus blacodes, Macruronus magellanicus,* and other teleosts. They were also found in such sharks as *Isurus oxyrhynchus, Prionace glauca,* and *Squalus acanthias.*

The larvae reach 84 mm in length and 6-9 mm in width, the 20-30 mm long larvae being found most commonly. The scolex, 7-9 mm in length, is indistinctly separated from the strobila. Two bothridia are embedded in the scolex rather than sticking out above it and externally form two grooves (dorsal and ventral) with slightly thickened edges. Four short, almost spherical proboscids are covered with strong, 0.18-0.26 mm long hooks.

The larvae, mostly non-encysted, settle in the fish abdominal cavity, their proboscids adhering to the intestinal peritoneum, or are surrounded by connective tissue cysts. They can also be found in the liver, gonads, and muscles.

Presumably, *Isurus glaucus* is the principal definitive host of the tapeworm, as adults were also found in other sharks (e.g. *Carcarias glaucus, Pristiurus catulus, Somniosus microcephalus, Lamna cornubica,* and *Squalus acanthias.*

The larval *Tetrarhynchidea* are particularly common in commercially caught fish species off the African shelf.

Order: Proteocephalidea

Synonym: Ichthyotaeniidea

Tapeworms in this order have a scolex with 4 circular lateral suckers; in some species the fifth sucker occurs on the top of the scolex. These cestodes are intestinal parasites of fish, reptiles, and amphibians.

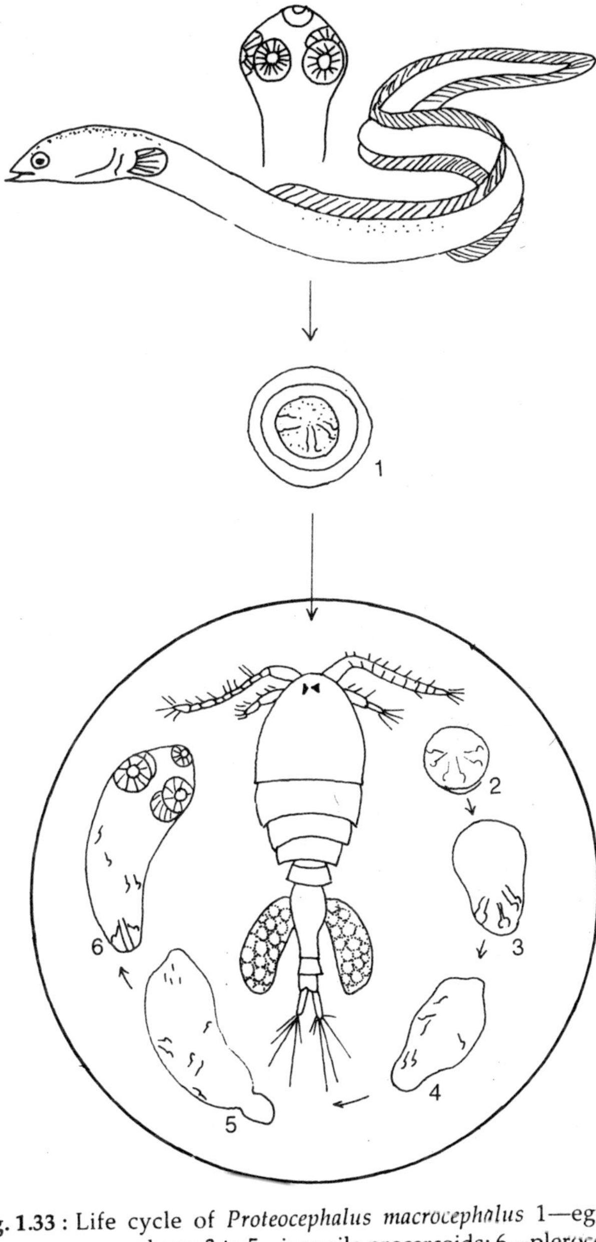

Fig. 1.33 : Life cycle of *Proteocephalus macrocephalus* 1—egg; 2—oncosphere; 3 to 5—juvenile procercoids; 6—plerocercoid.

Similarly to the *Diphylloidea,* the life cycle involves a single intermediate host, usually a copepod crustacean. The entire larval development, from the oncosphere to the fish-invading plerocercoid proceeds in the intermediate host.

Most of the *Proteocephalidea* parasitize freshwater fishes. They are highly host-specifc.

Both European and North American eels often contain *Proteocephalus mascrophthalmus,* a rather small (up to 20 cm in length), very delicate and easily disintegrating tapeworm.

Its intermediate hosts, as found experimentally by Doby and Jarecka (1966), are freshwater copepods *Cyclops strenuus, Cyclops* sp., *C. abyssorum, Acanthocyclops vernalis,* and *Diacyclops viridis.* The copepods infect eel during its fresh-water phase of life.

A few *prosobothrium* species only were found in sharks of the genera *Squalus, Zygaena, Carcharias,* and *Prionace* in the Mediterranean Sea and in the Sea of Japan.

The *Proteocephalidae* are of no economic importane.

CLASSIFICATION

Phylum platyhelminthes is divided into three classes. The classification followed here is from *Hyman* (1951).

I Class-Turbellaria
II Class-Trematoda
III Class-Cestoda

I CLASS - TURBELLARIA

Turbellaria is derived from Latin word turbella means a *stirring.* The animal produce turbulance in water by body cilia. They possess following characters:

1. The members of this class are broad, flat and leaf-like in shape.
2. They are lead a free living life in fresh or salt water, in moist soil, some members may be commensal or parasite.

3. Body unsegmented, dorso-ventrally flatenned and covered with cilia.
4. The anterior end is differentiated into head.
5. Epidermis cellular or syncytial contains rod-like structures called rhabdites. The rhabddites are curved structure capable of discharging a chemical secretion.
6. Epiderrmis and mesenchyma are provided with unicellular glands of two types (a) cyanophilous which secrete adhesive secretion and (b) eosinophilous.
7. Turbellarians are abundently provided with adhesive organs which are of two types (a) glandulo-epidermal adhesive organs and (b) glandulo-muscular adhesive organs.
8. Alimentary canal incomplete i.e. with single opening, the mouth Mouth is mid-ventral, sometimes near the centre of body. Intestine is blind or absent.
9. Excretory system with protonephridia.
10. Sense organs consist of tangoreceptors and chaemoreceptors.
11. Mostly bisexual or hermaphrodite or unisexnal with very few exceptions.
12. Reproduction by asexual, sexual and by regeneration.
13. Cleavage spiral. Development direct without larval stage.

Classification of Turbellaria

The members of this class are found almost all places. It includes more than 1500 species class Turbellaria is divided into 5 orders.

1. Acoela
2. Rhabdocoela
3. Allocoela
4. Tricladida
5. Polycladida

Order 1 - Acoela (Gr. a = without Koilos = cavity)

1. They are minute, delicate turbellarians with a ventral mouth.
2. Mouth and pharynx are present. Pharynx simple when present but usually absent and so the intestine is also absent.
3. Parenchyma syncytial not differentiated into mesoderm and endoderm.
4. Protonephridia absent.
5. Food is taken in solid form and digested by endodermal cells.
6. Gonoducts, gonads and yolk glands are absent.
7. Exclusively marine forms living under stores algae etc. at the sea bottom. Some live in the intestine of holothurians. Examples: Convoluta, Hoplodiscus

Order 2 - Rhabdocoela (Gr. rhabdos = rod; Koils = cavity)

1. Small Tubellarian usually less than 3mm in length.
2. Mostly free-living, marine of fresh water.

Some of them inhabit in the respiratory cavities of various fresh water crustaceans, *viz. Temnocephala.*

3. Body is elongated, cylindrical or compressed with simple pharynx and simple unbranched sack-like intestine.
4. Yolk-glands may or may not be present.
5. Yolk with two to many testes, whereas the female with 1 or 2 ovaries.

6. Nervous system consists of two longitudinal nerve cords. Eyes are present in most of them.

Order Rhabdocoela is divided into four sub-orders:

Sub-order (i) Natandropora

1. Exclusively fresh water forms.
2. Pharynx simple.
3. Single median protonephridium.
4. Single compact testis mass, penis unarmed.
5. Yolk-gland is absent.
6. Sexual fission with chains of zooids.

Examples: *Catenula, Rhycoscolex, Stenostomum*

Sub-order (ii) Opisthandropora

1. Fresh water or marine forms.
2. Pharynx is simple.
3. Excretory system consists of paired protonephridia.
4. Compact testes, penis armed with a stylet.
5. Yolk glands absent.
6. Asexual reproduction with chains of zooids.

Examples: *Microstomum, Macrostomum*

Sub-order (iii) Lecithophora

1. Fresh water, marine or terrestrial, commensel or parasitic.
2. Pharynx bulbose.
3. Protonephridia paired.
4. Separate ovaries and yolk glands.
5. Exclusively sexual reproduction.

Examples: *Graffilla, Anaplodium, Merostoma, Gyratrix*

Sub-order (iv) Temnocephalida

1. Freshwater ectocommensal forms.
2. Anterior end having 2-12 tentacles.
3. Posterior end with 1 or 2 adhesive discs.
4. Pharynx dolii form.
5. Gonopore simple.

Examples: *Temnocephala, Monodiscus*

Order 3-Allocoela (Gr. Alloios = different; Koilos = cavity)

1. Mostly marine, few freshwater and brakish water forms. Some are ectoparisite or ectocommensal in habit.
2. The pharynx is bulb-like or plicate.
3. Intestine is straight or branched.
4. Excretory system consists of paired protonephridia with two or three branches and nephridiopores.
5. Nervous system with 3 or 4 pairs of longitudinal nerve cords provided with transverse commissures.
6. Reproductive system consists of numerous testes and a pair of ovaries. Penis papilla mostly present.

Order Allocoela is divided into four Sub-orders.

Sub-order (i) Archiophora

1. Marine animals.
2. Pharynx plicate.
3. Female reproductive system primitive.

Examples: *Proporoplana*

Sub-order (ii) Lecithoepitheliata

1. Marine, freshwater or terrestrial forms.
2. pharynx simple or bulbose.

3. Penis with a cuticular stylet.
4. Female duct simple or none. Typical yolk glands absent, nutritive cells may surround the ova.

   Examples: *Hofstenia, Prorhynchus.*

Sub-order (iii) Cumulata

1. Fresh water or marine forms.
2. Pharynx plicate or bulbose.
3. Intestine mostly without diverticulae.
4. Penis unarmed.
5. Separate ovaries and yolk glands or *germoritellaria.* Examples: *Hydrolimax, Plagiostomum.*

Sub-order (iv) Seriata

1. Mostly marine and a few are fresh water forms.
2. Pharynx plicate.
3. Intestine with lateral diverticula.
4. Statocysts mainly present.
5. Ovaries and yolk glands are separate.

   Examples: *Monocelis, otoplana*

Order 4 Tricladila (Gr. treis = three; Klados = branch)

1. Marine, fresh water or terrestrial forms.
2. Large turbellarians, measuring 2-60 cm in length.
3. pharynx plicate usually backwardly directed.
4. Intestine three-branched sacs.
5. Eyespots usually present.
6. Two ovaries, 2 or numerous testes.
7. Female reproductive system provided with one or two copulatory bursa.

8. Penis with penis papilla, common genital aperture.

Order Tricladida is divided into three suborders.

Sub-order (i) Maricola

1. Exclusively marine forms.
2. A pair of eyes and auricular groores.
3. Typical penis papilla sometimes armed with stylet.
4. Rounded copulatory bursa is generally present.
5. Only sexual reproduction.

Examples: *Bdelloura, Ectoplana*

Sub-order (ii) paludicola

1. Mostly freshwater rarely brackish water forms.
2. Two to many eyes or eyeless.
3. Bursa with bursal canal, anterior to penis.
4. Reproduction mostly asexual.

Examples: *Dugesia, Kenkia, Dendrocoelum*

Sub-order (ii) Terricola

1. Terrestrial tropical forms.
2. Two or many eyes.
3. Bursa, if present, lies behind the penis.
4. Reprodution may be asexual.

Examples: *Geoplana, Dolicoplana*

Order 5-Polycladida (Gr. ply = many; klados = branch)

1. Size modrate, from 2-20 mm. Dorsoventrally flat leaf-like body; they are exclusively marine.
2. Intestine with many branches.
3. Eyes and nerve cords are present.
4. There are numerous ovaries and testes.

5. Genital apertures separate.

   It includes two suborders:

Sub-order (i) Acotylea

1. Pharynx usually vertical curtain like.
2. No suckers are present behind the female pore.
3. Tentacles nuchal type.
4. Eyes concentrated at anterior margin, not in pairs.

   Examples: *Shylochus, Notoplana*

Sub-order (ii) Cotylea

1. Sucker present behind the female pore.
2. Pharynxtubular.
3. A pair of marginal tentacles bearing eyes or with a cluster of eyes in place of tentacles.

   Examples: *Thysanozoon, Yungia*

CLASS II TREMATODA

1. They are ecto or endoparasites and commonly called 'flukes.'
2. Body is usually unsegmented, dorso-ventrally flatenned and leaflike.
3. Size varies from 0.5 mm to 1 or several cm. long.
4. Body is covered over by thick cuticle and no cilia.
5. Suckers and sometimes hooks are present. Oral sucker present at anterior end for feeding ventral sucker for attachment.
6. Epidermis absent. Rhabdites are absent.
7. Digestive system consists of mouth, muscular pharynx and intestine usually with two main branches. Anus is absent.

8. In some cases there is a system of parenchymal vessels showing the presence of primitive circulatory system.
9. Excretory organs consists of protonephridia.
10. There are three pairs of longitudinal nerve cords. Sense organs are poorly developed.
11. Bisexual or hermaphrodite except blood fluke (*Schistosoma*).
12. Single ovary and many testes. Cross fertilization takes place. Development direct or indirect.
13. Life-cycle simple or complicated with more than one hosts.

There are about 6000 species in class Trematoda. This class is divided into three orders:

Order 1-Monogenea
Order 2-Aspidobothrea
Order 3-Digenea

Order 1–Monogenea

1. They are ectoparasite mostly in cold-blooded animals (fishes).
2. The life-cycle is simple and passed in one host only, monogenetic.
3. Oral sucker absent, when present is poorly developed.
4. A pair of adhesive structures are present at anterior end.
5. A large disc-shaped structure, the opisthaptor, is found on the posterior end. It contains hooks and is lined with chitin.
6. Paired excrecory pores are present in the anterior part of the body.

7. Male and female genital pores are separate.
8. One or two vaginae are present. The uterus is small and contains few eggs.
9. Development is indirect through a larval stage called onchomiracidium.

Order Monogenea is divided in two suborders:

Sub-order (i) Monopisthocotylea

1. Opisthaptors are present without suckers but with hooks.
2. Genito-intestinal tract is absent.
3. Posterior adhesive organ is single or divided by septa.
4. Eye-spots usually present.

Examples: *Gyrodactylus, Monocotyle, Heterocotyle, Amphibdella*

Sub-order (ii) Polyopisthocotyle

1. Opisthaptor contains many suckers and also hooks.
2. Oral suckers two, opening into oral cavity.
3. Genito-intestinal tract present.
4. Eye-spots are absent.

Examples: *Polystoma, Diplozoon, Microcotyle.*

Order 2–Aspidobothrea

1. Endoparasite in the digestive tract of fishes and reptiles.
2. Oral sucker is absent.
3. The ventral sucker is divided by septa into many small parts. The hooks are absent in it.
4. Single excretory pore.
5. Only one testis is present.

6. Development direct, without alternation of hosts.

Order Aspidobothrea is divided into two families:

Family (a) Aspidogastridae

1. Sucker enormous circular, oval or elongated, divided by septa into one, three or four longitudinal rows of depressions termed alveloli.
2. Protonepliridia have separate bladders and separate or a common nephridiopore situated near the posterior end.
3. Testis single.
4. Laurer's canal may be present.
5. Endoparasite in the mantle, pericardial and renal cavities of clams and snail and in digestive system of fish and turtle.

   Examples: *Aspidogaster, Cotylaspis, Cotylogaster*

Family (b) Stichocotylidae

1. It consists of only one genus which lives in bile passage or spiral valve of fishes.
2. The life cycle is complex one, involving at least two hosts.

   Example: *Stichocotyle.*

Order 3–Digenea

1. They are endoparasite in invertebrate and vertebrates.
2. Two to four hosts in the life-cycle.
3. Two suckers, without hooks, mouth within angerior sucker.
4. There is only one excretory pore present in the posterior region of the body.

5. There is a common gonopore for male and female reproductive systems. Vagina is absent.
6. Uterus is long branched with many eggs.
7. Their larval forms reproduce asexually before metamorphosis.

Order Digenea is divided into many families; some important ones are as:

Family (a) Bucephalidae

1. They are commonly called gastrostomes. The mouth is located in the middle of the body.
2. Anterior end bears a simple or lobate oral sucker or has a generally lobed hood-like extension.
3. Intestine sacciform.
4. Acetabulum absent.
5. Genital pore posterior.
6. The cercaria larvae known as oxhead, have a pair of long tails.
7. The primary host is fish and secondary host is clams.

   Examples: *Bucephalus, Bucephalopsis.*

Family (b) Paramphistomidae

1. Body thick and fleshy with a powerful posterior sucker.
2. Cirrus pouch is present.
3. Oral sucker is absent.
4. Gonopore, situated in the anterior body half and provided with in some genera with a genital sucker.
5. Nephridiopore dorsal infront of acetabulum.
6. Uterus long having ascending limb. Vitelline glands follicular.
7. These are the endoparasites of gut and bile ducts.

   Examples: *Paramphistomum, Gastrothylax*

Family (c) Opisthorchidae

1. These are distomes of small to large size.
2. They live in the bile passage of fish-eating reptiles, birds and mamnials.
3. Body flat and semitransparent.
4. Excretory bladder 'V' shaped.
5. Testes posterior Genital pore infront of acetabulum.

   Examples: *Opisthorchis, Amphimerus, Metorchis*

Family (d) Fasciolidae

1. Commonly known as liver flukes.
2. Body leaf-like and acetabulum near the oral sucker.
3. Gonopore just infront of the acetabulum.
4. Testes, ovary, vitelline glands all are branched.

   Examples: *Fasciola hepatica, Fasciplopsis buski*

Family (e) Echinostomatidae

1. Body slender, elongated with spiny or scaly cuticle.
2. Acetabulum large behind the oral sucker.
3. Collar of spines is present in cercarial and metacercarial stages.
4. It is commonly found in the intestine of birds.

   Examples: *Echinostoma, Echinoparyphium, Parochis, Himasthla*

Family (f) Schistosomatidae

1. These are blood flukes that inhabit the hepattic portal system and pelivc veins of birds and mammals.
2. Sexes are separate.
3. Female lodges in a gynaecophoric canal of male.
4. Testes numerous and ovary tubular.

   Example: *Schistosoma*

Besides above families the others are:

Family - Zoognidae.	Examples: *Zoogonus, Nassa.*
Family - Allocreadidae.	Examples: *Bunodera, Plagioporus.*
Family - Azygiidae.	Examples: *Azygia, Otodistomum.*
Family - Hemiuridae.	Examples: *Halipegus, Derogenes.*
Family - Didymozoonide.	Examples: *Didymocystis*
Family - Gorgoderidae.	Examples: *Gorgoderina, Gorgodera*
Family - Plagiorchidae.	Examples: *Plagiorchis, Styphlodora.*
Family - Dicrocoelidae.	Examples: *Dircrocoelium, Eurytrema.*
Family - Heterophyidae.	Examples: *Stictodera, Heterophyes.*
Family - Microphallidae.	Examples: *Spelotrema, Maritrema.*
Family - Troglotrematidae.	Examples: *Paragonimus, Troglotrema.*
Family - Cyclocoelidae.	Examples: *Cyclocoelus, Typhlocoelum.*
Family - Clinostomatidae.	Example: *Clinostomum.*
Family - Notocotylidae.	Examples: *Notocotylus, Catatropis.*
Family - Strigeidae.	Examples: *Ophiosoma, Cyanthocotyle*
Family - Brachylaimidae.	Examples: *Brachylaima.*
Family - Sanguinicolidae.	Example: *Sanguinicola.*
Family - Spirorchidae.	Examples: *Spirorchis, Hapalorhynchus*
Family - Heronimidae.	Example: *Heronimus.*

CLASS III CESTODA

1. They are found as endoparasites in the intestine of vertebrates and are called tapeworms.
2. Body is flat, elongated and ribbon-like, measures from 1 mm to 10 metres in length.
3. Body is without epidermis, rhabdites, cilia but a cuticle is present.
4. Body is usually divided into a scolex or head, neck and few to many proglottids.
5. Scolex is provided with hooks or suckers or both.
6. Scolex is followed by neck which forms new

proglottids by transverse budding called strobilization.

7. Mouth and digestive tract are absent.
8. Excretory system consists of protonephridia with flame - cells.
9. Nervous system consists of ganglionated nerve ring in scolex and two pairs of lateral nerve cords in proglottids.
10. Each proglottid contains one or two sets of male and female reproductive organs. Thus they are hermaphrodite.
11. Several testes in the form of follicles and single-lobed ovary.
12. Uterus presents a great variety of structures.
13. Fertilization internal and self.
14. Life-cycle is complicated with hooked embryo, and is passed in more than one host.

Class Cestoda has about 3400 species. The class is divided into two sub-classes:

Sub-class (A) Cestodaria or Monozoa
Sub-class (B) Eucestoda or Merozoa

Sub-class (A) Cestodaria

1. Endoparasites in the intestine and coelomic cavities of fishes and reptiles.
2. Body is formed by one segment and without a scolex.
3. No alimentary canal.
4. Single set of reproductive organs.
5. Larva with ten hooks, known as Pycophore.

Sub-class Cestodaria is divided into two orders.

Order 1 - Amphilinida

1. They are endoparasites in the body cavity of ganoid fishes.
2. Frontal gland and protrusible pharynx are present in the anterior part of the body.
3. Suckers are absent.
4. A proboscis is present.
5. Male and female pores present in the posterior part of body.
6. Uterus is coiled and opens near the anterior end.

   Example: *Amphilina*

Order 2 -Gyrocotylidea

1. They are endoparasites of chimaeroid fishes.
2. Anterior end has cup-like sucker.
3. Eversible proboscis at anterior end.
4. Male and female pores are situated at the anterior half of the body.
5. Uterus not coiled.

   Example: *Gyrocotylus*.

Sub-class (B) Eucestoda

1. Elongated, flat body is divided into many proglottids.
2. Body is divisible into scolex, neck and strobila.
3. Scolex with suckers, spines and hooks for attachment to host.
4. Each proglottid with male and female reproductive organs.
5. Larva with six hooks.

   Sub-class Eucestoda is divided into following orders.

Order 1 - Tetraphyllidea

1. Endoparasite, exclusively in the intestine of elasmobranch fishes.
2. Scolex provided with four bothria (sessile suckers) often with hooks.
3. Testes lie infront of ovaries.
4. Cirrus armed with hairs, spines or hooks.
5. Common genital atrium marginal.

   Examples: *Phyllobothrium, Acanthobothrium.*

Order 2 - Lecanicephaloidea

1. Intestinal parasites of elasmobranch fishes.
2. Scolex divided by a transverse groove in two parts upper with disc and lower with four suckers.
3. Vitellaria in two lateral bands.

   Examples: *Lecanicephalum, Polypocephalus*

Order 3 - Proteocephalidea

1. These are endoparasites of freshwater fishes, amphibians and reptiles.
2. Scolex mobile with four lateral sucker, an apical sucker in some.
3. Ovary bilobed, Uterus with many branches and pores.
4. Vitellaria scattered.
5. Common genital atrium marginal.

   Examples: *Proterocephalus, Ophistaenia*

Order 4 - Diphyllidea

1. Parasitic in the intestine of elasmobranch fishes.
2. Proglottids only 20 or less.
3. Scolex with two bothria and a spiny head stalk.
4. Vitellaria scattered.

   Example: *Echinobothrium*

Order 5 - Trypanorhyncha

1. Parasitic in the spiral valve of digestive tract of elasmobranch fishes.
2. Strobila is segmented and is of modrate size.
3. Scolex with two or four sessile bothria and four protrusible spiny proboscis.
4. Testes extend beyond ovary posteriorly.

   Examples: *Haplobothrium, Tetrarhynchus.*

Order 6 - Pseudophyllidea

1. Endoparasites in the intestine of fishes, birds and mammals.
2. The larva lives in crustaceans.
3. Generally six shallow groove-like suckers and hookless bothrior are found on the scolex.
4. Proglottids may be absent or present.
5. Neck is not developed.
6. Many yolk glands are present.

   Examples: *Amphicotyle, Dibothriocephalus*

Order 7 - Nippotaeniidea

1. Endoparasite in the intestine of Japanese fresh water fishes.
2. Scolex with single terminal sucker.
3. Proglottids few with, each with one set of genitalia.
4. Vitellaria compect.

   Examples: *Nippotaenia, Amurotaenia*

Order 8 - Taenioidea

1. Endoparasites in the intestine of birds and mammals.
2. Elongated body, with proglottids.
3. Scolex bears four suckers often with an apical rostellum armed with hooks.

4. Excretory system with four longitudinal vessels and flame cells.
5. Complete sets of reproductive organs in each proglottids. Single compact yolk gland.

   Examples: *Taenia, Echinococcus.*

Order 9 - Aporidea

1. It is endoparasite of swan (birds).
2. Scolex with four suckers, rostellum armed.
3. No segmentation, ootype, yolk glands and sex-ducts or pores.

   Examples: *Nematoparataenia, Gastrotaenia*

POLYCHOERUS

Phylum	-	**Platyhelminthes**
Class	-	**Turbellaria**
Order	-	**Acoela**
Genus	-	***Polychoerus***

1. It is found in littoral zone of ocean shores in excessive winter.

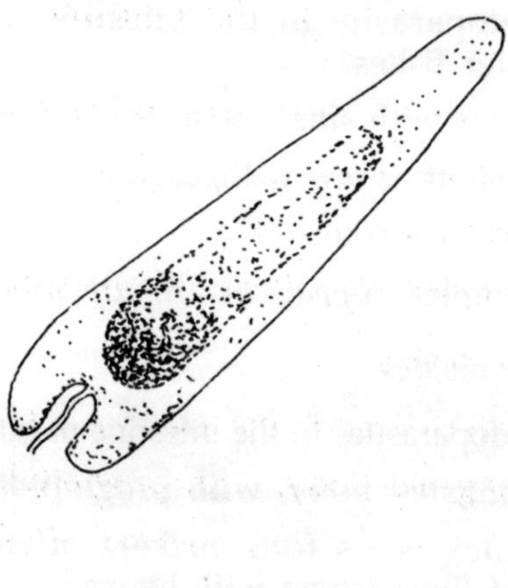

Fig. 1.34 : *Polychoerus.*

2. It is bringht orange or orange red in colour.
3. Body flat, oval, changeable shape. The anterior end is round while posterior end is not ched bearing tail filament.
4. Entire body is covered with cilia.
5. Statocyst is present in median like.
6. Digestive tract is absent.
7. Definite gonads are wanting. Yolk glands are absent. Definite oviduct absent.
8. The seminal bursa opens ventrally by female gonopore.
9. Penis is present.

CONVOLUTA

Phylum	-	**Platyhelminthes**
Class	-	**Turrbellaria**
Order	-	**Acoela**
Genus	-	***Convoluta***

1. Convoluta is found in sea water.
2. Living under stones, among algae and most commonly on the bottom mud.
3. Small worm with side of the body are held curved ventrally without any projections.
4. But the body has pigment cells.
5. The anterior of cephalic end has two eyes, a statocyst and cluster of frontal glands.
6. The mouth is located near the anterior half of the ventral surface.
7. A digestive cavity is lacking.
8. The excretory system is completely absent.
9. Hermaphrodite but protandry is common.

10. The eggs are enclosed in a thin gelatinous capsule and are struck to some objects.
11. The eggs are developed directly into young worms that escape the jelly.

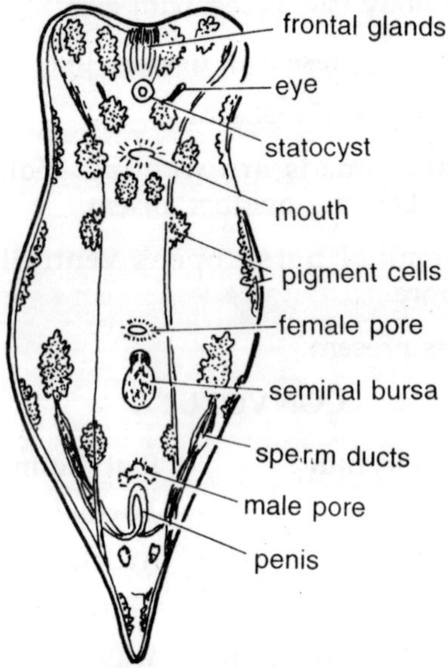

Fig. 1.35 : *Convoluta*

CENTULA

Phylum	-	**Platyhelminthes**
Class	-	**Turbellaria**
Order	-	**Rhabdocoela**
Suborder	-	**Notandropora**
Genus	-	***Centula***

1. It is a freshwater Turbellaria.
2. The pharynx is simple.
3. Single median protonephridium.
4. Nervous system consists of four pairs of longitudinal nerve cords and a statocyst.
5. Asexual reproduction by fission with chains of zooids.

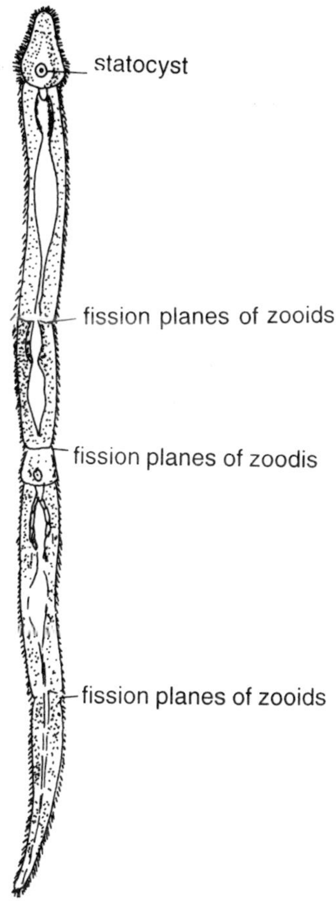

Fig. 1.36 : *Centula.*

6. No sexual reproduction.

MICROSTOMUM

Phylum	-	**Platyhelminthes**
Class	-	**Turbellaria**
Order	-	**Rhabdocoela**
Genus	-	***Microstomum***

1. It is a microscopic, freshwater form.

Fig. 1.37 : *Microstomum.*

2. It swims by means of cilia present on the body surface.
3. Its epidermis contains nematocysts derived from hydroid food.
4. Its is found that Microstomum attack Hydra only when it need of nematocysts.
5. The digestive cavity is simple, unbranched and straight.
6. It reproduce asexually by repeated transverse fission forming a linear chain of 4, 8 or 16 zooids. Each zooid separates to become an individual.

TEMNOCEPHALA

Phylum	-	**Platyhelminthes**
Class	-	**Turbellaria**
Order	-	**Rhabdocoela**
Genus	-	***Temnocephala***

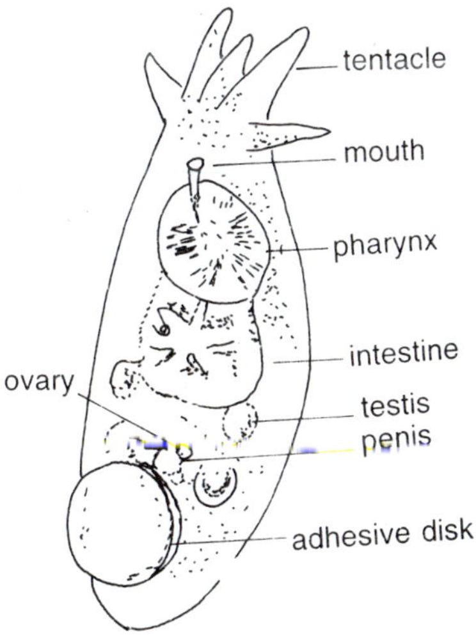

Fig. 1.38 : *Temnocephala.*

1. It is found as an ecto commensal on cray fish, prawns and other crustaceans.
2. It contains 12 finger-like projection or *tentaces* at anterior end and an adhesive disc at the posterior end.
3. The border of syncytial epidermis is apparently cuticle devoid of cilia.
4. Mouth anterio-ventral. Pharynx thick walled and simple lobulated intestine present.
5. It feeds upon diatoms, rotifers and other small animals.

6. A pair of nephridiopores is present on the dorsal surface of the anterior part of the body.
7. Cerebral ganglion bears two dorsal eyes.
8. There are four testes and single ovary. Their gonoducts leading into a common genital atrium.

PLAGIOSTOMUM

Phylum	-	**Platyhelminthes**
Class	-	**Turbellaria**
Order	-	**Allocoela**
Genus	-	**Plasgiostomum**

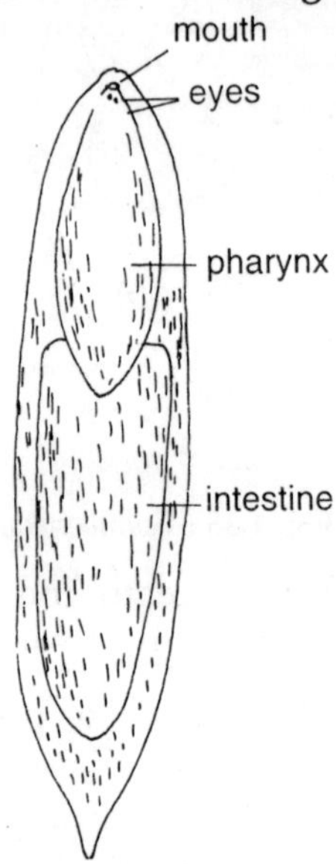

Fig. 1.39 : *Plagiostomum.*

1. It is present in both fresh water and marine.
2. Mouth is anterior.
3. Pharynx is bulbose. Intestine with short diverticula.
4. Gonopore posterior.

5. Sperm is characteristic in having witng-like lateral expansions.

BIPALIUM

Phylum	-	**Platyhelminthes**
Class	-	**Turbellaria**
Order	-	**Tricladida**
Genus	-	**Bipalium**

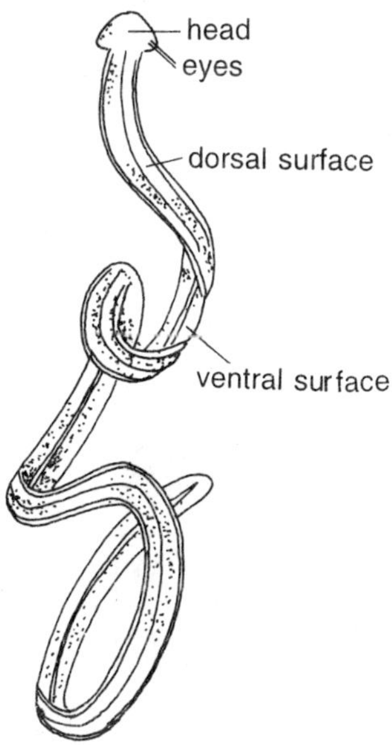

Fig. 1.40 : *Bipalium.*

1. Bipalium is a long terrestrial turbellarian living the humid soil on the floor of tropical forests.
2. It is cosmopolitan in distribution.
3. The animal measures 20-50 cm in length.
4. It consists of an expanded lunate head and cylindrical long body.
5. Numerous eyes are present on the margin of the head and sides of body.

6. There are stripes on the dorsal surface and a creeping sole on the ventral surface.
7. The stripes are purple, black, yellow, olive and grey in colour.
8. Reproduction generally asexual. It never becomes sexual in temperate climate and it propogates b fragmentation.
9. Bipalium adventium breeds sexually.

THYSANOZOON

Phylum	-	**Platyhelminthes**
Class	-	**Turbellaria**
Order	-	**Polycladida**
Genus	-	**Thysanozoon**

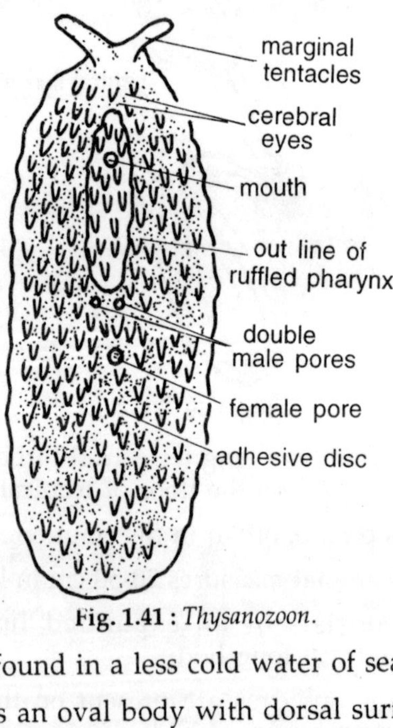

Fig. 1.41 : *Thysanozoon.*

1. It is found in a less cold water of sea.
2. It has an oval body with dorsal surface beset with papillae, each receiving a branch form intestine.

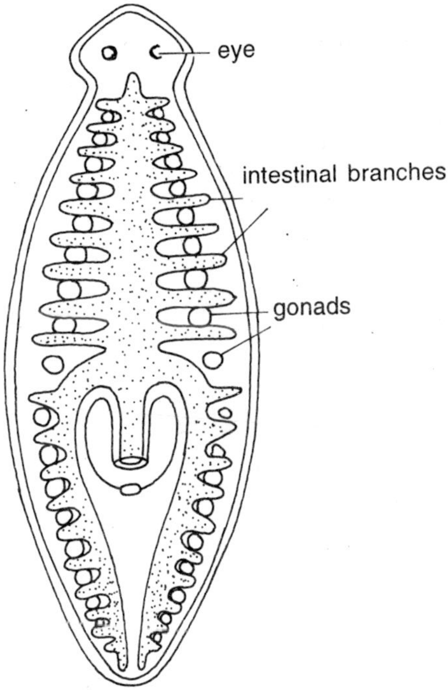

Fig. 1.42 : *Gunda segmentina.*

3. Anterior end bears a pair of marginal tentacles and numerous cerebral eyes.
4. Pharynx is tubular.
5. Glandulo-muscular adhesive organ is present on the dorsal surface behind the female gonopores.
6. A single female genital pore lies behind the male genital pores.
7. Seminal bursa is absent.

GUNDA

Phylum	-	**Platyhelminthes**
Class	-	**Turbellaria**
Order	-	**Tricladida**
Genus	-	***Gunda (Procerodes)***

1. It is a marine animal.
2. The auricle and a pair of eyes or ocelli are present in the anterior end.
3. Intestine has regularly arranged intestinal diverticulae. The gonads alternate with the diverticulae giving the appearance of segmentation.
4. Incapable of sexual reproduction.
5. *Lang* used the feature of intestine as a classical example in support the gonocoel theory of coelom formation by Haeckel.

APIDOGASTER

Phylum	-	**Platyhelminthes**
Class	-	**Trematoda**
Order	-	**Monogenea**
Genus	-	***Apidogaster***

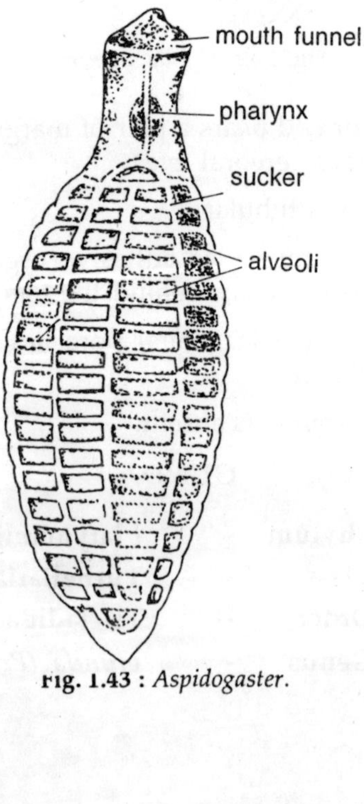

Fig. 1.43 : *Aspidogaster.*

1. It is an endoparasite in the pericardial and renal cavities of fresh water mussel and in the gut of fishes and turtles.
2. Body is elongated and dorsoventrally flatenned with anterior narrow end.
3. The narrow anterior end has a subterminal mouth which is devoid of oral sucker.
4. A large sucker appears which occupies great space in the posterior region of body. The sucker is sub-divided into four longitudinal rows of sucking cups or alveoli.
5. Gut is simple and straight.
6. Excretory system consists of protonephridia with excretory bladders.
7. Bisexual life-cycle simple without change of host.

POLYSTOMUM

Phylum - **Platyhelminthes**

Class - **Monogenea**

Order - **Monogenea**

Genusm - ***Polystomum***

Characters:

1. It is an endoparasite in the urinary bladdar of frogs, and turtles.
2. Body is leaf like and dorso-ventally flattened.
3. The posterior region has a disc-shaped opisthaptor with six cuplike suckers and 2 to 4 large chitinous hooks.
4. Mouth is surrounded by an oral sucker.
5. Intestine is bifurcated.
6. Hermaphrodite with single ovary and numerous tests.

Order **Monogenea**
Genus ***Polystomum***

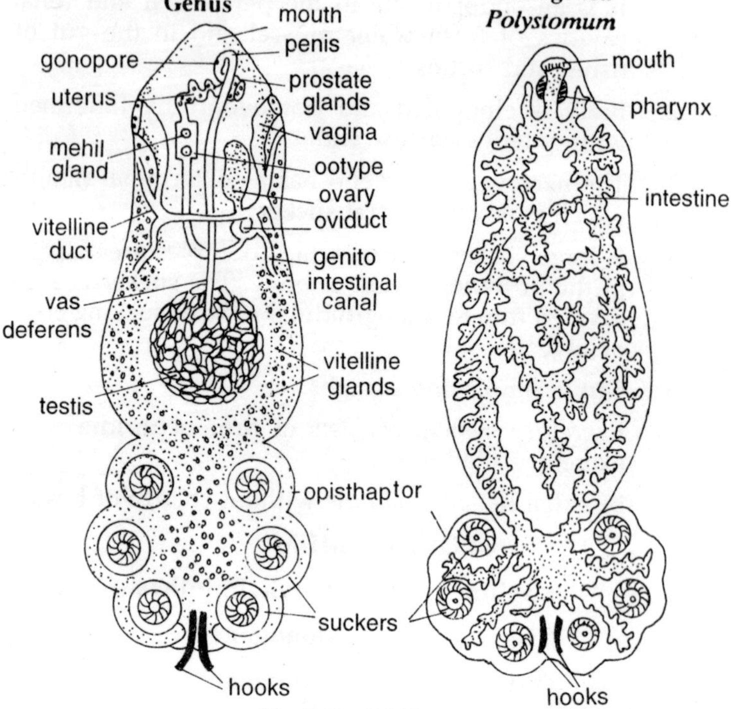

Fig. 1.44 : *Polystomum.*

7. Breeding season starts in spring. Eggs are produced in the urinary bladder of frong and discharged into the water along with the urine of the host.

8. Hatching takes place after 4 to 5 weeks and larva attaches to the gills of tadpole and later develop into adult in the alimentary canal of metamorphosed tadpole.

DIPLOZOON

Phylum	-	**Platyhelminthes**
Class	-	**Trematoda**
Order	-	**Monogenea**
Suborder	-	**Polyopisthocotyla**
Genus	-	***Diplozoon***

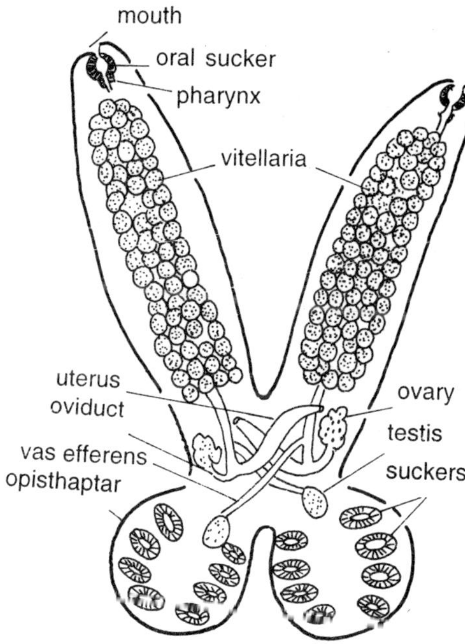

Fig. 1.45 : *Diplozoom.* Two individuals.

Characters:

1. Diplozoon is an ectoparasite, found attached to the gills of freshwater fishes, feeding on their blood.
2. Two individuals are always found attached permanently together in the form of 'X'.
3. Opisthaptor bears eight suckers arranged in two rows of four each.
4. Digestive system consists of mouth surrounded by oral sucker, pharynx and intestine. Intestine is not bifurcated.
5. Reproductive system consists of testis and ovary along with their usual ducts.
6. The sperm duct, uterus and vagina of one individual

cross with those or other and thus ensures cross fertilization. Uterus surrounded by mehlis' glands and contains one egg.

7. Egg differenciates into diporpa larva, two of which unite and metamorphosed into adult individual. Larva has two eyes digestive system & a pair of suckers.

GYRODACTYLUS

Phylum	**-**	**Platyhelminthes**
Class	**-**	**Trematoda**
Order	**-**	**Monogenea**
Suborder	**-**	**Monopisthocotyla**
Genus	**-**	***Gyrodactylus***

1. It is an ectoparasite on the skin and gills of freshwater fishes.
2. Body is minute and elongate.
3. Anterior end is provided with adhesive glands and organs. Oral sucker is absent.
4. Eyes are absent.
5. Opisthaptor is disc or wedge-shaped and bears one pair of anchors and eight pairs of hooklets.
6. Intestine is sac-like forked into two branches without diverticula.
7. Genito-intestinal canal is absent. Genital pore is median.
8. Development is intra-uterine *i.e.,* viviparous.
9. A remarkable feature is that a second embryo is produced within the first one, a third within the second and a fourth within the third. The first embryo, on completing its development passes out, still enclosing the other embryos, and attaches directly to a host fish.

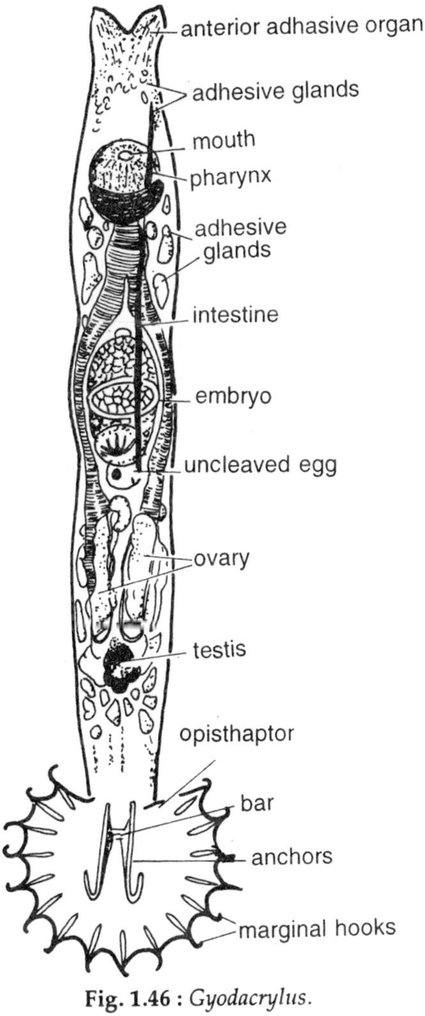

Fig. 1.46 : *Gyodacrylus.*

GIGANTOCOTYLE (PARAMPHISTOMUM)

Phylum	-	**Platyhelminthes**
Class	-	**Trematoda**
Order	-	**Digenea**
Family	-	**Paramphistomidae**
Genus	-	***Gigantocotyle***

1. It is an endoparasite in the rumen of cattles and bile-duct of sheep, goat etc.
2. Body is dorsoventrally flatenned and measuring 12-14 mm in length.
3. Mouth lics at the anterior end and without an oral sucker.
4. Acetabulum is very prominant and lies at the posterior end.
5. Digestive system consists of mouth, highly musculr pharynx and bifurcated, unbranched intestine.
6. Parenchyma has a sort of delicate tubes forming much branched lymphatic system.
7. Testes are lobed and occur near the middle of the body.
8. Ovary is small, lobed and lies posterior to testes.
9. Vitellaria are scattered on the sides of animal.
10. Genital pore is just behind the bifurcation of intestine.
11. Uterus is slightly folded and eggs are very large.

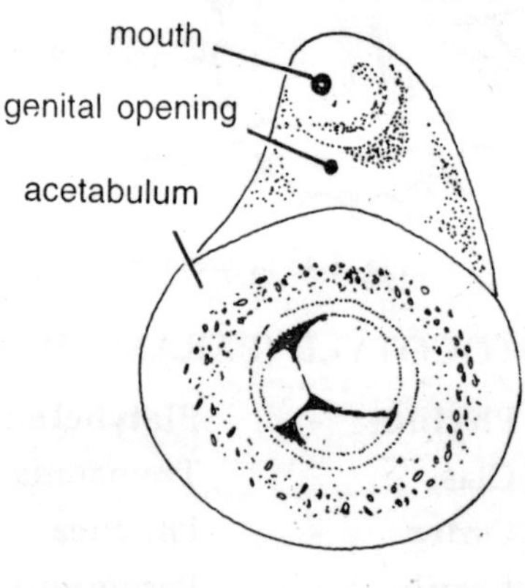

Fig. 1.47 : *Gigantocotyle.*

GASTRODISCOIDES HOMINIS

Phylum	-	**Platyhelminthes**
Class	-	**Trematoda**
Order	-	**Digenea**
Family	-	**Paramphistomidae**
Genus	-	***Gastrodiscoides***
Species	-	***hominis***

1. It is an endoparasite in the colon and caecum of man.
2. If is found in India and Indo-China.
3. It measures 5-7 mm in length.
4. Body is differentiated into anterior and posterior part.
5. Terminal mouth is present an anterior part.

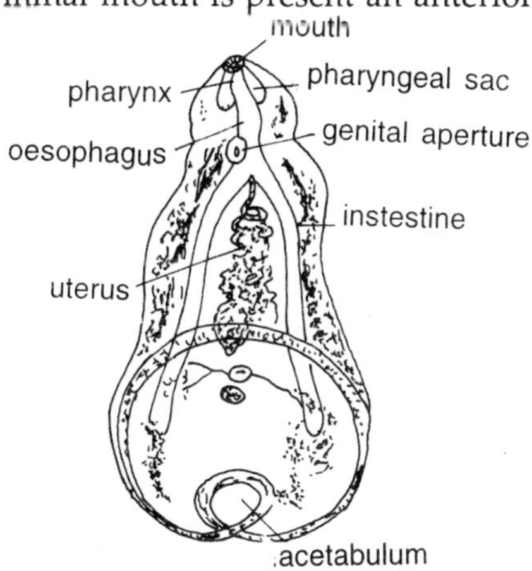

Fig. 1.48 : *Gastrodiscoides*

6. Oral sucker is absent.
7. Posterior part bears a prominant acetabulum on ventral side.
8. Gonopore is in the anterior half of the body on ventral side.

9. Excretory pore lies on the dorsal side infront of acetabulum.
10. The secondary host is snail.
11. It causes inflammation and diarrhoea.

GASTROTHYLAX

Phylum	-	**Platyhelminthes**
Class	-	**Trematoda**
Order	-	**Digenea**
Family	-	**Paramphistomidae**
Genus	-	***Gastrothylax***

1. It is an endoparasite in the rumen of cattles.
2. It is elongated animal possessing a deep ventral pouch which extends upto the ventral sucker and it is supposed to be elongated genital chamber.
3. Digestive system includes mouth, pharynx, oesophegus and simple non-diverticulated intestine.
4. Two rounded testes near the base of the pouch.
5. Female reproductive organs consist of single ovary, uterus, Ootype, vagina and seminal recepticle.

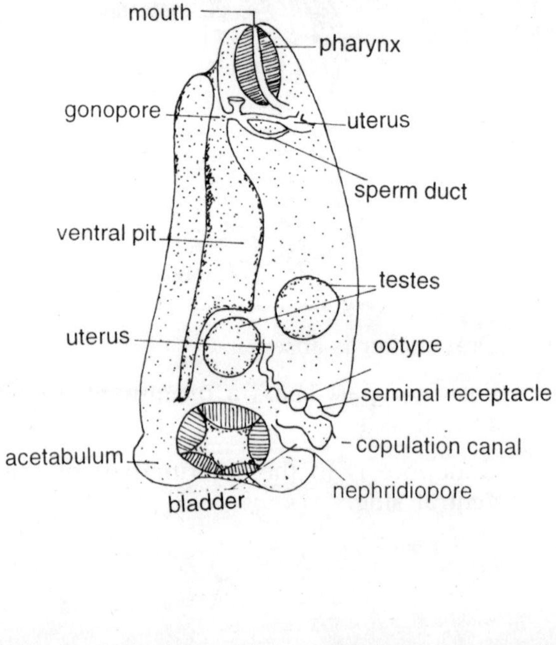

Fig. 1.49 : *Gastrothylax*

6. Single excretory pore is present.
7. Posterior end is divided with large acetabulum or adhesive sucker.

FASCIOLOPSIS BUSKI

Phylum	-	**Platyhelminthes**
Class	-	**Trematoda**
Order	-	**Digenea**
Family	-	**Fasciolidae**
Genus	-	***Fasciolopsis***
Species	-	***buski***

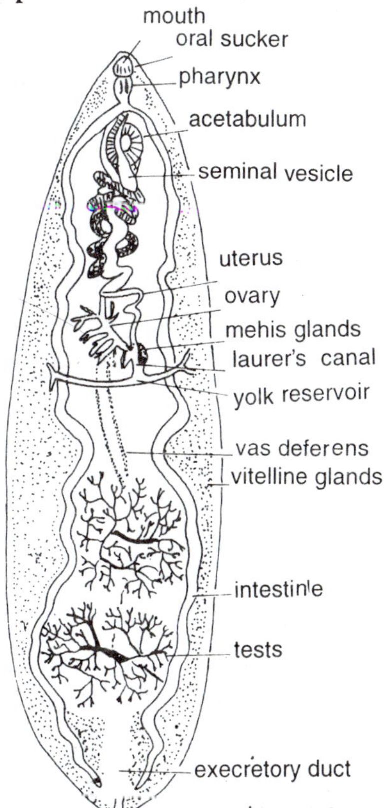

Fig. 1.50 : Fasciolopsis buski.

1. It is a large intestinal fluke of man, dogs, pigs etc.
2. It is found in India and China.
3. It measures about 70 mm in length and 20 mm in width.
4. Oral sucker and acetabulum both are well developed.
5. Mouth surrounded by oral sucker leads into small pharynx. Oesophegus is short and leads into non-diverticulated caeca.
6. Gonads are present in posterior region of body.
7. Intermediate host is snail.
8. Cercaria larval encyst on tubers of water caltrop or singhara. The singhara is eaten raw and thus infection occurs.
9. It causes inflammation and hemorrhage of the intestine.

PARAGONIMUS WESTERMANNI

Phylum - Platyhelminthes
Class - Trematoda
Order - Digenea
Genus - *Paragonimus*
Species - *westermanni*

1. *Paragonimus westermanni* or Lung fluke is parasite in the lungs of mammal including man.'
2. The body of the fluke is oval with well developed oral sucker surrounding the mouth and a ventral sucker in the centre of the body.
3. The internal organs are very much like that of Fasciola hepatica.
4. The cyst like pockets are formed around the parasite which rupture and liberate the eggs into bronchial tubes to be executed with sputum.

4. The cyst like pockets are formed around the parasite which rupture and liberate the eggs into bronchial tubes to be executed with sputum.

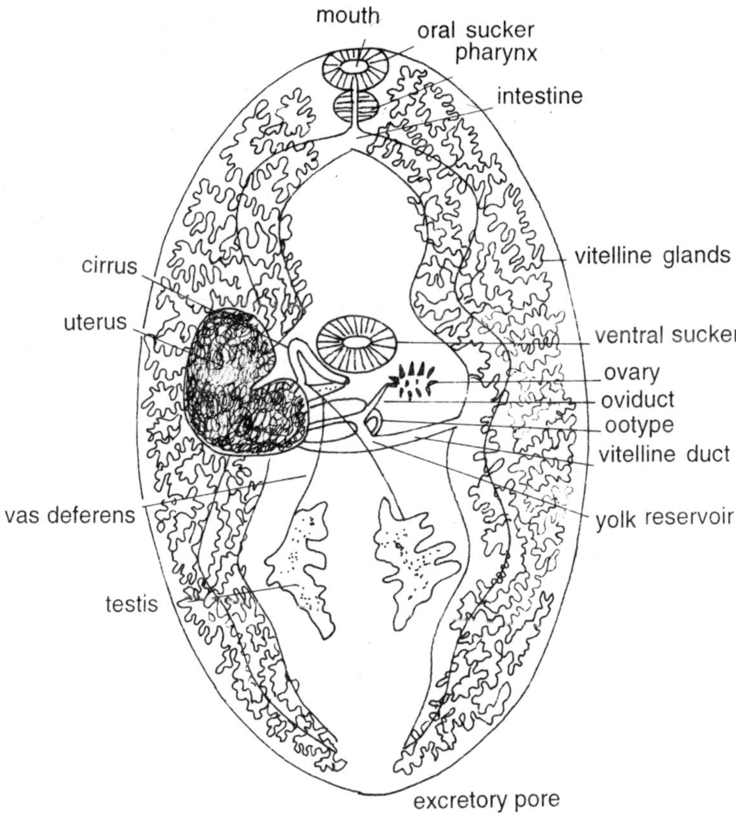

Fig. 1.51 : *Paragonius.*

5. The miracidia live only few hours after hatching.
6. The cercariae are long, have small knob like tail and spiny cuticle.
7. They do not swim but creep in leech-like manner.
8. They pierce the cuticle of crayfish where they become encysted inside the tissues and gradually develop into mature infective metacercariae.
9. The infection of the new host usually takes place by eating infected crab or crayfish.
10. The cause or help in tuberculosis, cough and blood stained sputum.

AMPHILINA

Phylum	-	**Platyhelminthes**
Class	-	**Cestoda**
Subclass	-	**Cestodaria**
Order	-	**Amphilinida**
Genus	-	***Amphilina***

1. It is an endoparasite found in the coelom of fish, Acipenser.
2. Body flat and leaf-like. Scolex is absent.
3. Anterior end is provided with weakly developed protrusible proboscis.
4. Testes are scattered and cirrus is armed.

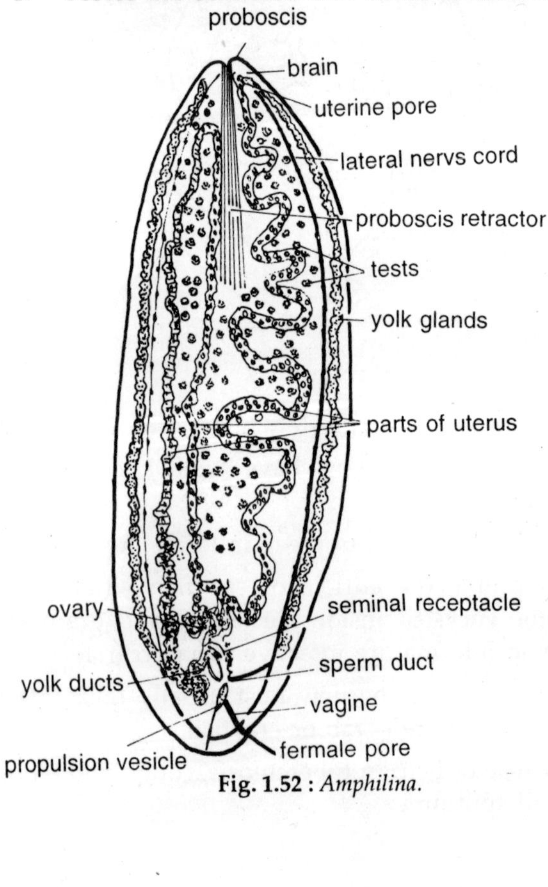

Fig. 1.52 : *Amphilina.*

5. Vagina lies behind the ovary and opens on the left side. Uterus is coiled.
6. The intermediate host is a crustacean, Gammarus, which is eaten by Acipenser.

GYROCOTYLE

Phylum	-	**Platyhelminthes**
Class	-	**Cestoda**
Subclass	-	**Cestodaria**
Order	-	**Gyrocotylidea**
Genus	-	***Gyrocotyle***

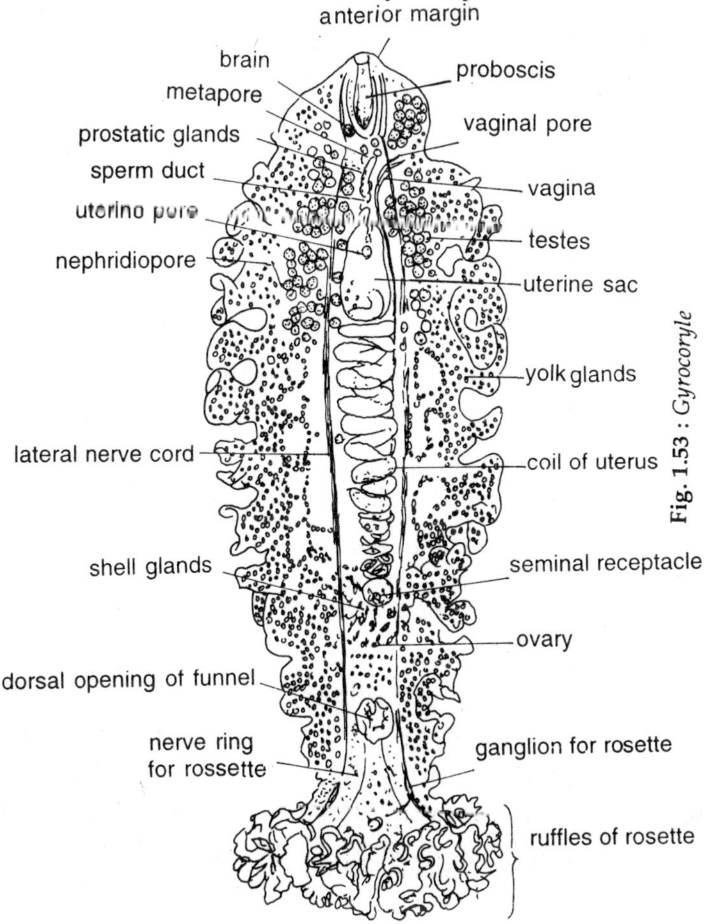

Fig. 1.53 : *Gyrocoryle*

1. These are endoparasite in the intestine of chimaeroid fishes.
2. Body is elongated and dorsoventrally flatenned.
3. Anterior end bears a large opening that leads into a highly muscular protrusible mass, the proboscis.
4. Posterior end bears a rosette-shaped adhesive organ, surrounding a funnel shaped depression.
5. The margins of body are generally ruffled.
6. Bisexual Male reproductive system comprise scattered testes, sperm duct and penis papilla. Female reproductive system comprises ovary, oviduct and uterus.
7. Follicles of yolk gland scattered throughout.

TETRARHYNCHUS

Phylum	-	**Platyhelminthes**
Class	-	**Cestoda**
Subclass	-	**Eucestoda**
Order	-	**Trypanorhyncha**
Genus	-	***Tetrarhynchus***

1. It is an endoparasite in the intestine of elasmobranch fishes.
2. Body is long divisible into scolex and proglottids.
3. Scolex is long and differentiated into a long proximal part containing the proboscis apparatus and a distal part bearing the bothria or sucker.
4. Scolexbears four suckers each with a eversible proboscide armed with spines. The proboscides are retractile in muscular sheaths which end in muscular bulbs.
5. Each proglottid contains a complete set of reproductive organs.

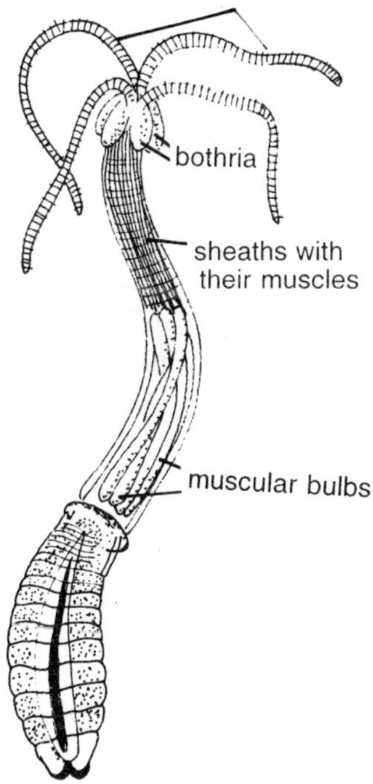

Fig. 1.54 : *Terrarhynchus.*

DIPHYLLOBOTHRIUM (DIBOTHRIOCEPHALUS)

Phylum	-	**Platyhelminthes**
Class	-	**Cestoda**
Subclass	-	**Eucestoda**
Order	-	**Pseudophyllidea**
Genus	-	***Diphyllobothrium***
Species	-	***latum***

1. It is found in the intestines of man. It is world wide in distribution.
2. It is the largest and most pathogenic cestoda of human beings, measuring 20 metres in length.

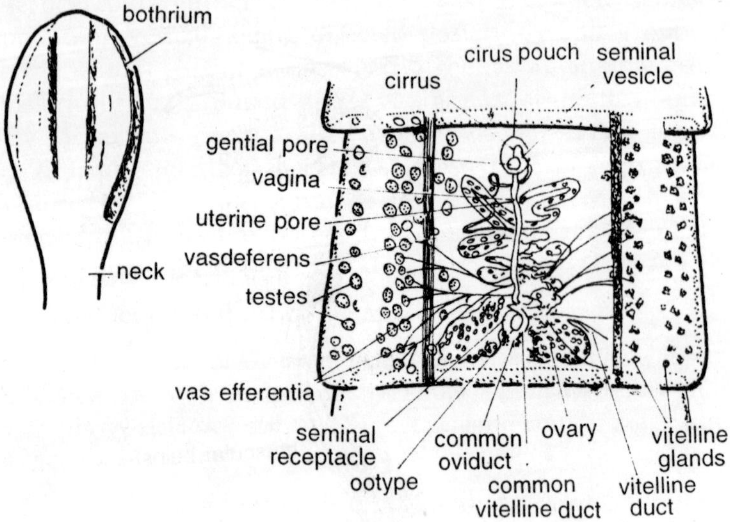

Fig. 1.55 : *Diphyllobothrium* A–Scolex; B–A mature proglottid.

3. Body is differentiated into scolex, neck and proglottids. The number of proglottids is 3000-4000.
4. Scolex is fusiform and bears two slit-like suckers or bothria.
5. Mature proglottid is broader than long.
6. Each proglottid is hermaphrodite and contains complete sets of male and female reproductive organs.
7. Testes are numerous. Vasa efferentia unite to form coiled sperm duct.
8. Ovary is trilobed. Vagina runs posteriorly forming seminal vesicle. Uterus very much twisted.
9. Gonopores ventral.
10. Life-cycle involves two intermediate hosts, one is *cyclops* (a copipod) and other a fish.
11. It causes bothriocephalus anaemia, erythropenia and haemorrhage.

MONIEZIA

Phylum	-	**Platyhelminthes**
Class	-	**Cestoda**
Subclass	-	**Eucestoda**
Order	-	**Taenioidea**
Genus	-	***Moniezia***
Species	-	***expansa***

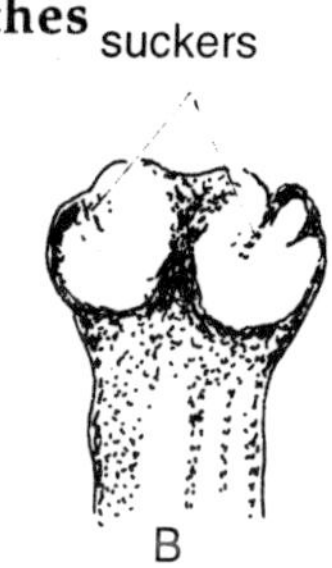

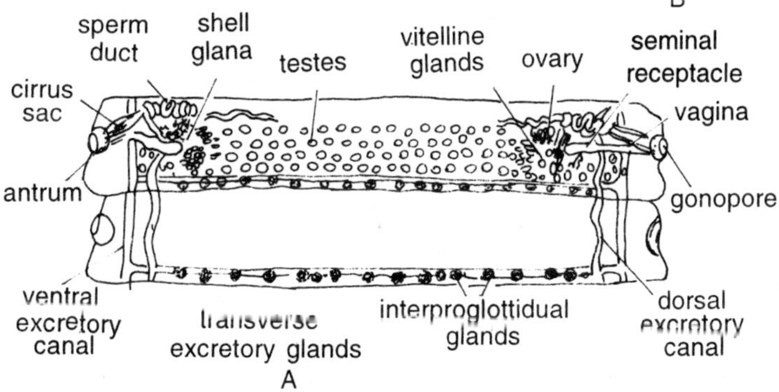

Fig. 1.56 : *Moniezia.* A-Two proglottids. B–Scolex.

1. It is found in sheep, cattles and other ruminants.
2. Body is divisible into scolex, neck and proglottids.
3. The scolex is small with prominant suckers. The rostellum and hooks are absent.
4. Each proglottid is broader than long and contains a double set of reproductive organs.
5. Genital ducts are dorsal to osmoretgulatory canals.
6. Testes are numerous scattered throughout the centre of proglottid. Cirrus in cirrus pouch.
7. Ovaries are in the form of an open fan. Uteri form networks.
8. On one side vagina is dorsal to cirrus pouch and on other ventral.

9. The intermediate host is mite (*Galumna*).
10. It is not so pathogenic but sometimes may cause diarrhoea or anaemia.

HYMENOLEPLS NANA

Phylum	-	**Platyhelminthes**
Class	-	**Cestoda**
Subclass	-	**Eucestoda**
Order	-	**Taenoidea**
Genus	-	***Hymenolepis***
Species	-	***nana***

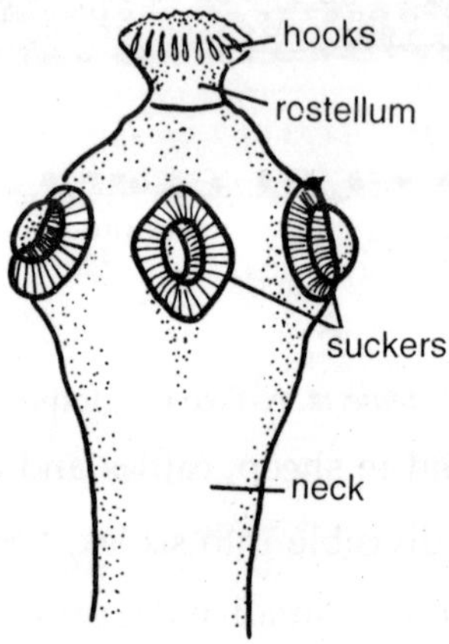

Fig. 1.57 : Scolex of *Hymenolepis*.

1. It is commonly called the dwarf tapeworm of man. It lives in the intestine.
2. It is world-wide in distribution.
3. It measures 25-40 mm in lengths and has about 100-200 proglottids.
4. The rostellum is well-developed and retractile. It bears a single circlets of 20-30 hooks.
5. Neck is long and slender.

6. There are three testes with varying arrangement. Cirrus pouch is large with external and internal seminal vesicle and with an accessory sac.
7. Life-history is simple without any intermediate host.
8. The inner membrane of the hexacanth bears long, wavy filaments.
9. Larvae develop in the intestinal villi while the adult lives in the intestine.
10. It causes abdominal pain and diarrhoea.

DIPYLIDIUM CANINUM

Phylum	-	**Platyhelminthes**
Class	-	**Cestoda**
Subclass	-	**Eucestoda**
Order	-	**Taenioidea**
Genus	-	***Dipylidium***
Species	-	***caninum***

1. It is commonly called dog-tapeworm.
2. It has world wide distribution.
3. It lives in the intestine.
4. It measures 15-40 cm in lengths and has under 200 proglottids.
5. The scolex is somewhat triangular and bears four deeply cupped suckers.
6. Rostellum is retractile into a rostellar sac in scolex. It bears about 60 hooks in 3-7 rings.
7. Mature proglottids are long and barrel shaped having a double set of male and female sex organs. The genital pores at the middle of both lateral margins.

8. Numerous testes are present before or behind the female genitalia.
9. The uterus first develops as a network of canales, but latter breaks up into a large number of uterine capsules, each enclosing 3-30 embryos.
10. Intermediate host is dog-fleas, *ctenocephalus canis* and *pulex irritans*.

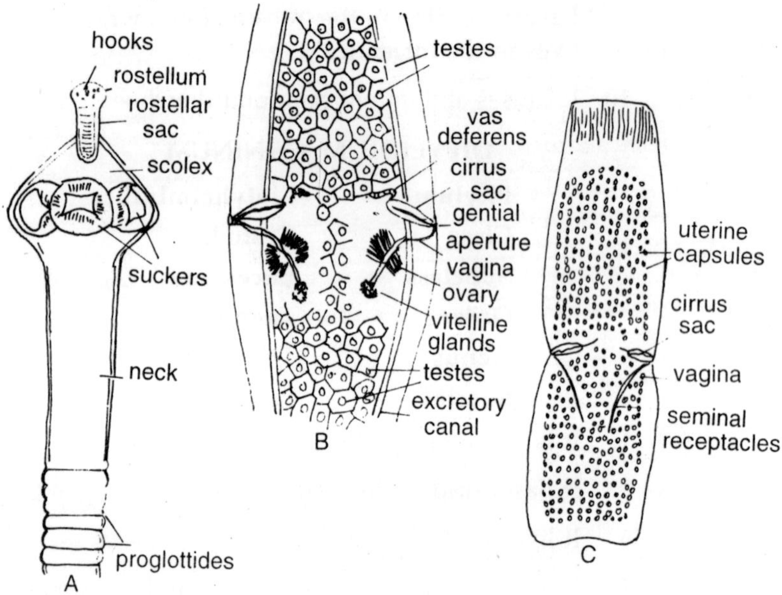

Fig. 1.58 : *Dipylidium.* A–Anterior part. B–Mature proglottis. C–Ripe progloltis.

TAENIA SAGINATA

Phylum	-	**Platyhelminthes**
Class	-	**Cestoda**
Subclass	-	**Eucestoda**
Order	-	**Taenoidea**
Genus	-	***Taenia***
Species	-	***saginata***

1. It is commonly called the beef tapeworm. It lives in the intestines of man where beef is eaten.

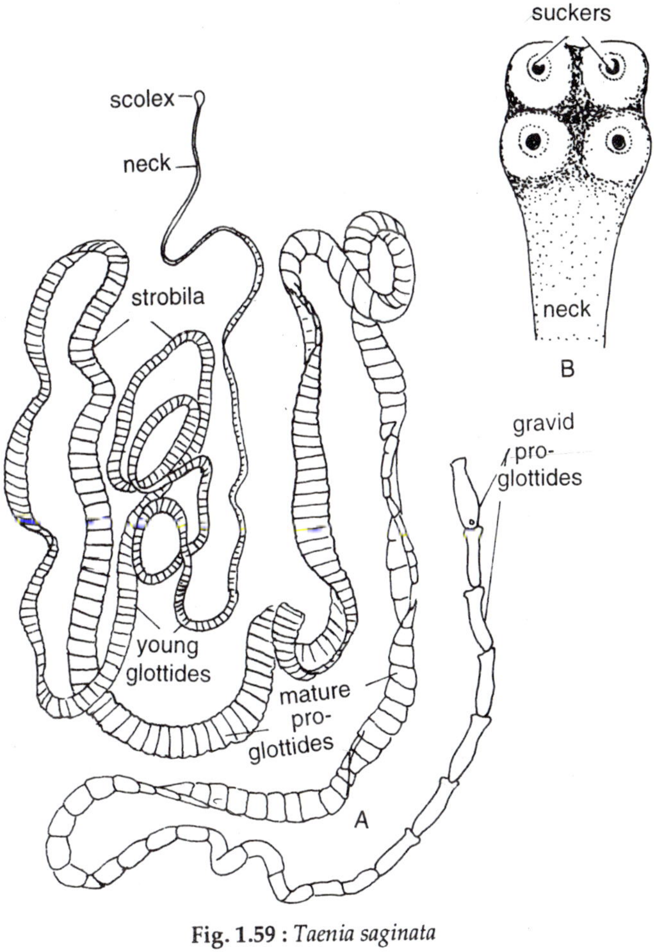

Fig. 1.59 : *Taenia saginata*

2. The body is dorso-ventrally flatenned and much larger than taenia solium. It measures 15-20 feet in length.
3. Body is divisible into scolex, neck and proglottids.
4. Scolex bears four large suckers, adhesion and devoid of rostellar hooks.

5. About 1000 proglottids are present.
6. Testes are about 300-400. Ovary in the form of two lobes. Ootype is present surrounded by Mehli's gland.
7. Life-history is similar to *T. solium*.
8. Intermediate hosts are cattles.

Chapter—2

Origin of the Bilateria

Since the first Metazoa were almost certainly radial animals, the Bilateria must have sprung from a radial ancestor, and there must have been an alteration from radial to bilateral symmetry. This change constitutes a most difficult gap for phylogeneticists to bridge, and various highly speculative conjectures have been made. It is generally believed that bilateral symmetry originated in consequence of the assumption by a radial ancestor of creeping mode of life. We have already traced this ancestor to a planuloid form resembling the planula larva of the Cnidaria. This planuloid ancestor was an elongated, radially symmetrical organism, without mouth or archenteron, and consisted of a ciliated or flagellated epidermis, probably composed of epithelio-muscular cells, and a solid interior mass of digestive cells. There was probably a nerve net under the epidermis with an accumulation of neurosensory cells at the anterior pole, and indifferent mesenchyme cells, capable of differentiating into sex cells and other types of cells, were scattered through the interior mass. Food catching and digestion remained wholly of the protozoan type so that the organism had no need of a mouth or archenteron. The creature was polarized with definite anterior or aboral and posterior or oral ends but lacked dorsoventral differentiation.

This planuloid organism may be supposed to have given rise to the coelenterate line by formation of a mouth at the site of the closed blastopore, hollowing out of the interior to form a primitive gut, and sprouting of tentacles, thus becoming a simple medusa similar to an actinula. It is her assumed also to have given rise to the Bilateria by adopting a creeping mode of life, then flattening out and developing bilateral symmetry and dorsoventral differentiation.

There are three chief theories of the steps from the radiate of to the bilateral condition: the *ctenophore-polyclad,* the *ctenophore-trochophore,* and the *planuloid-acoeloid* theories. The ctenophore-polyclad theory was first suggested by Kowalevsky in 1880 as a result of his finding of the peculiar ctenophore *Coeloplana* but has been supported and elaborated chiefly by Arnold Lang, the eminent student of the polyclad flatworms, who based his conclusions on his studies of the anatomy and embryology of various turbellarians, and on the peculiar features of the aberrant ctenophores *Ctenoplana* and *Coeloplana.* These ctenophores resemble polyclads in their oval flattened shape with dorsoventral differentiation; in the presence of two dorsal tentacles; in creeping upon the entire ventral surface; in the centrally located ventral mouth and branched, blindly ending digestive canals; and in the radiating, anastomosing nervous system. There is a certain amount of resemblance in the embryology, as both ctenophores and polyclads have determinate cleavage with the formation of micromeres and macromeres, and in both there is a large stomodaeal invagination which in both contributes to the ventral surface. Many polyclads have a swimming larval stage with eight ciliated lobes which seem comparable to the eight comb rows of ctenophores. However, as Lang fully realized, *Coeloplana* and *Ctenoplana* have no trace of real bilateral symmetry, being in fact biradially symmetrical, and their nervous center is in the middle of the dorsal surface, whereas it is characteristic of the Bilateria that the nervous center is anteriorly located. To bridge these differences, Lang postulated a series of stages from the platyctene ctenophores to the polyclads by a forward shift of the nerve center and tentacles and extensive branching of the digestive canals.

Lang's theory of the ctenophores as intermediate forms between the radiates and the polyclad flatworms had a wide and extended acceptance but is to be rejected on the following grounds. First, it is agreed by all students of ctenophores that the Platyctenea are simply highly aberrant ctenophores without phylogenetic significance. Second, the embryology of the two groups is in fact very different and the cleavage patterns follow very diverse plans, that of the ctenophores being biradial, and that of the polyclads spiral. Third, in the theory of Lang, the polyclads are necessarily regarded as the most primitive existing Bilateria; this is a mistaken idea, since it is now clear that the order Acoela occupies that position and the polyclads stem from acoel ancestors.

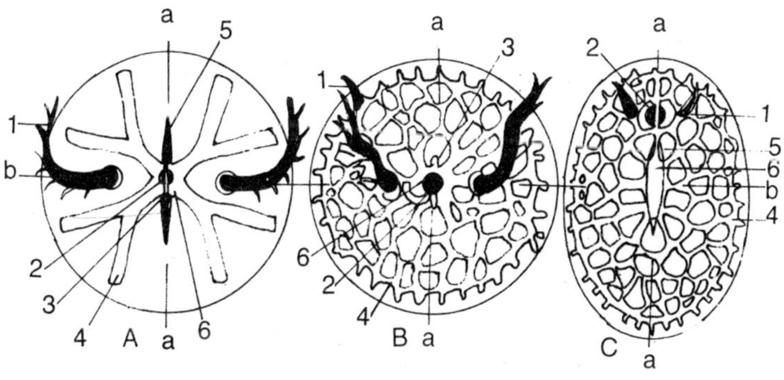

Fig. 2.1 : Diagram to show change from a ctenophore to a polyclad. *A.* Ctenophore. *B.* Hypothetical intermediate stage. *C.* Polyclad. 1, tentacle; 2, nervous center; 3, polar field; 4, gastrovascular branches; 5, pharynx; 6, main canal of gastrovascular system. *aa,* sagittal axis; *bb,* transverse axis.

The ctenophore-trochophore theory is a variant of the trochophore theory, discussed below, and is based on alleged resemblances between ctenophores and the trochophore larva. These resemblances are best stated after the anatomy of the trochophore larva has been described. Here it may be pointed out that this theory proceeds from the ctenophores to the annelids and other higher Bilateria and fails to account for the origin of the lower Bilateria. Supporters of the theory have realized this difficulty and take one or the other of the two possible explanations. Some regard the larvae of the acoelomate Bilateria (*i.e.,* the flatworms and the nemertines) as earlier stages of the trochophore larva. But then such stages would necessarily fall between the ctenophore and the trochophore and should more nearly resemble the ctenophore than does the trochophore larva. This, however, is not the case for in fact the ctenophore similarities are manifested only in the fully developed typical trochophore larva. The second explanation is to the effect that the acoelomate Bilateria are degenerated from the higher Bilateria. This cannot possibly be admitted. It may be regarded as firmly established that the free-living flatworms, especially the order Acoela, are of exceedingly primitive structure, positively not degenerated from or modified from a higher anatomical type. The ctenophore-trochophore theory therefore appears to the author unacceptable, although it has been widely supported.

The planuloid-acoeloid theory is based upon the remarkable features of the flatworm order Acoela, whose primitive structure was first recognized in 1882 by Ludwig von Graff, the distinguished pioneer in the study of the free-living flatworms. The Acoela were nevertheless long regarded as degenerate forms and hence of no phylogenetic importance until recent years when their primitive nature has again been recognised by all students of the Turbellaria. The Acoela are group of very small flatworms with the following primitive characteristics. The ciliated epidermis is often syncytial, a in many coelenterates, and in several members has basal muscle fibers, as also in coelenterates. There is no distinct basement membrane beneath the epidermis. The interior is a solid mass of cells as in the planula larva without a digestive cavity, and comparison of the embryonic development with that of other flatworms indicates that this interior mass represents entoderm and ecto- and entomesoderm. There is a mouth in the center of the ventral surface, and this leads into the interior mass either directly or by way of a simple tubular pharynx.

These worms feed by ingesting small animals through the mouth into the interior mass where intracellular digestion occurs; thus it is directly proved that an archenteron is not a necessity for the early bilateral holozoic animal, as supposed by the Haeckelian gastrula theory. The nervous system is of primitive construction, consisting of a plexus beneath the epidermis with several more or less distinct longitudinal strands radiating from an anterior brain mass; but in some acoels the nerve plexus is situated in the basal part of the epidermis and the brain is a mere thickening also epidermally located. Most of the Acoela have a statocyst near or embedded in the brain; this structurally is more like the statocyst of medusa than of a ctenophore. There are no distinct gonads, but the sex cells differentiate out of the cells of the interior mass. Female ducts are wanting, and the eggs discharge either through the mouth or by rupture of the body surface. There are usually indistinct male ducts which lead to a copulatory organ. Although the animals are hermaphroditic, cross-fertilization is the rule and is internal. The Acoela lack an excretory system.

It is evident that there is no difficulty in passing from a planuloid type of ancestor to an acoeloid form. Epidermis, muscular system, interior mass, nervous system, and sex cells are, even in present Acoela, distinctly at a coelenterate stage or but slightly advanced from this stage. Consequently we accept the theory of the origin of the Bilateria from a planuloid ancestor by way of an acoeloid form. The main changes necessary are the alteration from radial to bilateral symmetry and the forward shifting of the nervous center. As no intermediate forms are known, the steps by which these changes occurred may be inferred from the embryology of the lower Turbellaria. As the mouth of the Acoela occupies the site of the blastopore, it may be taken as a fixed point. It is then evident that, as postulated by Lang, the planuloid form must have flattened down in the oral-aboral axis and at the same time there must have occurred a forward migration of the sensory center, as actually happens in the development of polyclads, through processes of differential growth. The result is a simple flatworm with bilateral symmetry, ventral mouth, and anteriorly located brain and sense organs. A later stage, represented by the higher Turbellaria, is attained by the hollowing out of the central part of the interior mass to form an archenteron and the greater differentiation of the anterior region as a head.

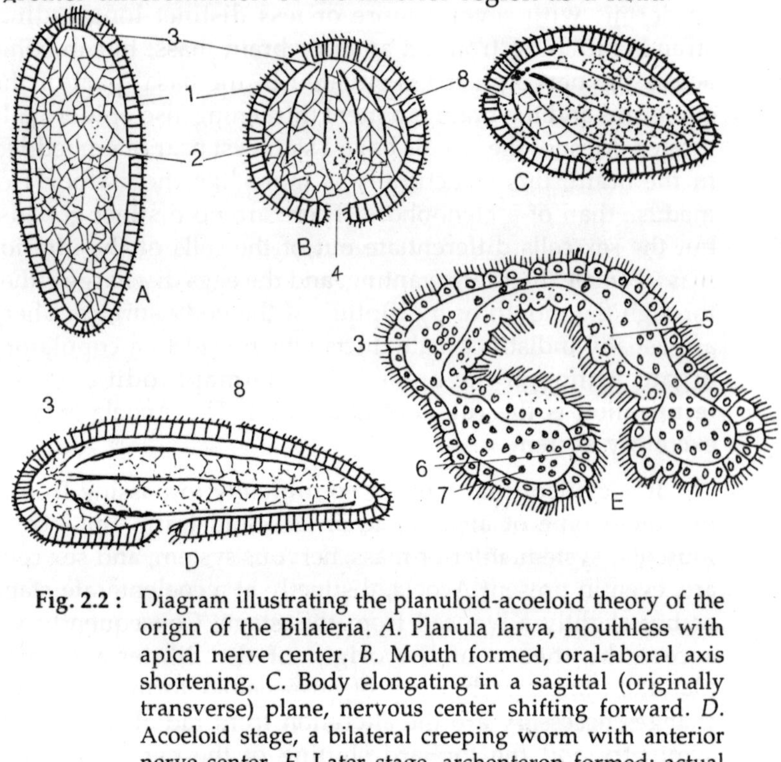

Fig. 2.2 : Diagram illustrating the planuloid-acoeloid theory of the origin of the Bilateria. *A.* Planula larva, mouthless with apical nerve center. *B.* Mouth formed, oral-aboral axis shortening. *C.* Body elongating in a sagittal (originally transverse) plane, nervous center shifting forward. *D.* Acoeloid stage, a bilateral creeping worm with anterior nerve center. *E.* Later stage, archenteron formed; actual embryonic stage of a polyclad. 1, ectoderm; 2, entodermal mass; 3, nervous center; 4, mouth; 5, archenteron; 6, stomodaeum; 7, mesoderm; 8, nerve cords.

It is thus seen that the oral surface of the Radiata becomes the ventral surface of the Bilateria while their sagittal plane is retained as the sagittal plane of the latter. But the dorsoventral flattening is not a mere shortening of the oral-aboral axis, for this axis has been curved forward by the anterior shifting of the nervous center. Thus, whereas in the Radiata the oral-aboral axis of the larva is retained as the oral-aboral axis of the adult, in the Bilateria the gastrular axis has no direct relation to any of the adult axes.

Spiral Cleavage and Determinate Development

Determinate or *mosaic* development, characteristic of most of the Protostomia, is usually associated with the spiral type of cleavage, found in polyclad flatworms, nemertines, annelids, and mollusks. Among these groups the cleavage pattern and the fate of the various blastomeres are so nearly identical that a common descent is scarcely to be doubted. In spiral cleavage, the spindle axes are oriented obliquely with respect to the polar axis of the egg. As a result successive tiers of blastomeres alternate, the cells of one tier resting in the angles between the cells below them. This displacement may be either clockwise (dextrotropic) or counterclockwise (levotropic). Spiral cleavage is commonly holoblastic but unequal, and this inequality is often evidenced even at the first cleavage. In spiral cleavage, the fate of each blastomere can be determined, a type of study known as *cell lineage* and pursued with brilliant results by American embryologists. For clarity in describing the course of events, each blastomere is named. The four blastomeres formed by the first two meridional cleavages are labeled *A, B, C, D.* The third division is transverse and latitudinal and results in eight cells, four small animal ones called the *first quartet of micromeres* and four large cells below, called *macromeres*. The micromeres are named 1*a*, 1*b*, 1*c*, 1*d*, and the macromeres 1*A*, 1*B*, 1*C*, 1*D*. At the next cleavage, the macromeres give off above a *second quartet* of micromeres, called 2*a*, 2*b*, 2*c*, 2*d*, and are themselves then designated as 2*A*, 2*B*, 2*C*, 2*D*; and at the next cleavage the latter again divide, giving off a *third quartet*, 3*a*, etc., themselves becoming 3*A*, etc. Meantime the first and second quartets have cleaved, and their offspring are also numbered by a definite system. Thus the two daughter cells of 1*a* become $1a^1$ and $1a^2$; the exponent 1 is applied to the cell nearer the animal pole. When these divide again, exponents are added; thus the daughter cells of $1a^1$ are numbered $1a^{11}$ and $1a^{12}$,

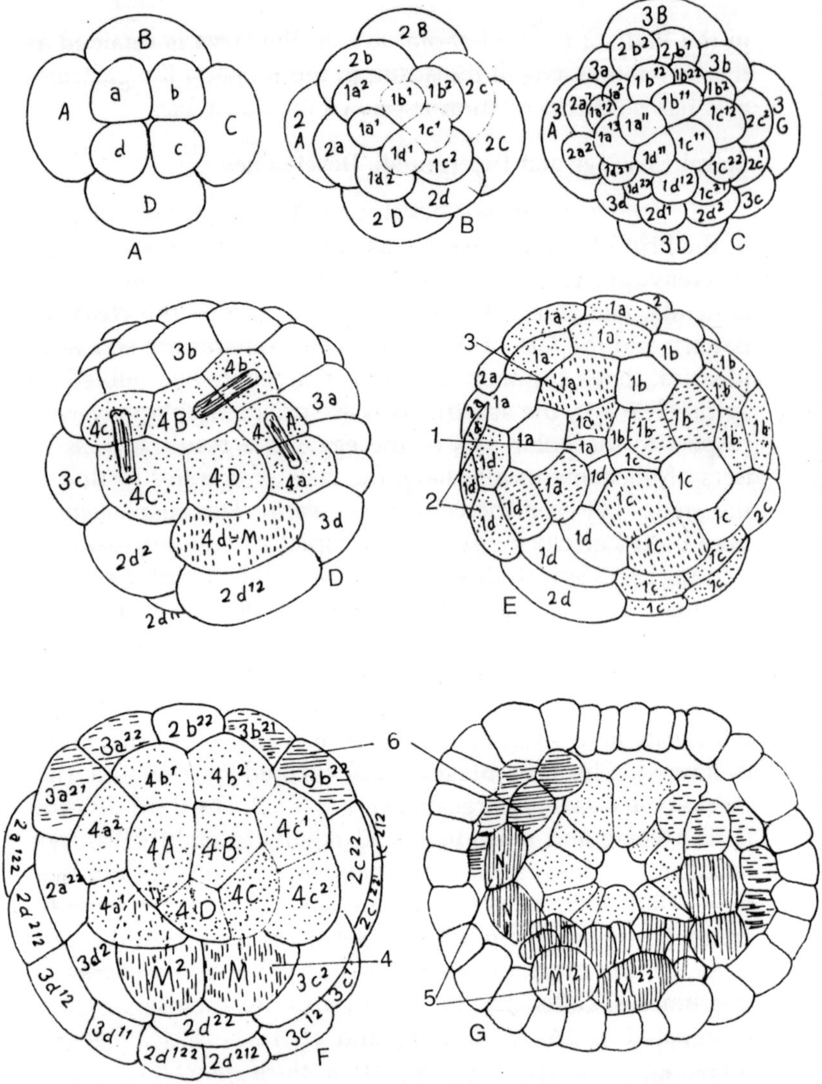

Fig. 2.3 : Spiral cleavage. *A*. Third cleavage, division into micromeres and macromeres. *B*. Formation of the second quartet of micromeres; first quartet has divided. *C*. Formation of the third quartet of micromeres; second quartet has divided again. (*A, B, C,* based on development of polyclads.) *D*. View from vegetative pole, showing the macromeres giving off the fourth quartet of micromeres;

and of $1a^2$, $1a^{21}$ and $1a^{22}$, respectively. This system obviously is capable of indefinite expansion and serves to designate exactly the origin of any particular cell. All the cleavages are oblique and are alternately dextrotropic and levotropic. All the blastomeres bearing one letter, as *a*, are offspring of one of the original four blastomeres, in this case *A*, and hence more or less occupy one quadrant of the embryo.

When the third quartet of micromeres has been given off, the germ layers are already fixed (usually 32-cell stage, as the first and second quartets have cleaved). The three quartets and their offsprings form the whole of the ectoderm (and also the larval mesoderm, i.e., the ectomesoderm) and hence are often called *ectomeres*. The four macromeres, 3*A*, 3*B*, 3*C*, 3*D*, form the whole of the entoderm and the true or entomesoderm. At their next division, they give off at the vegetal pole a fourth quartet of micromeres, 4*a*—4*d*, three of which are also purely entodermal. The fourth cell, 4*d*, is larger than the corresponding macromere, 4*D*, is called the *mesentoblast* cell, or *M*, and is the entire source of the entomesoderm. It was originally supposed that all its offsprings are mesodermal, but later researches have shown that a few of these remain entodermal and contribute to the gut wall. The main offspring of 4*d*, however, in annelids and mollusks,

---

entoderm cells stippled, ento-mesodermal cell, 4*d* = *M*, vertically hatched. *E*. View from the animal pole, showing the rosette and the annelid cross (cells of cross hatched); primary trochoblasts stippled; only the series to which each cell belongs is indicated. *F*. View from vegetal pole of embryo of the snail *Physa*, showing entoderm (stippled), the two teloblasts (vertically hatched), and the cells that give rise to the ectomesoderm (crosshatched). *G*. Later stage of *Physa*, looking into interior from animal pole, showing the teloblasts giving off the mesoderm bands (vertically hatched), of which the cells marked *N* become the metanephridia; entoderm stippled, ectomesoderm crosshatched. (the cleavage of *Physa* spirals in the reverse direction from that of the other forms figured.) 1, rosette; 2, primary trochoblasts; 3, cross; 4, teloblasts; 5, mesoderm bands; 6, ectomesoderm.

are two cells known as the *teloblasts, pole cells,* or *primordial mesoderm cells.* These take up a bilateral position just in front of the future anus and each proliferates a band of cells, the *mesoderm band,* the sole source of the entomesoderm.

The embryo becomes an inequal coeloblastula or, in the case of very yolky eggs, a stereoblastula. The ectomeres spread and grow down over the macromeres, which also invaginate. A typical invaginate gastrula thus arises partly by epiboly and partly by emboly. The mouth forms at the site of the blastopore by way of stomodaeal invagination. In forms with an anus, the blastopore elongates and closes together medially so as to leave its anterior end as mouth and its posterior end as anus. The anal end usually closes temporarily but later reopens as the permanent anus. A stomodaeum is universal throughout the Protostomia. The mouth is usually shifted forward by growth processes elsewhere in the embryo. The embryo develops an apical tuft of cilia and an equatorial girdle of cilia and swims about as a free-living larva.

The fate of the various cells of the cleavage pattern may be stated briefly. The micromeres of the first quartet form the ectoderm of the aboral part of the larva. The $1a^{111}$–$1d^{111}$ series occupy the animal pole where they make a four-rayed pattern termed the rosette which becomes the aboral sensory plate with its tuft of cilia. Their sister cells, the $1a^{112}$–$1d^{112}$ series, in annelids alternate with them as a cross-shaped figure called the *annelid cross.* A similar *molluscan cross* occurs in molluscan development but there is formed of the cells which lie between the arms of the annelid cross, namely, the $1a^{12}$–$1d^{12}$ cells, with some assistance from second quartet cells. Thus the arms of the cross are interradial in annelids, radial in mollusks. The cross, very evident in early stages, eventually becomes obliterated and plays no special role in development as such. It is often regarded as a ctenophore reminiscence, and the ctenophore resemblance is enhanced in some cases by the bifurcation of the tips of the arms to form an eight-rayed figure. The $1a^2$–$1d^2$ descendants of the first quartet are termed the *primary trochoblasts;* they form four groups of four cells each, one group in each quadrant of the embryo near the equator. Each cell develops a transverse tuft of cilia reminiscent of a ctenophore comb. These four groups of ciliary tufts constitute the primitive ciliated girdle, or *prototroch,* of the larva. Later the prototroch is completed by the formation of cilia on the cells between the primary trochoblasts, termed *secondary trochoblasts,* which come from the first or second quartet or both.

The cells of the second and third quartets contribute to the surface ectoderm, and certain ones, termed *stomatoblasts,* invaginate as the stomodaeum. In annelids, the cell 2*d*, known as the first *somatoblast,* gives rise to practically the entire ectoderm of the adult trunk. Cells are also given off from the second and third quartets into the blastocoel as mesenchyme which differentiates into larval muscles. This mesenchyme constitutes the larval mesoderm or ectomesoderm and is undoubtedly a reminiscence of the original ectomesoderm of the ancestral stem form of the Bilateria. Descendants of $3c^2$ and $3d^2$ give rise in annelids to the larval excretory organs, the *archinephridia.* In the most primitive case, that of the annelid *Polygordius,* the archinephridium comes from two cells, one forming the tubule, and the other the solenocyte.

The entoderm originates almost wholly from seven cells, namely, 4*a*, 4*b*, 4*c*, 4*A*, 4*B*, 4*C*, and 4*D*, which invaginate and produce the entodermal wall of the stomach and intestine of the larva. As already noted, a few cells descended from 4*d* become incorporated into the intestinal wall. The intestine (in the higher Protostomia) connects to the exterior by an anus formed at the site of the rear end of the blastopore. In producing the anus, the ectoderm invaginates slightly, and this ectodermal termination of the gut is called the *proctodaeum.*

The two teloblasts, offspring of 4*d*, lie in the blastocoel just in front of the anus. By repeated divisions each gives rise to a band of cells which extends forward into the blastocoel. These bilateral *mesoderm bands* are the whole of the entomesoderm and give rise to muscle, connective tissues, gonads, excretory system at least in part, blood vessels, etc. The formation of typical mesoderm bands from teloblasts is confined to the phyla Annelida and Mollusca, but the bands are well evidenced in the Arthropoda.

In the annelids and mollusks, spiral cleavage often results in a characteristic larva, known as the *trochophore,*[1] and somewhat similar larvae occur in other protostomous phyla. As great phylogenetic significance has been attributed to the trochophore larva, it becomes necessary to describe its characters before proceeding further.

The Trochophore Larva, The Trochophore Theory, And The General Phylogenetic Significance of Larval Types

The trochophore larva is a somewhat biconical creature with a protruding equator. Its external surface consists of a one-layered epithelium (ectoderm) thickened at the apical pole into a sensory plate which bears a tuft of cilia. Around the equator there is a girdle of cilia termed the *prototroch* which passes above the mouth; and there may be present a second equatorial girdle, the *metatroch,* passing below the mouth, and sometimes also a ciliated circlet around the anus, the *paratroch.* A complete digestive tube is present extending in an L-shape from the mouth at the equator to the anus at the lower pole; it consists of a ciliated stomodaeum leading into an expanded rounded stomach from which the short narrowed intestine proceeds to the anus. Between the digestive tube and the ectoderm there is a spacious cavity which is the blastular cavity or blastocoel. This is more or less occupied by mesenchyme cells and well-developed muscle cells, all derived from the ectomesoderm. The muscular system is often quite complicated consisting not only of bands acting on the digestive tube and ciliary girdles but also of bands under the ectoderm accompanying the nerves. The nervous system according to the best description may attain an astonishing degree of complexity, consisting of a ganglionic mass under the apical plate, a variable number of longitudinal nerves radiating from this mass, and one to several nerve rings connecting the radial nerves. The main nerve ring underlies the prototroch. Various sense organs, including eyes and statocysts, may be present. To either side of the intestine lies a mesoderm band derived from a teloblast, and near this there is on each side a nephridium consisting of a tubule whose inner end is closed by one or more solenocytes. A solenocyte is a long tubular cells having a flagellum playing in the tube. Solenocytes occur in several invertebrate groups and also in *Amphioxus*, and nephridia with solenocytes are undoubtedly only a variant of nephridia with flame bulbs; the latter are general throughout the acoelomate and pseudocoelomate Bilateria.

The *trochophore theory*, elaborately developed by Hatschek (1857), is to the effect that the trochophore is the

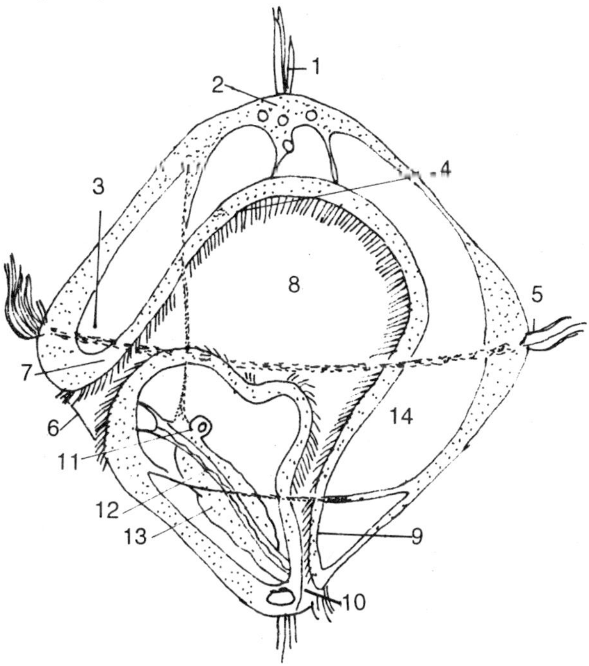

Fig. 2.4 : Trochophore larva of an annelid. 1, apical tuft of cilia; 2, apical sensory plate; 3, ectomesoderm; 4, eye; 5, prototroch; 6, mouth; 7, stomodaeum; 8, stomach; 9, proctodaeum; 10, anus; 11, statocyst; 12, archinephridium, consisting of one solenocyte; 13, mesoderm bands; 14, blastocoel.

larva of an ancestral form, the *trochozoon*, which was the common ancestor of most, if not all, the bilateral phyla, and which, of living forms, most nearly resembled a rotifer. It is to be noted that the trochophore is regarded as recapitulating the *larva* of the ancestor, not as being itself a replica of the ancestor. The rotifer resemblances usually cited are the general pseudocoelomate grade of structure, the ciliary girdles, the structure and location of the brain and the nature of its attached sense organs, the muscle bands, the nephridia, the form of the digestive tract, and the presence of an anus. The rotifer-like nature of the hypothetical trochozoon was

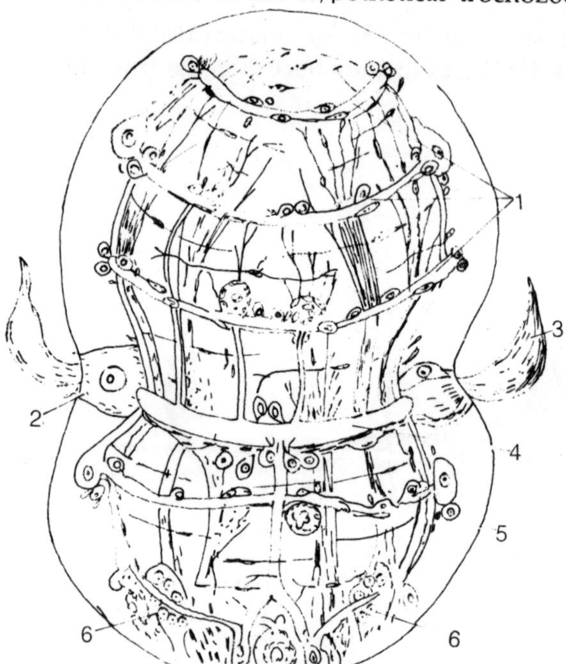

Fig. 2.5 : Trochophore of the annelid *Lopcdorhynchus*, showing nervous system (blank) and muscular system (striated). The orientation of the larva is such that the apical sensory center does not show. 1, upper three nerve rings; 2, ciliary cell of the prototroch; 3, prototroch; 4, main nerve ring underlying the prototroch; 5, lower nerve ring; 6, mesoderm bands.

considered strongly supported by the discovery by Semper in 1859 in the rice fields of Philippines of a rotifer which he named *Trochosphaera* and which through its inflated spherical form and equatorial ciliary girdle bears considerable superficial resemblance to a trochophore larva. Time has detracted from the rotifer part of the trochophore theory, for those rotifers whose ciliary arrangements most suggest the girdles of the trochophore are now regarded as specialized types and the primitive ciliary apparatus of the rotifers has not much likeness to the trochophore girdles. Further the rotifer *Trochosphaera,* which has been refound and restudied in various localities, is now known to be merely an aberrant type of no especial evolutionary significance. Nevertheless, while discarding the rotifer part of the trochophore theory, one must admit the general soundness of Hatschek's view that the trochophore is about at the structural level of a rotifer, *i.e.,* is of the pseudocoelomate grade of structure. The fact further that nearly identical trochophores occur in two phyla in which the adults are anatomically very different, *i.e.,* the annelids and the mollusks, suggests that the trochophore is indeed a reminiscence of the common ancestor of the eucoelomate Protostomia and perhaps also of the pseudocoelomate groups. However, the origin of the acoelomate Bilateria remains unaccounted for on this theory.

The ctenophore-trochophore theory, mentioned above, is a more recent and widely accepted version of the trochopore theory. It proposes to pass directly from a ctenophore-like ancestor to the trochozoon and so to bridge the gap between the radiate and bilateral animals. The chief ctenophore resemblances of the trochophore are: the apical nerve center with its attached sense organs, the radiating subectodermal nerves, often (but not always) eight in number, the occurrence of the "cross" in the cleavage pattern, and the origin of the prototroch from four groups of ciliated cells. The apical nerve plate and sense organs are, however, also a feature of the planuloid ancestor, and the other resemblances are a necessary consequence of the quadripartite pattern characteristic of spiral cleavage. This quadripartite pattern is the result of the two meridional cleavages producing four blastomeres, each of which roughly forms one-fourth of the embryo. The argument therefore boils down to the question whether cleavage into four blastomeres is in fact reminiscent of a ctenophore ancestry. And this does not appear to be the case since the ctenophore egg goes through three meridional cleavages producing a wreath of eight blastomeres, and the details of its cleavage are not similar to the course of spiral cleavage. Further the nervous system of the developed trochophore with its longitudinal nerves connected by circular nerves is more like that of the acoel flatworms (which often also have eight longitudinal nerve trunks) than it is like that of a ctenophore. The ctenophore-trochophore theory is therefore rejected on these grounds and those previously mentioned.

A main difficulty in the path of the trochophore theory and its variants has been to allocate the acoelomate groups, for these can no longer be disregarded as degraded from higher bilateral phyla. This difficulty is here met by the interpolation in the phylogeny of an acoeloid grade of ancestor from which the flatworms are readily derived. The phylogenetic series here proposed is then: planuloid ancestor-acoeloid ancestor-trochozoon (pseudocoelomate grade). The gap from the acoeloid ancestor to the trochozoon is probably too large and may need to be filled by a type with a digestive tract but no anus, resembling stages seen in the development of polyclads.

An attitude of conservatism should always be maintained in regard to the evaluation of larval types. A larva is a developmental stage that leads a free existence. Larvae are practically limited to marine animals. The main reasons given for this are: that larvae are too delicate to withstand the much wider range of environmental factors in fresh water than in the sea; that they would be swept away by currents; and that fresh water is poor in the salts needed by many larvae. It results that animals, to reproduce successfully in fresh water, must produce larger and fewer eggs, better provided with food and salts than their marine relatives; the eggs must be protected with shells, jelly, etc.; must be fastened to objects or develop inside the mother; and the young animals must hatch in a fully developed condition. Consequently, the development of fresh-water animals is generally highly modified and furnishes poor material for phylogenetic speculations. It is therefore difficult to work out the relationships of groups of animals which are predominantly fresh-water. Even among related marine groups, some may have free larvae and others not, without apparent reason. In such cases the question arises which type of development is the more primitive. Again as larvae must fend for themselves and find food, they are very apt to display adaptive modifications which are without phylogenetic significance. Pelagic larvae tend to present common characters adapted to pelagic life, as transparency and lightness of body, swimming apparatus mostly in the form of ciliated bands, lobes and projections for buoyancy, and a food-catching mechanism, mostly of the mucous-ciliary type. Evidently great caution must be exercised in the interpretation of larvae, and most weight should be placed on characters not common to pelagic animals. In the case of the trochophore larva, the most salient features are the form of the nervous system, the presence of an anus, and the occurrence of primitive nephridia. These suggest an ancestral form higher than a flatworm and lower than an annelid.

The foregoing analysis carries the line of evolution from the radiates to an through the protostomous Bilateria. The relation of the Deuterostomia to this line of ascent remains very dubious, and this situation casts much doubt on the generally accepted ideas of the original modes of origin of mesoderm and coelom.

Modes or Origin of the Mesoderm

The term mesoderm is rather loosely applied to all cells, cell layers, or cell masses which occur in the embryo between the ectoderm and the entoderm. The mesoderm is also spoken of as the third germ layer. As already noted, there are two kinds of mesoderm which seem to be phylogenetically distinct, the ectomesoderm and the entomesoderm. The ectomesoderm exists in both the Radiata and the Protostomia and is undoubtedly the oldest form of mesoderm. It is always mesenchymal and typically consists of inwandered ectodermal cells, in spiral cleavage of cells descended from the second and third quartets of micromeres. Ectomesoderm appears to be absent in the Deuterostomia. The Protostomia presumably inherited the ectomesoderm from a radiate ancestor, and their entomesoderm is a later development originating in the evolution of the Protostomia from a radiate ancestor.

The entomesoderm, the "true" or definite mesoderm, includes all mesoderm which arises from or with the entoderm. It may form as mesenchyme or as bands, plates, or sacs, more or less epithelial in character. The attempt of the Hertwigs (1882) to draw a sharp distinction between entomesoderm of mesenchymal origin and that of epithelial origin (mesothelium) was mistaken. The entomesoderm may arise by either method, and mesoderm which begins as epithelium may later become mesenchymal or vice versa. The time of origin of the entomesoderm is also variable. The following modes of formation of the entomesoderm occur among the Bilateria.

Chapter—5

Dugesia

Planaria or *Dugesia* belongs to class Turbellaria. The Turbellarians are mostly free-living, unsegmented flatworms that are clothed with a cellular epidermis, which is usually ciliated, sometimes only in parts. Turbellarians are primarily aquatic, and the great majority are marine. Although these are a few pelagic species, most of them are bottom dwellers than live in mud or sand, under stones and shells, or on sea weeds, *Degusia* is among them. The following description mostly belongs to the common fresh water genus, the *Degusia* formarly called *Planaria*. It is a well-known representative of the Turbellaria.

SYSTEMATIC POSITION

Phylum	—	**Platyhelminthes**
Class	—	**Turbellaria**
Order	—	**Tricladida**
Family	—	**Planariidae**
Genus	—	***Dugesia***

Habits and habitat. Dugesia is the commonest free-living platyhelminthes. They are usually found on the vegetation or under the stones in ponds, lakes, marshes, rivers and small streams. They generally inhabit pure, cool and clear water. Some species remain in currents, while others like to remain in quiet water. When active, they glide over the surface but when at rest they crawl down to the bottom or under other smooth objects, where light is not bright. *Dugesia* is found under the stone generally attached in great numbers, if pieces of meat are dropped in water near the bank.

It starved they become smaller, but if well-fed they lay cocoons containing eggs which hatch to give young. They are carnivorous and extremely cannibalistic.

MORPHOLOGY

Shape, Size and Colour

These are small, thin, dorsoventrally flattened, bilaterally symmetrical animals measuring about 12 mm long. Colouration is due to the dissolved pigments present beneath the epidermis. The colour may be dark yellow or olive, brown or blackish brown and may have stripes of white, orange or yellow colour. The colouration is light or whitish on ventral side.

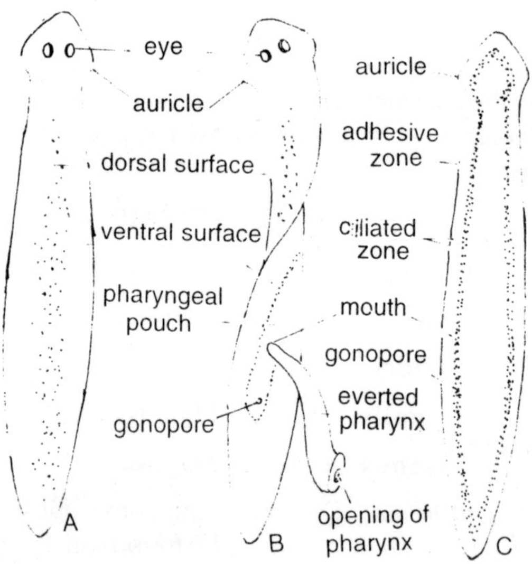

Fig. 3.1 : *Dugesia*. External features. A = Dorsal view; B = Lateral view; C= Ventral View

Structure

There are two ends in the body. Anterior end is blunt, while the posterior end is tapering. Anterior end is almost square or triangular with its apex pointing forward. This portion of the body is known as head, because with it animal moves forward. There are two lateral projections, called '*acuricles or ears*' at the base of the head. But these ears are not sensory organs. On the dorsal side of the head, near the anterior end, there are two black *eye-spots*. Behind the auricles the body is somewhat constructed to form a so-called *neck*. On the ventral side of body, near its middle, is found the mouth through which *proboscis* or *pharynx* protrudes. Thus, in Dugesia mouth is not situated on the head. The muscular walls of the pharynx serve for capturing the food. Behind the mouth in mature worms is situated a small *genital pore*. There are also excretory openings, the *nephridiopores*, on the dorsal surface situated laterally, but they are very minute and are, therefore, invisible. The ventral surface of the body is covered with cilia, which help in locomotion.

Body-wall

The body of a planarian is covered by a ciliated epidermis. The epidermal cell on its free surface also show

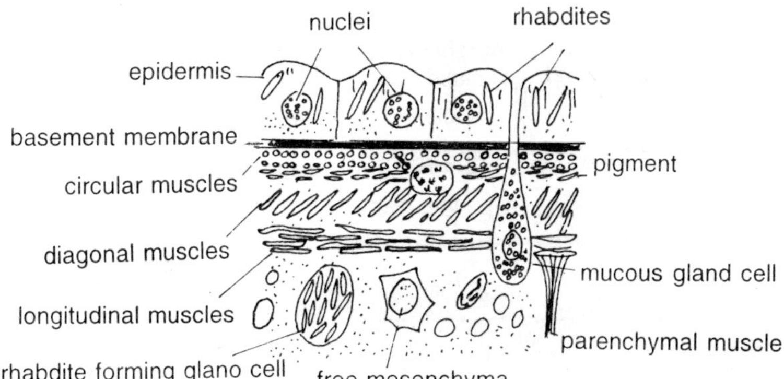

Fig. 3.2 : *Dugesia*. T. S. of body wall to show ciliated epidermis and muscie layers.

a number of microvilli. Cilia are more prominent on the ventral surface. The epidermal cells rest on a basement membrane. Among the epidermal cells the following other types of cells are also present.

(i) *Gland cell* secreting adhesive substances.

(ii) *Rhabditogen cell* which secretes and contains rhabdites.

(iii) *Sensory cell.*

Rhabdites are arranged at right angles to the body surface. They are hyaline elliptical structures secreted by rhabditogen cells. When they are expelled out they swell on contact with water forming a sticky froth around the body. This may be protective, help in gliding or even in capture of food. Some workers like *Reisinger* and *Kelbetz* (1964) suggest that the rhabdites and nematocysts are evolutionarily related.

Below the epidermis are granules and rods of pigments. The gland cells are unicellular, some occur in the epidermis but most of them are in the mesenchyme. They have long necks opening on the surface and secrete mucous.

Below the epidermis is a thin elastic basement membrane at which the epidermal cells rest. It maintains the general form of the body and provides surface for the attachment of underlying muscles.

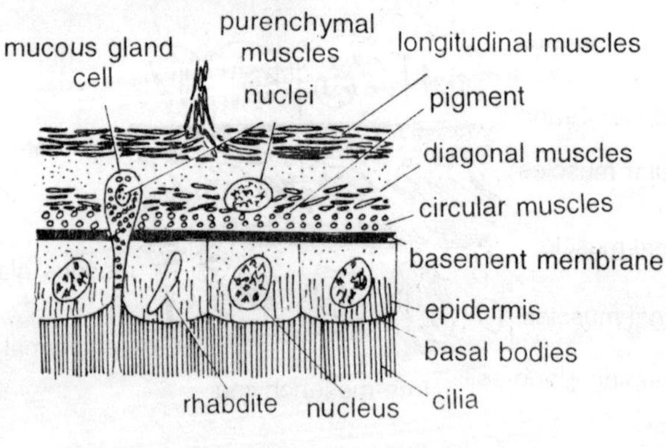

Fig. 3.3 : *Dugesia.* V.I.S. Ventral body wall.

The muscle layer consists of outermost this sheet of *circular muscle fibres,* middle broad *diagonal muscles* and innermost is the longitudinal muscles. On the ventral body wall the *longitudinal muscles* are more developed as this surface is important in the locomotion of the animal. Some muscle fibres run dorsoventrally in mesenchyme-*Mesenchyme.* The bulk of body-wall and the interior is filled with mesenchyme, which consists of connective tissue (net-like syncytium containing nuclei), free wandering cells and fluid filled spaces. The wandering cells multiply by mitosis and differentiated into other types of cells. These cells help in regeneration in the *Dugesia.* Parenchyma or mesenchyma helps in transport of food, assimilation and in the excretion of waste products.

PHYSIOLOGY

Locomotion

Dugesia is aquatic but it does not swim. The animal exhibits two types of movements:

1. Gliding movement
2. Crawling movement

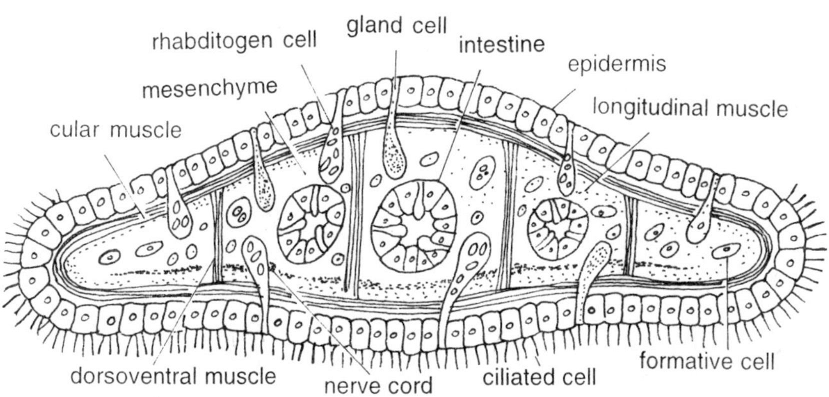

Fig. 3.4 : *Dugesia.* T.S. body showing the disposition of muscles, nerve cord etc.

1. *Gliding movement.* In gliding movement the cilia present on the ventral surface beat against a track of mucus secreted by the mucous glands. As a result of ciliary action, the animal is propelled forward in a gliding manner. The wave of ciliary action proceeds from the anterior to the posterior end and the head of the animal is slightly elevated.

2. *Crawling movement.* It is brought about by the contraction of muscles. The wave of contraction extends from anterior to the posterior end, contraction of circular and dorsoventral muscle elongate the body, the anterior is held fixed to the substratum by mucous, and the posterior end is then pulled forwards by contraction of longitudinal muscles.

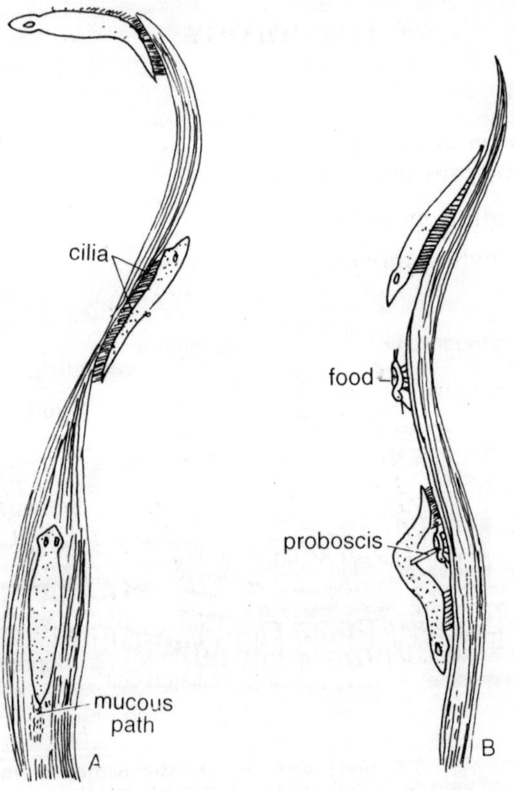

Fig. 3.5 : *Dugesia.* A = Locomotion, B = Feeding.

Sometimes transverse waves of contraction sweep the length of the body causing its undulations in the vertical plane and enabling the planaria to crawl forward. Differential action of local muscle groups produces turning or twisting movements.

Digestive System

1. *Alimentary canal.* The alimentary canal is incomplete *i.e.,* there is only one opening for ingestion and egestion. The alimentary canal comprises: mouth, plarynx and intestine.

Mouth. The mouth is an oval or rounded aperture situated mid-ventrally a little behind the middle of the body. It is surrounded by radial and circular muscle fibres from the subepidermal musculature. The muscle fibres regulate

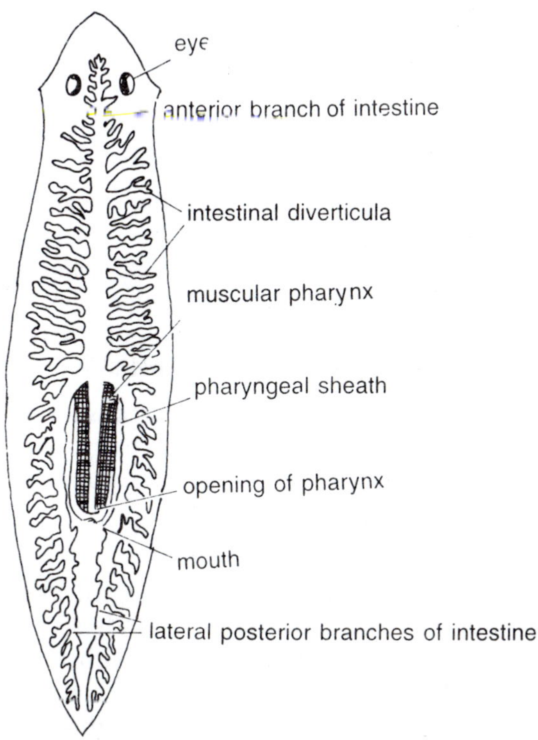

Fig. 3.6 : *Dugesia*. Alimentary canal.

the opening or closing of mouth. It leads into a large cylindrical elongated chamber. The *pharyngeal pouch*.

Pharynx. The pharynx is a tubular organ situated within the pharyngeal pouch projects posteriorly. The pouch is bounded by a muscular sheath, the *pharyngeal sheath*. A space sometimes called buccal cavity is present between mouth and the opening of pharynx. The free end of the pharynx projected out of the mouth during feeding. The projecting pharynx is called *proboscis*. This is brought about by the action of muscular wall. Pharynx opens into the intestine by way of a short *oesophagus*.

Intestine. The intestine immediately divides into three branches. The median branch extends forwards upto the head while the other two lateral branches extend backwards upto the posterior end, one on either side of pharyngeal cavity. All the three branches give off numerous branching and blind diverticulal which end into the mesenchyme. The much branching intestine is a means of increasing the surface area for digestion, absorption and distribution of food.

Histology of alimentary canal. The pharyngeal wall shows a complicated histological structure. It consists of nine layers from the surface to the lumen: epithelial cells, longitudinal muscle layer, circular muscle layer, outer gland cells, nerve plexuses, inner gland cells, longitudinal muscle layer, circular muscle layer and an endodermal epithelial lining.

The intestine is simple and thin consists of a single-layered epithelium, the gastrodermis. The gastrodermal cells are of two types phagocytic cells and granular cells.

Food and feeding. Normally *Dugesia* is *holozoic* and canivorous subsisting for the most part on crustaceans, insect larvae and small worms. During feeding, it glides through the action of cilia and the adhesive glands secrete adhesive secretion. When prey comes in contact with the secretion it becomes firmly adherent. *Dugesia* then folds the anterior end of its body over the prey and thus immobilizes it. The pharynx is then produced out through the mouth and through its peristaltic action sucks up small particles of the prey into its intestine.

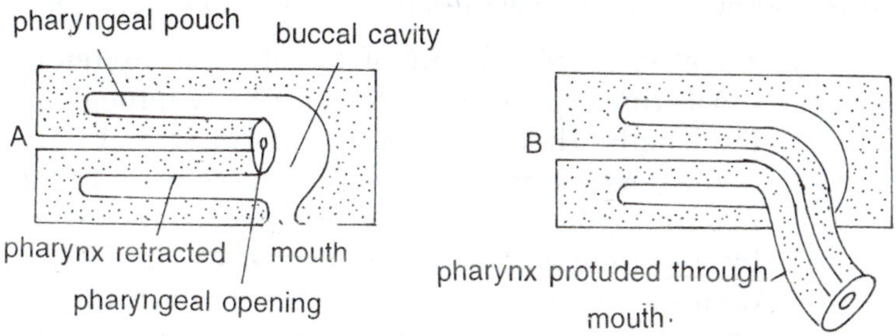

Fig. 3.7 : *Dugesia.* Diagrammatic representation of A = Retracted pharynx, B = Protruded pharynx.

Digestion. Digestion is initially extracellular. Food is broken down by pharyngeal proteolytic enzymes and endopeptidases secreted by the gland cells lining the intestinal wall. The broken pieces of partially digested food are then digested by intracellular method. The phagocytic cells of the intestine engulf these particles which are digested by intracellular endopeptidases inside cytoplasmic vacuoles in acidic medium. After about 8 hours the vacuolar content becomes alkaline and digestion in this medium is completed by exopeptidases, lipases and carbohydrases. The excess of food is stored as fat and proteins in the intestinal cells and mesenchyma.

Freshwater planarians are able to withstand prolonged experimental starvation. In extreme cases, they utilize part of enteron and all the mesenchyma and reproductive system. In fact, the body volume may be reduced to as little as 1/300 of the original.

Egestion. Since there is no anus, the egestion of undigested food material takes place through mouth. It is brought about by the contraction of the muscles of body-wall.

Respiration. Mode of respiration is aerobic but there are no special organs for this. Oxygen from the surrounding water is directly into the body and carbon dioxide is diffused out through the entire body surface. Flattened body is an admirable adaptation for this type of respiration.

Excretory system. The excretory system or *protonephridial system* of *Dugesia* consists of a pair of *longitudinal excretory canals* and numerous tiny enlargements known as flame cells.

(i) *Excretory canals.* A pair of longitudinal excretory canals running on each side of the body throughout the length. They are highly inter-coiled and open to the dorsal surface by several minute poses called nephridiopores. Anteriorly two canals are united by a *transverse vessel,* in front of eyes. Each canal divides into a number of branches which further divides into very fine blind *capillaries* which bear flagella at the inner end (*Wilson* and *Webster,* 1974). The terminal structure is composed of flame cells or cryptocytes.

(ii) *Flame cells.* Each flame cell represents the unit of excretory system. It is a large, tubular, protoplasmic body. It is produced into numerous protoplasmic

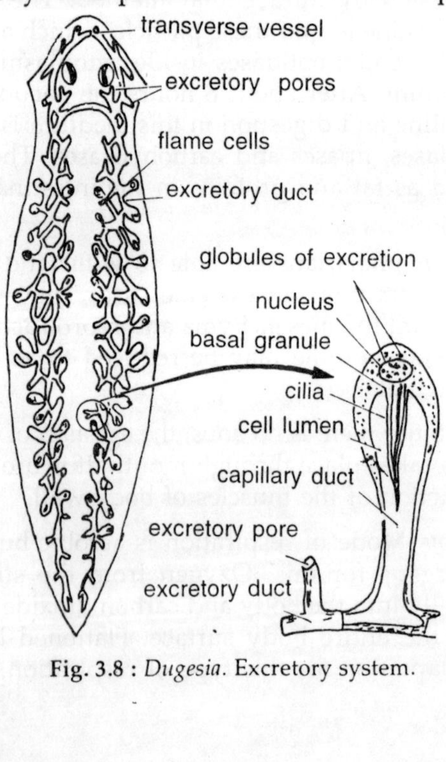

Fig. 3.8 : *Dugesia*. Excretory system.

processes reaching the mesenchyme. Its terminal part is swollen into a bulb and contains a conspicuous nucleus. The flame cell has an intracellular space which is continued into capillaries. The space encloses a bunch of cilia which vibrate giving the appearance of flickering candle flame.

Besides the flame cells, certain glandular cells called the *arthrocytes* or *paranephroytes* occur in close contact with the excretory canals and also help removal of excretory products.

Physiology of excretion. Excretory substance is collected from the mesenchyme and is transferred into the cavities of flame cells. The beating of cilia of flame cells causes hydrostatic pressure by which the fluid waste passes into the longitudinal trunks and goes out of nephridiopores.

According to Storer, Wolcott, Guyer, Hickman etc. the flame cells simply help in excretion of nitrogenous wastes. But according to some workers like Mckanna, (1968); *Prusch,* (1976) etc. the excretory system in *Dugesia* is largely osmoregulatory bailing out excess of water. As osmoregulatory organs, the protonephridia might function in the following manner. The beating of cilia of flame cells would produce a negative pressure within the end of the tubule, causing filteration of fluid across the membrane covering the fenestrations. Selective reabsorption of ions, especially K+ and Cl-, by the tubule would yield a hyposmotic fluid.

Nervous system. The nervous system shows a much better development than what we found in the coelenterates and can be differentiated into the *central* and *peripheral systems.*

Just below the epidermis and behind the eyespots, there is a bilobed *brain* or *cerebral ganglia* in the form of an inverted 'V'. Both the ganglia are connected by several transverse fibres. From the brain, several thin nerves are given off anteriorly and to the eyes. From the brain two *lateral longitudinal nerve cords* extend posteriorly along the ventral side. The two nerve cords are connected together by *transverse connectives.*

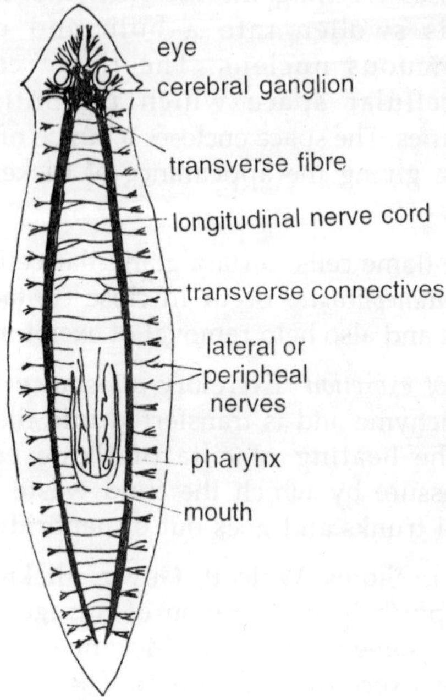

Fig. 3.9 : *Dugesia*. Nervous system.

The peripheral system is present in the form of a *sub-epidermal plexus* (just below the epidermis) and a *sub-muscular plexus* below the muscle layers in the mesenchyme.

The nerve-cells or neurons may be unipolar, bipolar or multipolar. All the neurons possessing neurites are similar. These neurites cannot be differentiated into axons and dendrons.

Sense organs. Dugesia possesses special sensory organs, called receptons. These are photoreceptors or eyes, auricular organs, tangoreceptors and rheoreceptors.

(i) *Photoreceptors or eyes.* The eyes are two dark spots on the dorsal surface of the anterior end, one on either side of the median line. Each eye has the form of a cup whose opening faces anterio-posteriorly.

Structure of each *eye* is very simple. It is made up of highly pigmented *retina*. Inside the cup are situated numerous nerve-cells, the *fibrillae* of which are in contact with the retina and their opposite ends are continued into an optic nerve, which is joined with *brain*. The nerve cells enter the cup through the opening. Ectoderm over the eye is not pigmented. *Lens* is absent in eye.

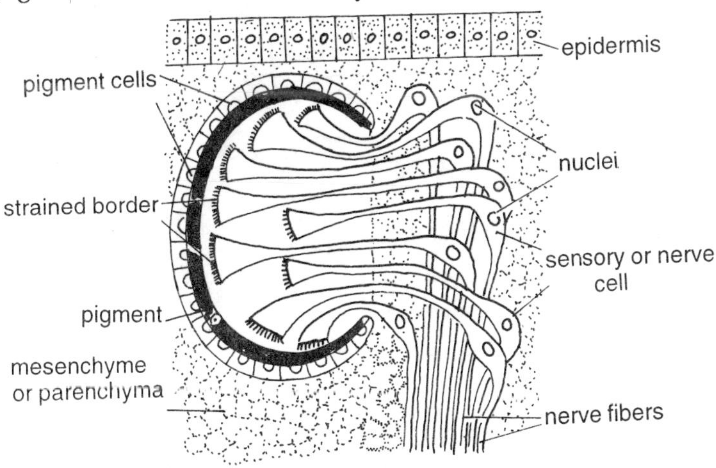

Fig. 3.10 : *Dugesia*. V.S. of eye.

Eye are capable of a crude discrimination of the direction of light. The pigment cup serves as a shield and light can enter only through its opening to stimulate the photosensitive expanded ends of retinal cells, thus the animal can detect the direction light. The animal is negatively phototactic and is most active at night. It the eyes are removed it can still react to light. However the reaction is much slower. This shows that it possesses some light sensitive cells over the general body surface.

(ii) *Ciliated pits*. Besides eyes, there are *ciliated pits* which are also sense-organs. They are situated on either side of the head, having special sensory cells with long cilia. These ciliated pits are connected with nervous network. Tentacles, which are found in some turbellaria with long cilia, are also supposed to be organs of chemical sense, which help in finding food.

(iii) *Auricular organs.* These are sensitive to chemical and are chemoreceptors. These are in the form of a pair of whitish ciliated groores, lying one on either side of the head near the base of the auricles. If the auricles are removed the planarian cannot locate food.

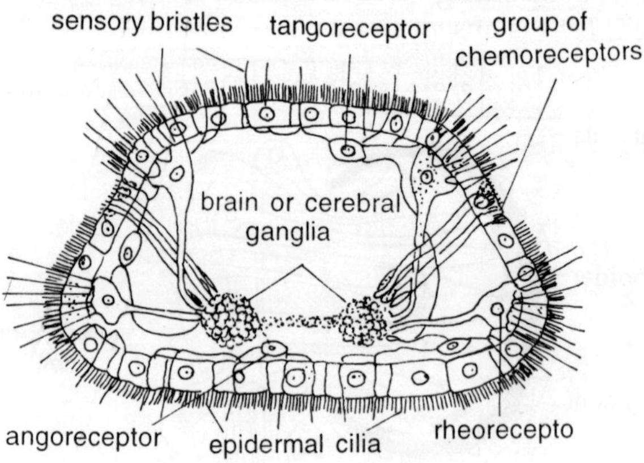

Fig. 3.11 : *Dugesia*. Digrammatic T.S. Showing various receptors.

(iv) *Tango receptors.* These are sensitive to touch temperature etc. A tangoreceptor is a small cell with fine sensory bristles that project over the surface except one that poses to the brain. They are particularly concentrated on the ventral surface around the mouth, auricles and the body margins. Thus ventral surface is positively thigmotactile while the dorsal surface is negatively thigmotactic.

(v) *Rheoreceptor.* These are sensitive to water current. They are of general occurrence and their sensory processes or bristles project much beyond the level of cilia.

Reproductive System

Dugesia reproduces with asexually as well as sexually.

(1) Asexual Reproduction

Asexual reproduction occur by the transverse binary fission. The plane of fission usually forms posterior to the pharynx. The posterior part attaches to substratum by mucus, the anterior part stretches forward until the animal snap into two. Each half regenerates the wanting parts resulting in the formation of two complete planarians. Some planarians fragment into two or more pieces each growing into a complete individual. Asexual reproduction is restricted to some species only. Some species reproduce asexually in one season, and may develop sex organs in the other, but these are some species that reproduce exclusively asexually and rarely develop sex organs.

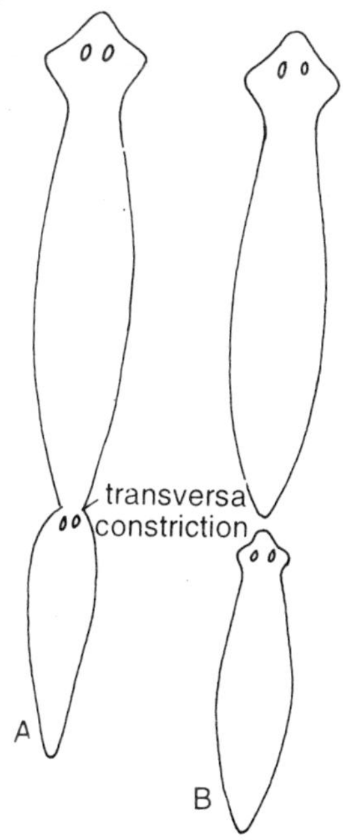

Fig. 3.12 : *Dugesia*. Transverse binary fission.

(2) Sexual Reproduction

The reproduction organs are more complicated in these animals. The gonads *i.e.*, testes and or aries are derived from the meroderm and there is a system of tubules and chambers in which fertilisation occurs.

The animal reproduces in early summer and the gonads are developed temporarily in the breeding season. The *Dugesia* is bisexual or hermaphrodite but self-fertilisation does not take place and the cross fertilisation is the rule. After breeding season the gonads degenerate and disappear and the worm reproduces asexually.

Male reproductive system consists of a large number of small spherical testes, which are situated on both the sides of the body. Each testis is connected with minute tubes, called *vasa effertia*. There form two large ducts called *vasa deferentia*. The two vasa deferentia run parallel to one another towards the posterior end, where these bend mesially towards the penis. Each vas deferens dilated to form the seminal or spermlducal vesicle with thick muscular wall. The sperms are stored here. The penis shows two regions, the anterior large spherical *penis bulb* and the posterior narrow conical *penis papilla* with muscular wall. The lumen of penis

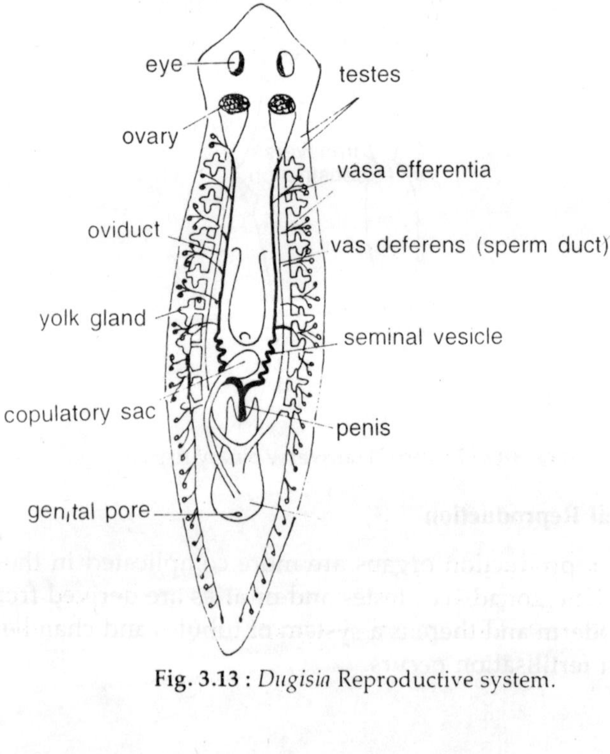

Fig. 3.13 : *Dugisia* Reproductive system.

is very narrow and called ejaculatory duct. The penis papilla opens into the genital atrium which terminates in the common gonopore, situated mid-ventrally behind the opening of month. The lumen of penis bulb is formed by *bulbar cavities*. The bulb contains numerous uni *cellular glands,* the prostate glands which open into it lumen.

Female reproductive organs. The female reproductive organs consist of a pair *ovaries,* which are small apherical bodies situated laterally just behind the head. From each ovary arises an oviduct which extends posteriorly through the parenchyma. Immediately after its emergence each oviduct presents a dilation, which is known as *seminal receptacle.* The two oviducts unite to form a common vagina which opens into the *genital artium.* The oviducts along their whole length are surrounded by numerous vitelline glands which open into oviducts by short *vitelline ducts.* Because of the close association of oviduct and vitelline glands, these are sometimes collectively referred to as the ovovitelline ducts. Opening into the genital chamber is found a muscular sac-like *copulatory sac* or *uterus* or *bursal canal.* The bursal canal opens into a large sac the *bursa conplatrix* which receives penis and temporarily stores the sperms received during copulation. It opens below into female genital atrium.

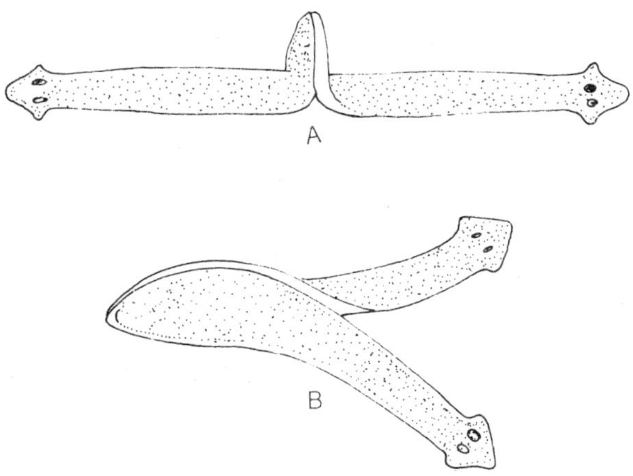

Fig. 3.14 : *Dugesia.* Copulation

Copulation

Although *Dugesia* is hermaphrodite, yet *cross-fertilization* is a rule. *Copulation* takes place between the two worms in which their posterior and ventral surface come together and the penis of each is inserted into the *genital atrium* of the other. In this way sperms are deposited in the *seminal receptacles* of the mating partner. The sperms are stored inside the copulatory sac for some time and then travel along the oviducts to reach the seminal receptacles. The eggs are deposited in the seminal receptacle and are fertilized. Thus the fertilization is internal. During breeding season, each worm may copulate many.

Cocoon Formation

The zygote thus formed, pass backwards through the ovovitelline ducts and reach the genital atrium of ovovitelline ducts and reach the genital atrium of female. In their way they are mingled with yolk cells which are discharged by yolk glands into the oviducts. The deposition of yolk outside

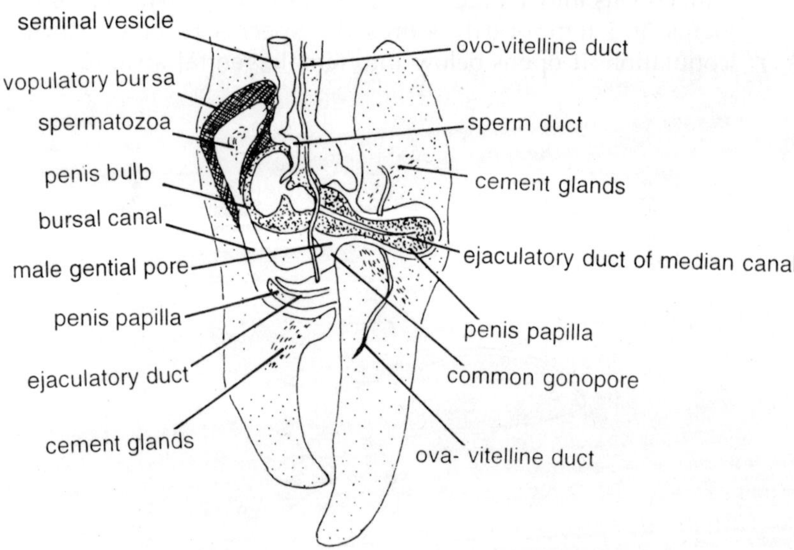

Fig. 3.15 : *Dugesia*. A pair of copulating animals in section.

the zygote is unique condition in platyhelminthes. Such eggs are called *ectolecithal*. When the eggs are reached in genital atrium 5-10 eggs along with yolk cells are enclosed in a protective, proteineous shell forming a cocoon or an egg-capsule. The capsule is coated by a sticky substance secreated by the cement glands that surround and open into the common genital atrium. The capsule is now extruded through the genital pore. As it is passing out, sticky coating is drawn out into a sticky stalk that fastens it to some object chiefly the under surface of stones. The capsules are laid in succession at intervals of a few days.

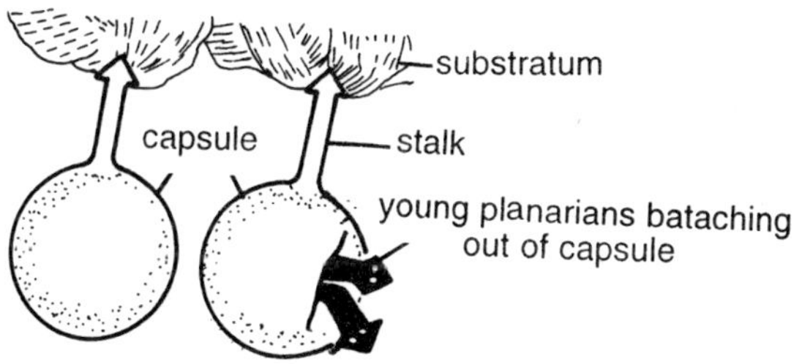

Fig. 3.16 : *Dugesia*. Young one haching from the cocoon.

Development

Development starts soon after the cocoon is laid and the hatching occurs in 2 to 3 weeks depending upon the temperature of water. The cleavage is *spiral* and *determinate* type. No larval stage is formed hence the development is *direct*. A young planaria or *juvenile* glides out of the capsule from each egg.

Regeneration and Transplantation

Those Turbellaria that reproduce asexually also possess high powers of regeneration and consequently have served as material for innumerable studies on regeneration. In other Turbellaria regeneration of a head is limited to pieces from anterior regions or to pieces containing the cerebral ganglia; or else the animals are incapable of regenerative processes

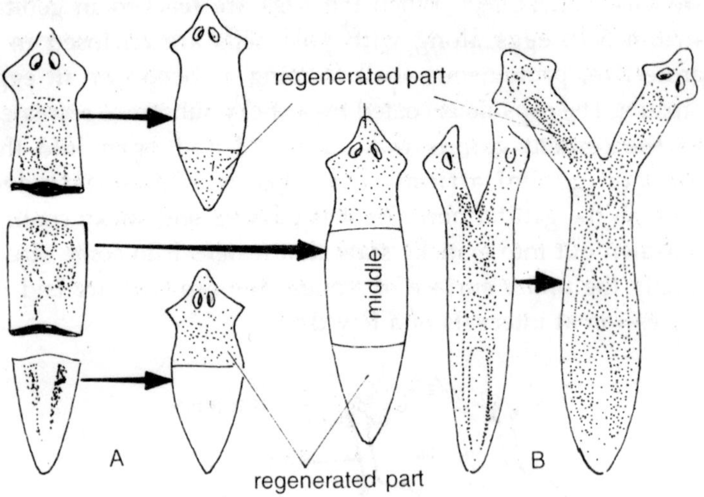

Fig. 3.17 : *Dugesia* Regeneration. A = Transverse, B = Longitudinal.

beyond wound healing. Many forms can replace posterior regions, including the pharynx and the copulatory apparatus when the cuts are made posterior to the cerebral ganglia. The region anterior to the ganglia is incapable of regeneration in any case. Polarity is strongly retained in pieces, *i.e.*, the anterior cut surface regenerates a head, the posterior cut surface, a tail; but alternations of polarity occur or may be induced or increased by experimental means. The capacity to regenerate, the rate of the process, the amount of new tissue produced, and often the type of anterior regenerate are markedly related to level, being greatest or most normal anteriorly and declining posteriorly. In short, the flatworm body like the coelenterate is a polarized system; the polarisation consists in some sort of physiological gradation along the anteroposterior axis. The characteristics of and evidence for the existence of such an axial gradation have recently been summarized by Child (1941). Regeneration involves two processes: regeneration proper, or *epimorphosis*, in which new tissue grows out from the cut surface in partial

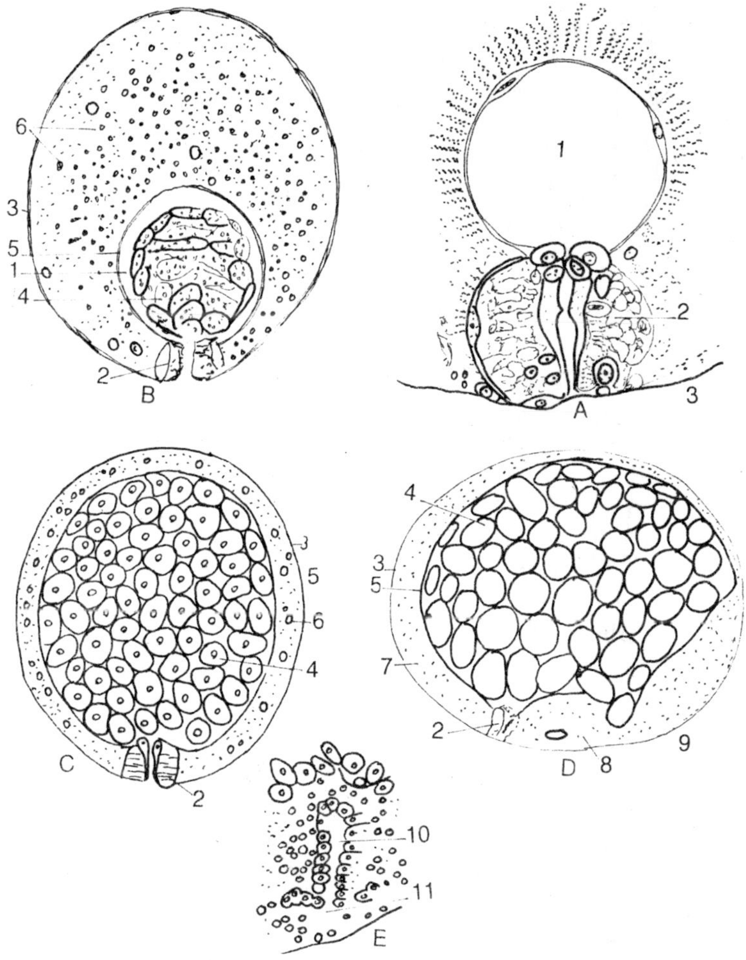

Fig. 3.18 : Development of triclads (continued). *A*. Embryonic pharynx and intestine completed. *B*. Embryo beginning to ingest the yolk (*after Metschnikaff*, 1883). *C*. Ingestion of yolk completed *D*. Three embryonic masses formed. *E*. Formation of definitive pharynx by appearance of spaces. 1, embryonic intestine; 2, embryonic pharynx; 3, bounding membrane; 4, mass of ingested yolk; 5, inner bounding membrane; 6, blastomeres; 7, anterior mass to form brain; 8, middle mass to form definitive pharynx; 9, posterior mass to form mesenchyme and reproductive system; 10, pharynx lumen; 11, pharyngeal cavity.

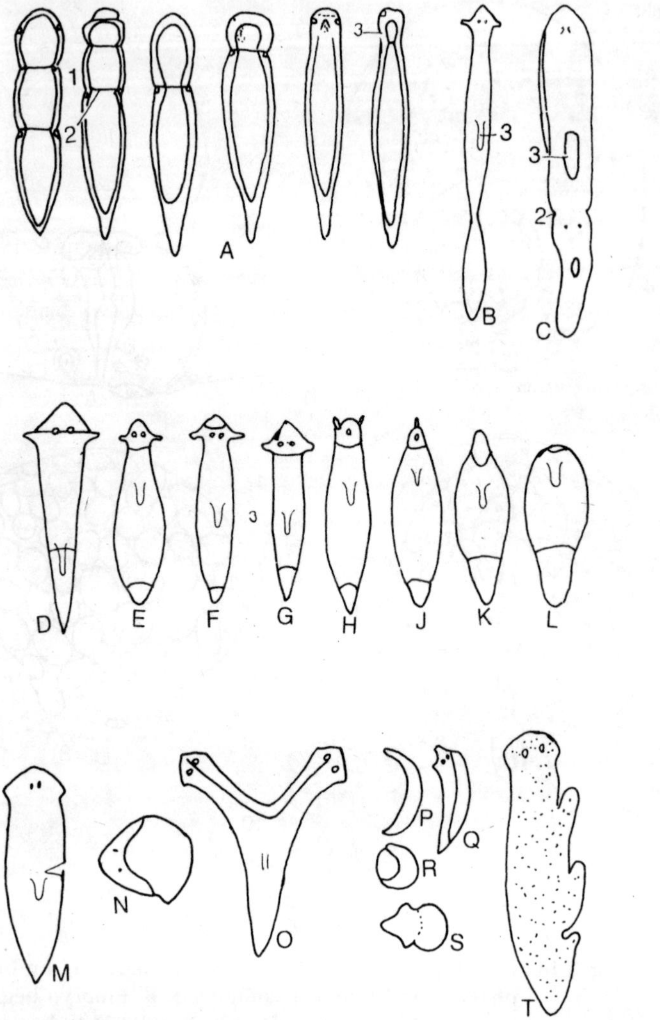

Fig. 3.19 : Asexual reproduction and regeneration. *A*. Six stages in the regeneration of a piece of *Stenostomum* containing fission planes; the best developed zooid absorbs parts anterior to it and becomes a new worm. *B*. *Dugesia* in fission. *C*. *Planaria fissipara*, Trinidad, triclad that forms fission planes prior to fission. *D*. *E*. Effect of level on regeneration. *D*. Regenerate from anterior piece; has large head, posteriorly located pharynx. *E*. Regenerate from

replacement of the part removed; and *morphallaxis*, or the working over of the old tissue and organs to fit into the new animal. The relative roles of the two processes vary with different species and with size and level of the piece; but in any event reorganisation of the old parts is necessary.

The Acoela have limited powers of regeneration, as illustrated by experiments on *Polychoerus* and *Aphanostoma*. Portions removed anterior to the cerebral ganglia or rear portions are replaced perfectly; in pieces cut posterior to the ganglia the amount of anterior tissue formed is less the more posterior the level, but in no case is a normal anterior end regenerated, for ganglia and statocyst are not replaced. Longitudinal halves replace the missing half.

Those rhabdocoels that form chains of zooids naturally regenerate readily, and the regeneration of *Stenostomum* has been much studied. Pieces from almost any region reorganise into complete new worms. In some species the anterior cut end forms the new head while in others preexisting fission planes or planes formed after cutting absorb the part anterior to them and become the head; or in cut pieces containing fission planes, these planes may continue to differentiate and fission may ensue, leaving a piece anterior to the fission

---

postpharyngeal piece; has small head, more anterior pharynx. *F-L*. Series to show decline of head regeneration in *Dugesia dorotocephala* with decreasing size and more posterior location of piece. *F*. Normal regenerate. *G*. Teratophthalmic regenerate with fused eyes. *H*. Teratomorphic regenerate with one eye and auricles shifted anteriorly. *J*. Extreme teratomorphic type with fused auricles. *K*. Anophthalmic regenerate without eyes or auricles. *L*. Headless, no anterior regeneration. *M, N*. Regeneration of lateral triangular piece at right angles to original polarity. *O*. Double heads produced by keeping open an anterior split. *P, Q*. Regeneration of a narrow side piece, polarity retained. *R, S*. Alternative method of regeneration of a narrow side piece, polarity at right angles to original. *T*. Head regenerated on each of three anteriorly directed oblique cuts. Line in each case separates new from old tissue. 1, ciliated pit; 2, fission plane; 3, pharynx.

plane that may or may not be able to regenerate a head. The regeneration of *Microstomum* is similar to that of *Stenostomum*. Some other rhabdocoels may regenerate the tail tip but in general rhabdocoels that lack asexual reproduction and alloeocoels are incapable of any regeneration, although wounds are healed by epidermal migration.

Fresh-water planarians of the genera *Dugesia, Phagocata, Polycelis,* etc., display remarkable powers of regeneration and have been the object of numerous researches. In general, pieces of moderate size from any level form a complete new worm. With reduction in size of piece, complete worms may regenerate from even very short pieces in some species, while in others short pieces tend to regenerate reduced or imperfect heads or may remain headless, and this tendency is greatest the more posterior the level. The size and normality of the regenerated head the rate of anterior regeneration, and the length of tail produced are generally greater the more anterior the level from which the piece was cut and decline in posterior levels. In *Dugesia dorotocephala* extensively studied by Child, a series of anterior regenerate ranging from normal heads through several types of reduced heads to headlessness occurs in relation to level and shortness of piece. This series shows increasing suppression of the middle region with consequent approximation and fusion of eyes and auricles. These facts constitute some of the evidence for the existence of the axial gradation mentioned above.

Reconstitution also involves the formation of a new pharynx. Pieces that do not regenerate anteriorly may form a pharynx if they were cut from postpharyngeal regions of the original worm; but pieces cut anterior to the pharynx will not regenerate a pharynx unless some sort of head first regenerates. The new pharynx may arise in either the new or old tissue and is nearer the anterior end the more posterior the level from which the piece was cut and the less normal the anterior regenerate. If a head has regenerated the old portions of the intestine are worked over so as to reproduce in the regenerated worm the typical three-branched pattern. In general, the ability of a piece to regenerate into a complete worm of typical morphology depends upon the regeneration of a head at the anterior cut surface. This means that the head controls the pattern of the morphology, *i.e.*, the spatial arrangement and proportions of parts. This control is an expression of the dominant position of the head in the axial gradation already mentioned.

Some other points in planarian regeneration may be touched on briefly. Very short pieces especially from anterior regions may regenerate biaxial heads, *i.e.*, form a head at each cut surface. Isolated heads cut off just behind the eyes also often regenerate a reversed head at the cut surface. In some species short posterior pieces produce biaxial tails. After oblique cuts, a head usually forms at the most anterior point, a tail at the most posterior point, of the cut surfaces. Lateral longitudinal pieces usually regenerate with reference to the original polarity; but short lateral strips, or those cut lateral to the ventral nerve cords, or small triangular lateral pieces tend to regenerate the new head at the cut surface, *i.e.*, at right angles to the original polarity (Morgan, 1900b; Olmsted, 1918; Beyer and Child, 1930). By means of appropriate splits, double or multiple heads or tails may be induced at will; heads will grow out from the body sides behind forwardly directed gashes, tails behind posteriorly directed ones one or more "crotch" heads may arise in the angle of posterior longitudinal splits if these are carried far enough forward (Beissenshirtz, 1928; Silber and Hamburger, 1939; Hull, 1940); heads will form along the edges of windows through the body and in fact an endless variety of bizarre forms can be produced by appropriate cuts[1]. Regeneration is also affected by external factors such as temperature, osmotic pressure, and chemical composition of the medium; many experiments have been performed on the effect of various salts, anesthetics, radiations, etc., on regeneration.

If a sexually mature planarian is cut in two between the pharynx and the copulatory apparatus, the latter degenerates and the piece regenerates into an asexual worm. The large anterior piece regenerates a new copulatory apparatus if active testes are present. The presence of ovaries is not pertinent. This result indicates that the formation and maintenance of the copulatory apparatus depend on the presence of testes.

The Dendrocoelidae are incapable of asexual reproduction and have limited powers of regeneration. Anterior pieces retaining the head regenerate the missing posterior parts, but decapitated pieces can reform the head only when taken from anterior levels, in our common dendrocoelid, *Procotyla fluviatilis*, only from the anterior third. Similar conditions obtain in marine triclads (*Procerodes, Cercyra*), some of which cannot regenerate a head when the cut lies behind the cerebral ganglia whereas in others prepharyngeal pieces may regenerate a head. Pieces from pharyngeal and postpharyngeal regions are incapable of head regeneration but may regenerate a tail at the anterior cut surface. In marine triclads slight dilution of the sea water accelerates regeneration; other osmotic changes retard. Some land planarians, as *Bipalium kewense*, regenerate very well (Morgan, 1900a); others, as *Geoplana* (Goetsch, 1933), regenerate perfectly when the pieces are large, less so in short pieces.

The polyclads in general have poor powers of regeneration (Child. 1904a, b, c, 1905a, b, c, d, 1910a; Levetzow, 1939; Olmsted, 1922b). Pieces containing the cerebral ganglia regenerate perfectly. Pieces cut behind the cerebral ganglia are incapable of anterior regeneration but can regenerate parts behind their level of section.

Grafting experiments may also be performed with suitable fresh-water planarians; species of *Dugesia* are generally used. When crosspieces are fused in normal or reversed orientation, the result is similar to that found with hydroids *i.e.*, regeneration occurs according to the original polarity of each piece except that the polarity of very short pieces fused to larger pieces may be altered. Triangular or rectangular pieces may be excised and grafted into windows made into other parts of worms (Miller, 1939, Okado and

Sugino, 1937; Santos, 1931) in normal or altered orientation. Such grafts from the cephalic region containing much nervous tissue are more effective in inducing a head-like outgrowth and causing reorganization of adjacent parts of the host with production of one or more pharynges, the more posteriorly they are grafted into the host. Similar grafts from tail regions implanted anteriorly are influenced by the host to develop into a tail regardless of their original polarity.

Heads regenerated from grafts or at a cut influence adjacent regions for a certain distance, reorganising them to a greater or less degree into postcephalic regions (with pharynx and intestinal branches), and prevent the formation of another head at a cut surface or otherwise within a certain distance.

The histology of the regenerative process has been repeatedly studied in planarians with not entirely concordant results; the differences are probably real, depending on the species employed. The wound is considerably closed by muscle contraction (this contraction is omitted if the cutting occurs in an isotonic medium) and the adjacent epidermal cells flatten, become syncytial, and creep over the cut surface. If the exposed surface is sufficiently large, regeneration cells may participate in the formation of this covering membrane, which later differentiates into the typical epidermis. A mass of cells, variously known as regeneration cells, formative cells, etc., accumulates beneath the covering membrane and differentiates into the various missing parts. In the case of anterior regeneration strands of these cells attach to the cut nerve cords and differentiate into brain and eyes. They also adhere to the cut intestinal branches and build on appropriate terminations, and differentiate into muscle cells and gland cells in appropriate locations. It appears, however, that in some species or with regard to some systems, the new parts may be formed by proliferation of the adjacent cells of the old parts, i.e., Bandier (1936) states that in the regeneration of land planarians both cut nerve ends and intestinal branches grow from their own cells. The source of these regeneration cells is still under

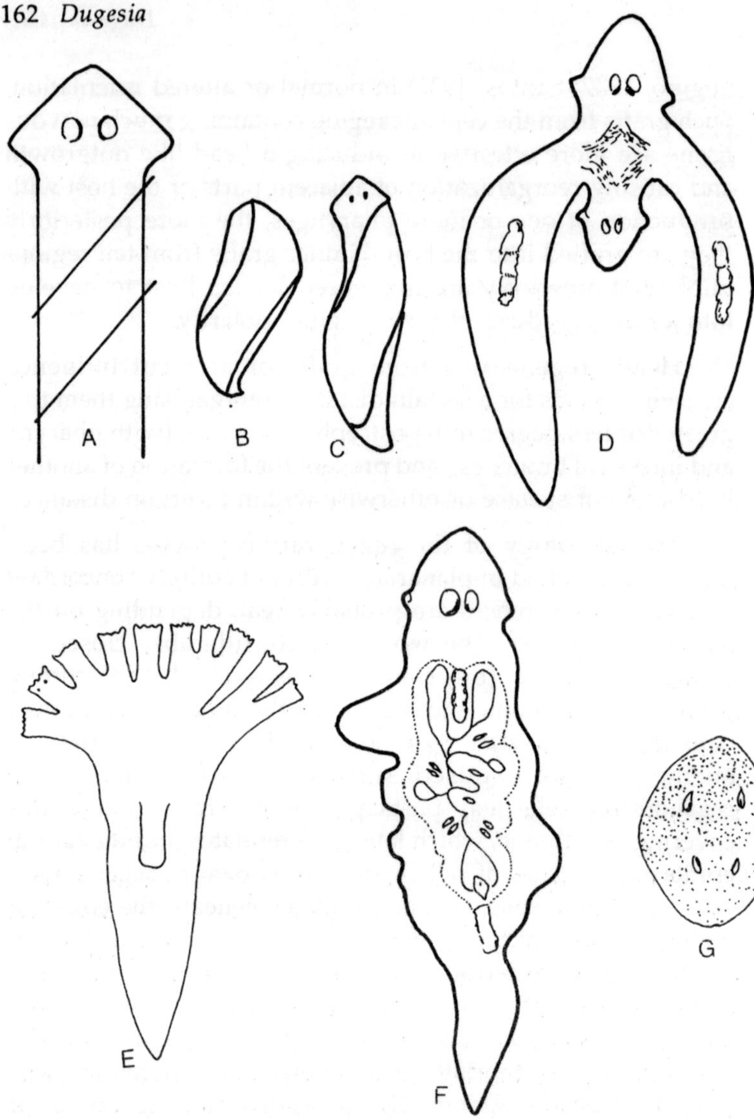

Fig. 3.20 : Regeneration of planarians (continued). *A-C*. Regeneration of an oblique piece. *D*. Crotch head, produced at anterior end of posterior split that extends far forward. *E*. Numerous heads produced on *Dendrocoelum lacteum* by repeated splits. *F*. Many heads regenerated along the sides of a central long window. *G*. Head regenerated at posterior cut surface of head piece of *Curtisia foremanii*.

dispute. There is little doubt that they comprise preexisting cells that migrate to the cut surface either with or without multiplying by mitosis. According to some authors, they consist entirely of the free cells of the mesenchyme; others state that, after cutting, various body cells, such as gastrodermis, yolk glands, gland cells, etc., undergo dedifferentiation into amoeboid cells and join the free mesenchymal cells to form the mass of regeneration cells. An extreme view is to the effect that the regeneration cells are persistent embryonic cells held as a reserve to function in regeneration and formation of the reproductive system. It appears to the author that there is no evidence in support of such a view and that some of the evidence cited above oppose it. Curtis and Schulze have found that there are conspicuously fewer free mesenchymal cells in planarians with poor powers of regeneration, as *Procotyla fluviatilis,* than in those with high powers, as species of *Dugesia;* but their tabulation fails to show any marked decline in number of free mesenchymal cells in those regions of *Procotyla* that cannot regenerate a head as compared with regions that can. Further regions unable to regenerate a head can regenerate a tail and also can produce (from free mesenchymal cells) the entire complicated copulatory apparatus.

Attempts to shed some light on these problems have been made by studying the effect of X rays and radium on regeneration. (Bardeen and Baetjer, 1904; Gianferrari, 1929; Meserve and Kenney, 1934; Schaper. 1904; Strandskov, 1937; Weiand, 1930). The subject has been reviewed by Curtis (1936) who has researched extensively on the matter. In general, exposure to sufficient dosage of X rays and radium inhibits regeneration altogether in most species, although *Dugesia dorotocephala* will regenerate after any exposures that do not kill.[2] The failure to regenerate is caused, according to some of the above workers, by the killing off of the free mesenchyme cells, according to others by the inability of these cells after radiation to migrate or to differentiate. The result therefore only proves what was already known, namely, that cells of the type of free mesenchymal cells are responsible for regeneration. It does not settle the questions whether these cells are in fact embryonic rests and whether they get contributions from dedifferentiated tissues. The difference in susceptibility to radiation between the mesenchyme cells and other body cells evidently varies with different species.

Chapter—4

Fasciola Hepatica

Fasciola belongs to class Trematoda. Most of the trematods living in other animals inhibit the digestive tract or its accessory tubes or cavities, and it seems probable that those species living outside gut are derived from gut dwelling forms. Probably the most important of these flukes, *Fasciola hepatica,* was the first digenetic fluke to have its life-cycle uncovered. *Thomas* (1883) described its life-cycle. *Reinhard* (1957) redescribed its life-cycle in an interesting way.

SYSTEMATIC POSITION

Phylum	-	**Platyhelminthes**
Class	-	**Trematoda**
Order	-	**Digenia**
Family	-	**Fascioliodae**
Genus	-	***Fasciola***
Species	-	***hepatica***

Geographical Distribution

World wide distribution occurs in areas where sheep, cattle and goats are raised and there is a niche for lymnaied snails. In U.S.A., organisms founding the gulf Coast, the North-West, rocky Mountains, Florida etc. in U.S.A., it is endemic.

Habits and Habitat

Fasciola hepatica is derived from Greek and Latin words - (*L., fasciola* = small bandage; Gr. *hepar,* = liver) is a *digenetic parasite, i.e.,* its life history is completed in two hosts - a primary vertebrate host and a *secondary* or intermediate invertebrate host. Adult *f. hepatica* occurs in the liver and bile passages of the primary host which is usually sheep; but, at times, it may occur in some other vertebrates, like goat, horse, dog, ass, ox, deer, antelope, rabbit, elephant, monkey, etc. Humans are occasional host for *F. hepatica,* and infections are most often obtained through eating watercress (*Nasturtium officinale*) in salid. A single host may harbour as many as about 200 of these flukes in its liver, which may consequently cease to function normally. This effect is known as liver-rot. *F. hepatica* spends a part of its life history in an intermediate host, a freshwater snail which is either *Limnea truncatula* or some specific species of *Planorbis, Fossaria modicella, Pseudosuccinea* etc.

Shape, Size and Colouration

The body is large, soft and fleshy. It is dorsoventrally flatenned and leaf-like having somewhat oval appearance. The size varies from 1.8-5.1 cm in length and about 0.4-1.5 cm in breadth near the midbody. The colour of body is generally pinkish. The margins are blackish or brownish due to the presence of vitelline glands. The alimentary canal is brownish is colour due to the presence of ingested bile of the host.

Structure

The *anterior end* of *F. hepatica* is triangular while *posterior end* is tapering. The anterior end terminates into a projection, which is called apical or cephalic cone or *head lobe.* At the apex of the head lobe is situated at small opening, the *mouth,* which is surrounded by an anterior muscular sucker, called oral or *anterior sucker.* At a short distance behind the head lobe is situated another saucer-shaped sucker known as *ventral sucker* or acetabulum, situated on the ventral side. The ventral sucker is superficial and has no connection with the interior of the body.

Fasciola hepatica was formerly known as *distomum hepaticum* and the name of the groups to which it belongs was known as *Distomidae*. These names were adopted as the mouth and anterior sucker were regarded as two suckers to two mouths (*di,* two; *stoma,* mouth). The word *stoma* was used in trematodes for suckers, which are characteristic structures. But now it is certain that the mouth cannot be claimed as a sucker.

One the dorsal surface there is a single aperture in the middle line at about one-third of the length of the animal from the anterior end. This aperture is the opening of *Laurer's canal*. This aperture appears during breeding season.

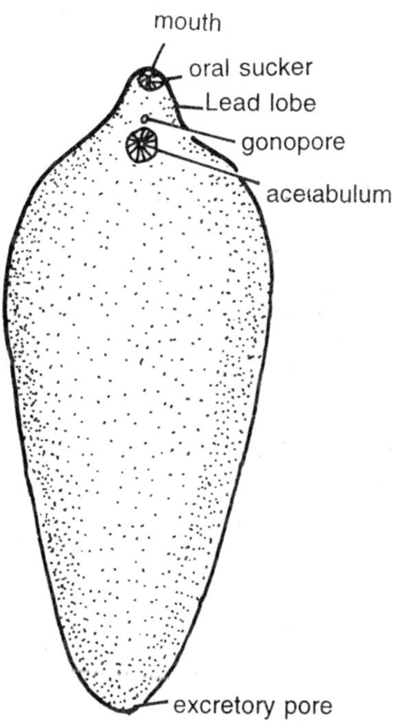

Fig. 4.1 : *Fasciola hepatica*. External features.

Body Wall

Body wall in Fasciola is characteristic and adapted to parasitic mode of life. It lacks a cellular epidermis and consists of the following layers.

1. *Cuticle.* It is the outermost layer having stout, rough, blunt scales which project over the general surface. Cuticle is a secretion of *mesenchyme,* either of special type of cells called gland cells formed between muscle-fibres or of general mesenchyme. Cuticle is definitely non-chitinous and is made up of *scleroteins.* It is situated over the body and protects the liver-fluke from the harmful effects of the different types of juices produces in the host. Below the cuticle is situated a thin transparent layer known as *basement membrane.*

2. *Musculature.* Immediately beneath the cuticle are three layers of muscles-the outermost circular layer, the medical longitudinal layer and the inner digonal layer. In addition to these muscles other make up the suckers and pharynx. Still others, the parenchymal muscles, are dorsoventrally oriented along the lateral regions of the body.

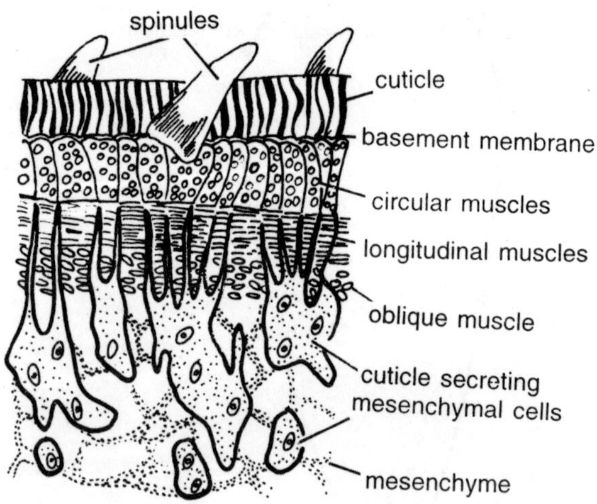

Fig. 4.2 : *Fasciola.* V.L.S. body wall seen under light microscope.

3. *Parenchyna or mesenchyma.* Immediately beneath the musculature are numerous cells, known as *beta cells* or *mesenchymal cells.* Some of the beta cells posses cytoplasmic tubules that penetrate through the muscular layers and reach the cuticle. These specialise beta cells are thought to secrete cuticular materials that reinforce the existing cuticle.

Electron microscopic studies by *Threadgold* (1968) have revealed that the so called *cuticle* of Fasciola is actively a thick layer of multinucleated protoplasm containing mitochondria, canals of endo-plasmic reticulum, vacuoles and vesicles. It is termed as *tegument.* It is continuous with *tegument secreting cells* lying embedded in the mesenchyme. The outer surface of tegument is thrown into numerous folds. These increase the surface area considerably for the absorption of fluids of the host. Further, tegument is perforated by fine *pore canals.* Through these canals fluids are absorbed into the mesenchyme.

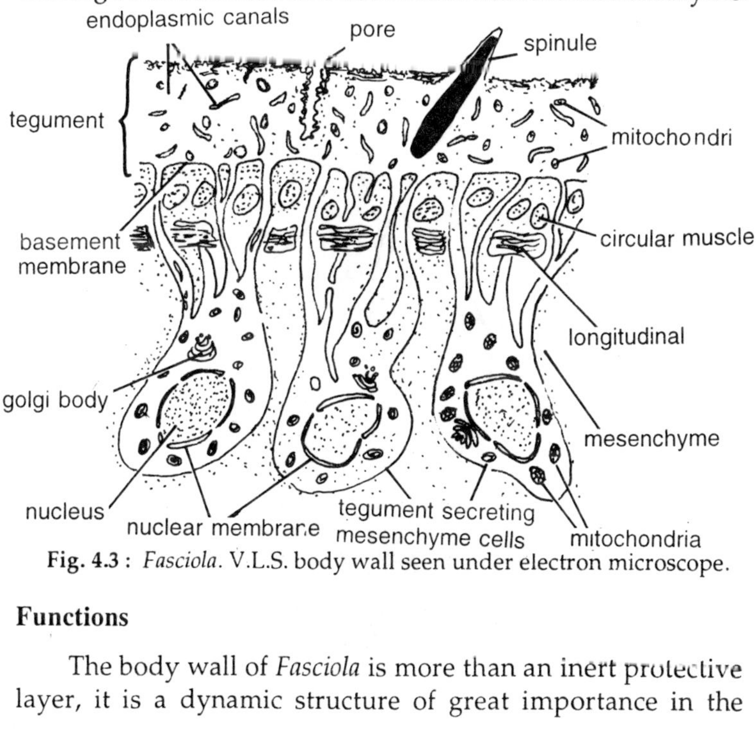

Fig. 4.3 : *Fasciola.* V.L.S. body wall seen under electron microscope.

Functions

The body wall of *Fasciola* is more than an inert protective layer, it is a dynamic structure of great importance in the metabolism of the organism. Although the animals derive a great deal of their nutrients through ingestion, certain chemicals such as certain amino acids, which are necessary as energy sources and as raw materials for the synthesis of complex body chemicals, are sbsorbed through the body wall. It is also possible that certain metabolic wastes are discharged through cuticle.

Digestive System

The digestive system is complex and highly branched with *moth* but no *anus*. It consists of the following parts:

Mouth. It is a small opening situated ventrally at the apex of head lobe and surrounded by oral sucker.

Oral chamber. Mouth opens into a sort but funnel-shaped oral chamber.

Pharynx. It leads into a short, rounded, muscular pharynx with thick wall and pharyngeal glands.

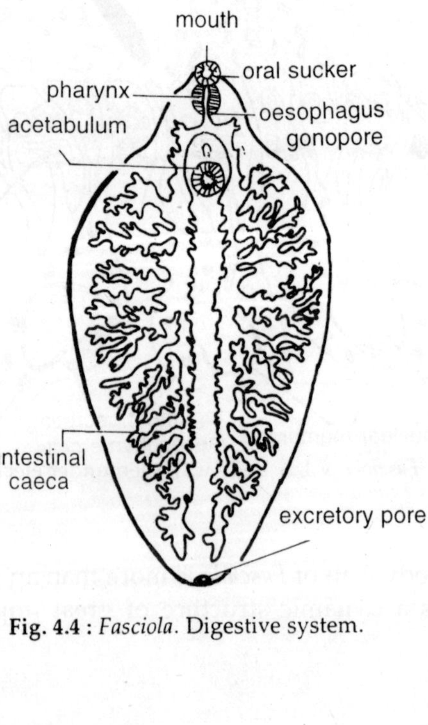

Fig. 4.4 : *Fasciola*. Digestive system.

Oesophagus. The pharynx communicates with a short-narrow oesophagus joining the intestine.

Intestine. The intestine immediately lbifurcates into two lateral branches, running one on either lateral side of the body upto the posterior end where these end blindly. A large number of caeca or diverticula are given out along either side of its whole length. The diverticula of the outer side are large and highly branched, while those of the inner side are short and unbranched.

Food and feeding. The food of pasciola is bite and blood as it lives in the bile passage of sheep and cattle. It takes about 0.18-0.2 ml of blood per day. Small microscopic debris is also taken as food. The food is taken by the sucking effect of the pharynx and is pumped into the intestine.

Digestion. The exect manner of food digestion is not well understood. Digestion is supposed to be extracellular, carried in the intestinal ceaca. Proteolytic enzymes are reported in the animal. The digested food diffuses into the surrounding mesenchyme. The mesenchyme distributes digested food to every part of body. As earlier mentioned the tegument also helps in the absorption of certain amino acids as well as monsaccharides (glucose and fructose). The reserve food material is in the form of glycogen and fat which are stored in mesenchymal cells and muscles.

Egestion. As there is not anus, the undigested food materials are egested out through mouth.

Respiration

In the *Fasciola,* the respiratory organs are absent so that the respiration takes place through general body surface by the process of diffusion. Since the animal is an endoparasite, the rate of respiration is very low and is *anaerobic*. In this process, there is no necessity of oxygen. Glycogen undergoes anaerobic glycolysis to form pyruvic acid. Pyruvic acid is decarboxylated to form carbon dioxide and an acetyl group. The latter combines with coenzymes to form acetyl coenzymes

A. Acetyl coenzyme A is finally condensed and reduced to form fatty acids. During the process energy is released.

Excretory System

The excretory system consists of flame cells or protonephridia connected with an intricate system of excretory ducts.

1. *Excretory ducts.* A single and median longitudinal excretory canal which runs just behind the ventral sucker and terminates into the *excretory pore*. It is remified throughout the body. In the anterior region it receives two ventral and two dorso-lateral excretory canals. The dorsal ducts before opening into the main longitudinal excretory canals are united so as to form a common duct. Main excretory canal as well as dorsal and ventral-lateral excretory canals divide into small capillaries or tubules, each of which ends internally into a *flame cell*.

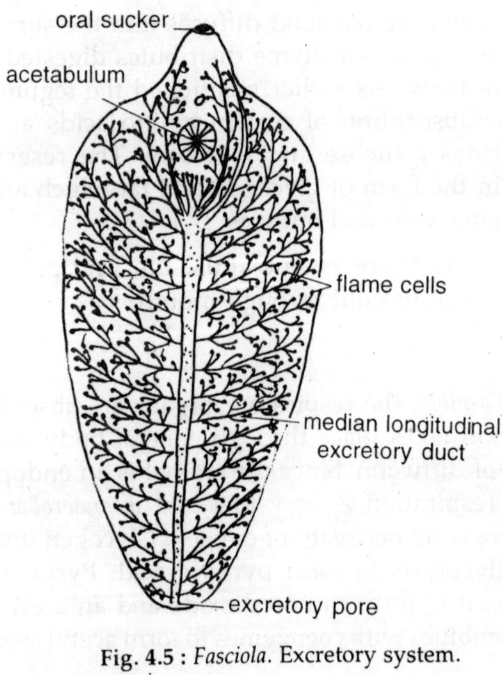

Fig. 4.5 : *Fasciola*. Excretory system.

2. *Flame cells.* The flame cells are characteristic, each has a thin elastic wall with a nucleus and a cavity containing many long cilia arising from basal granules or blepharoplasts. The cilia is a living animal vibrate like the flickering of the flame, hence named as flame cells. The cavity of flame cells is continued with the cavity or capillary. The shape of these cells is irregular due to preudopodial processes into the mesenchyma.

Physiology of Excretion

The excretory products are carbon dioxide fatty acids and ammonia. These are diffused into the flame cells from the surrounding mesenchyme. The fluid moves through the tubule by the action of cilia into the excretory ducts and finally passes out of the body through the excretory pore. All the ducts except the median longitudinal duct are lined with cilia. The course of the flow of excretory products is as follows:

Excretory products - -> mesenchyme - -> flame cells
↓
main longitudinal canal <- - vessels <- - twigs <- - Capillaries
↓
Excretory pore - -> outside.

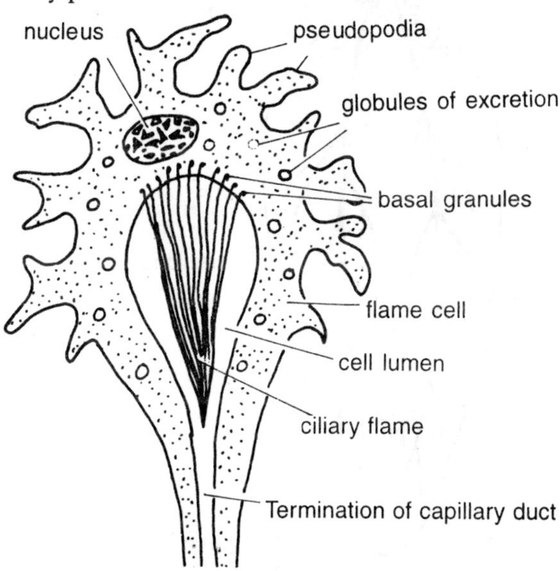

Fig. 4.6 : *Fasciola*. A flame cell.

Nervous System

The nervous system is well developed in spite of the fact that due to sluggish and parasitic habitats such as large correlation is not needed. The nervous system includes:

(a) *Central nervous system.* The cerebral ganglia are prominent masses of nervous material joined together by a nerve ring around the oesophagus. From each cerebral ganglion arise three longitudinal nerve cords, dorsal, ventral and lateral. Of these the lateral nerves are much longer than the other. They alone reach the posterior end of body. They are interconnected at intervals by transverse commissures.

(b) *Peripheral nervous system.* From each cerebral ganglion three nerves arise anteriorly and innervate the head lobe and the oral suckers. Fine nerves arise from nerve cords all along their course and supply the various parts of the body.

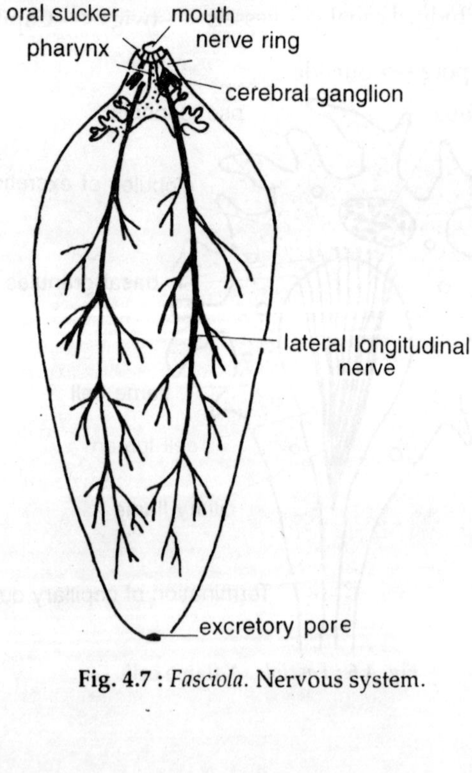

Fig. 4.7 : *Fasciola*. Nervous system.

Sense Organs

Due to parasitic mode of life, special sense organs are absent. It, however, possesses tectile organelles or *Tangoreceptors*. These are bulb-like nerve ending. These are abundant in the sucker region.

Hormones in Fasciola

In *Fasciola* serotonin is present and it seems to play a role in the tissues comparable to that played by adrenaline in mammalian tissue. It is not known whether seroctonin acts as a humoral component of the nervous systems, as it is reported to do in some other animals. The tissues of *Fasciola* contain an acetylcholine-like substance and cholinesterase; the specificity of cholinesterase for acetylcholine has not been established, but it has been suggested that these components are related to activity of nervous system.

Insulin has been reported to have an effect on the carbohydrate metabolism but thc evidence are lacking.

Reproductive System

Fasciola hepatica is hermaphrodite or disocious. The glands are well developed and the male and female genital ducts open into a common chamber, the genital atrium. It is situated anteriorly in the body and opens to the exterior though the common genital aperture or gonopore, located ventrally in front of the acetabulum.

The Male Reproductive Organs

The male reproductive organs are:

(i) A pair of testes

(ii) A pair of *vasa deferentia*

(iii) Seminal vesicle

(iv) An ejaculatory duct

(v) Cirrus and cirrus sac.

(i) *Testes*. These are greatly ramified and tubular structures situated in the middle part of the body, one behind the other (*tandem arrangement*).

(ii) *Vasa deferentia*. A narrow delicate duct, the *vas deferens*, emerges out from each testis and runs forward to met the fellow of opposite side forming the common sperm duct just below the ventral sucker.

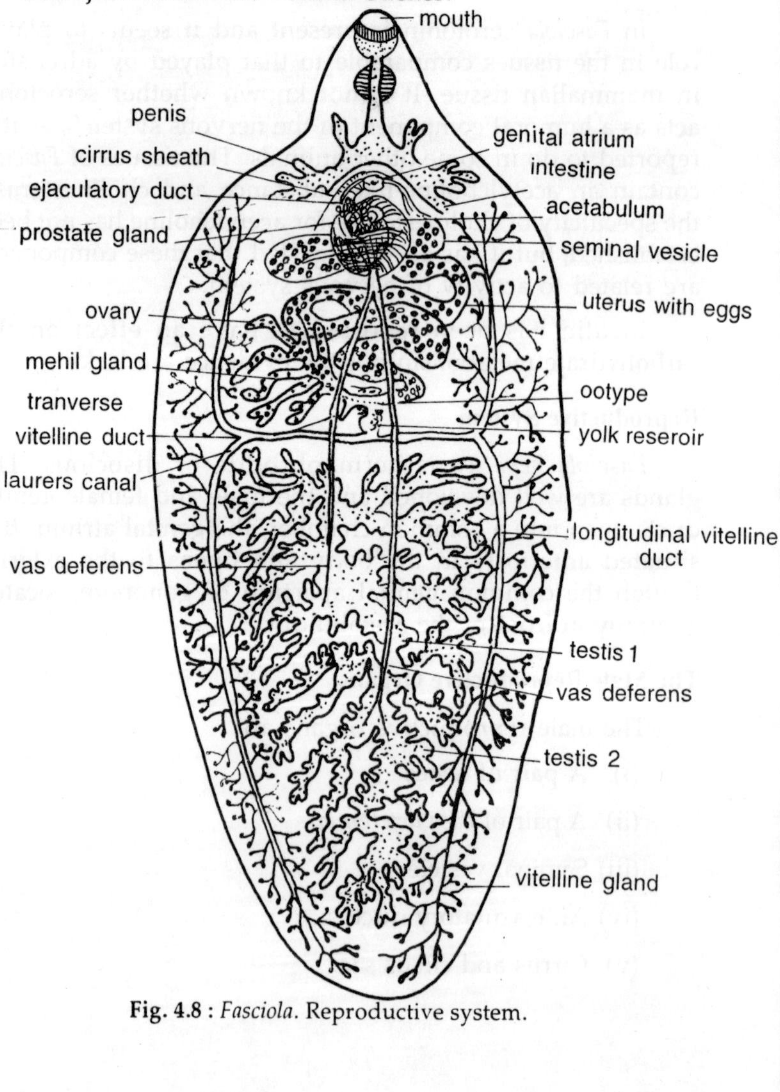

Fig. 4.8 : *Fasciola*. Reproductive system.

(iii) *Seminal vesicle.* The common sperm duct immediately leads into a large muscular pear-shaped sac, the *seminal vesicle,* which lies in front of ventral sucker.

(iv) *Ejaculatory duct.* The ejaculatory duct arises as a fine convoluted tube from the seminal vesicle and runs forward in a zigzag fashion through the cirrus to open into the *genital chamber* through the *male genital pore.*

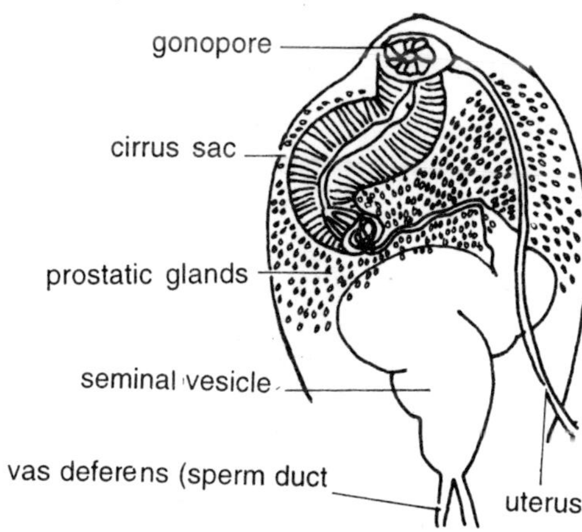

Fig. 4.9 : *Fasciola.* Details of male reproductive system.

(v) *Cirrus and cirrus sac.* The cirrus or penis is a muscular cylindrical structure traversed internally by the passage of ejaculatory duct. It can be everted out and drawn in through the genital pore and thus helps in

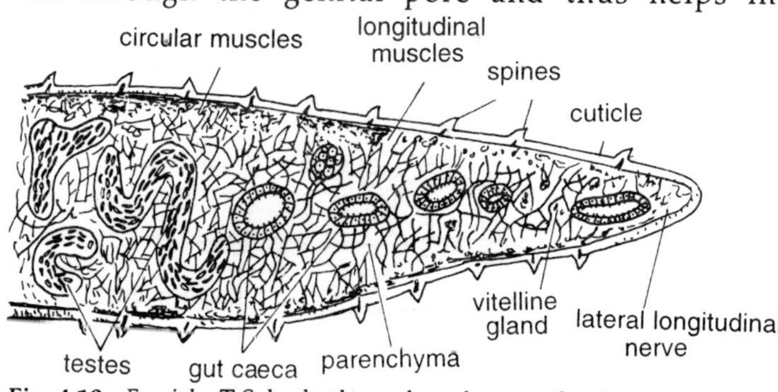

Fig. 4.10 : *Fasciola.* T.S. body through male reproductive system.

copulation. The cirrus and the seminal vesicle both are enclosed in a bag - like *cirrus sac.*

(vi) *Prostate glands.* Numerous unicellular prostate glands are found around the ejaculatory duct.

Female Reproductive Organs

The female reproductive organs are:

(i) Ovary

(ii) An oviduct

(iii) Uterus

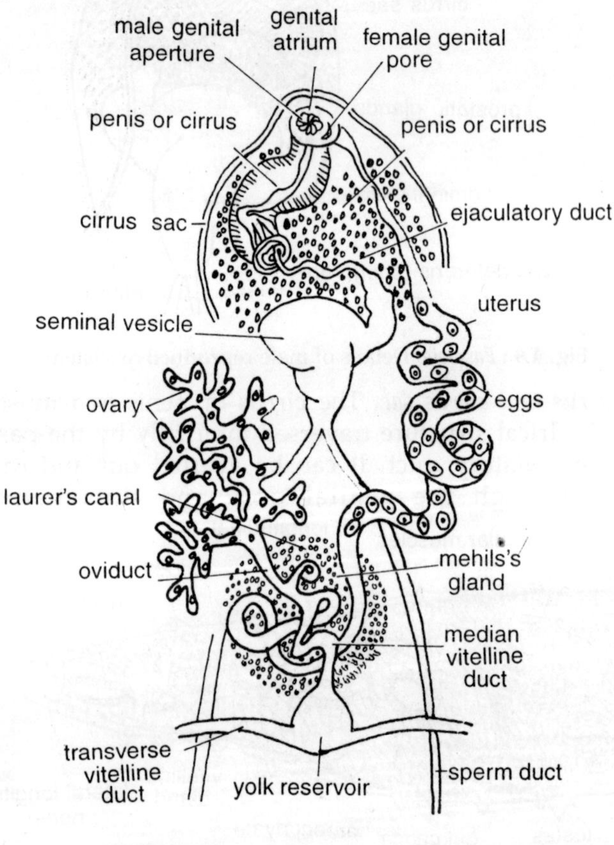

Fig. 4.11 : *Fasciola.* Details of male and female reproductive system.

(iv) Laurer's Canal

(v) Mehli's gland

(vi) Vitelline glands and vitelline ducts.

(i) *Ovary*. The ovary is a large, highly branched tubular structure which is situated on the right side in front of testes in the mesenchyme. It occupies the anterior one-third of the body.

(ii) *Oviduct*. From the inner side of ovary arises a short, narrow and convoluted duct, the *oviduct*. It runs down ward to join the uterus.

(iii) *Uterus*. It is a long wide and highly convoluted tube that extends upto the genital atrium, opening to it through the female genital aperture, close to male genital pore on the life-side. It contains a large number of capsule containing fertilized eggs.

(iv) *Laurer's canal*. The laurer's canal arises from the oviduct and acts as a sperm duct. It develops a temporary opening on the dorsal body surface during breeding season. It may serves as vagina.

(v) *Mehli's gland*. At the junction of oviduct, vitelline duct and uterus a mass of unicellular gland lies known as *Mehli's glands*. The function of these glands remains unclear. It was believed that these glands were merely responsible for the formation of shell but *Stephenson* (1947), *Rennison* (1953) and *Johri* and *Smyth* (1955) have shown definitely that this cannot be the case, because the shall of eggs is rich is various proteins, phenols and phenolases that are not present in these glands. Several view have been proposed to explain the function of Mehli's glands:

(a) These glands secrete a fluid that enhances the hardening or tanning process of newly formed eggs. (b) The secretion of these glands cause release of the shell globules from the vitelline glands. (c) the secretion forms a thin membrane around the cells forming eggs, and the shall globules than built up from within this membrane (d) The secretion lubricates the uterus facilitating passages of the eggs. (e) The secretion activates sperms which are passed down. All or some of these hypothesis may be valid.

(vi) *Vitelline glands and vitelline ducts.* The vitelline glands or the yolk glands are in the form of numerous minute vesicles or follicles arranged on either side of the body from one end to the other. A fine duct arises from each vesicle and after uniting with similar ducts from other vesicle opens into the *lateral longitudinal ducts.* The two longitudinal ducts are connected in the middle line by a *transverse vitelline duct.* A medium duct arises mesially from the transverse vitelline duct joins the oviduct.

The functions of vitelline glands is now know. In addition to contributing yolk material for incorporation within the egg, the vitelline glands also secrete large globules known as *shell globules* - which envelop the developing egg and eventually coalesce and become hardened to from the shall. This hardening process involves the tanning of protein present within the coalesced globules by quinone. Thus vitelline glands contribute both yolk and shell material.

Phenolase

Phenol ------------> quinone ---------------->Tanned protein

Protein ------------> Protein

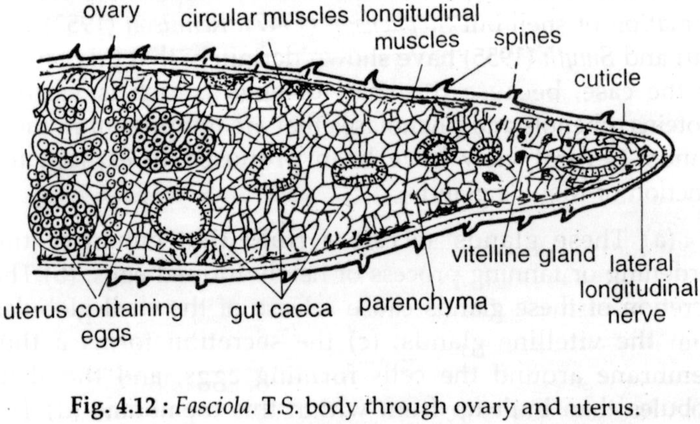

Fig. 4.12 : *Fasciola.* T.S. body through ovary and uterus.

Ootype. In some cases of the oviduct after receiving Laurer's canal and common vitelline duct swells to form ootype. In *F. hepatica* according to *Stephenson* (1947) the ootype is absent.

LIFE HISTORY

Copulation and fertilization. Cross-fertilization takes place by insertion of cirrus of one fluke into the Laurer's canal of another fluke. Transference to biflagellate sperms directly into the oviduct is possible. Secretion of prostate glands and as mentioned above, the secretion of Mehli's gland keep the sperms active for fertilisation. Laurer's canal become invisible after copulation. Fertilisation takes place in oviduct. Self-fertilisation is also seen here. The sperms enter the uterus of same fluke through female genital pore and pass down to reach oviduct where fertilisation takes place.

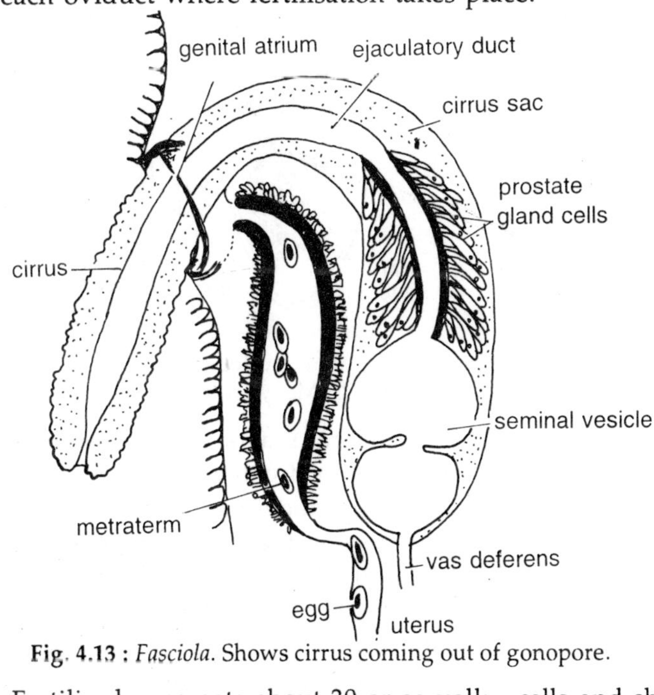

Fig. 4.13 : *Fasciola.* Shows cirrus coming out of gonopore.

Fertilised eggs gets about 30 or so yolk - cells and shell globules from vitelline glands by the way of various vitelline ducts. The shell is formed with a lid or *operculum* on one side. *Rowen* (1956,57) demonstrated that there is a viscous granular cushion at the opercular end of the egg, situated immediately beneath the operculum. The zygote and mass of yolk - cells enveloped by the shell constitute the capsule. When complete the *capsule* passes into the uterus.

Capsule formation occurs throughout the year and one fluke may produce about 50,000 capsules. A single fluke may produce a total of 2,00,000 capsules in about 11 years.

The capsules differ in shape and size in different animals. Hence, with the size and shape of capsules the parent flukes are recognized.

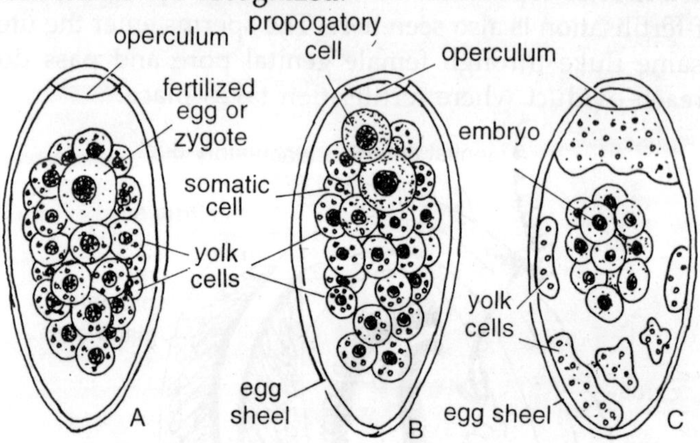

Fig. 4.14 : *Fasciola*. Early stages of development. A = Fertilized egg, B = Two cell-stage, C=Many cell stage.

Development. Cleavage starts immediately after fertilisation. The fertilised egg divides into a small granular *propagative cell* and a large ectodermal *somatic cell*. Somatic cell divides and gives rise to the body-wall of the larva. Propagative cell later on divides into a cell similar to *somatic cell*, which takes part in the development of mesoderm, endoderm, and a cell called *germ cell* afterwards reaches the posterior part of the body and forms germ-ball consisting of 14 to 20 germ cells.

Capsules then escape through the gonopore and are liberated into the bile capillaries from when they reach the intestine of the host and are passed out with the faeces.

The development remains arrested while the embryos remain in the faces, they may survive in wet faecal matter for several months. If washed free, the development of the embryo proceeds. The optimum temperature for development ranges from 10-30°C. At 30°C the encapsulated embryo differentiates into *miracidium larva* within eight days. The availability of oxygen is also a factor. The fully formed miracidium is a more or less conical little animal covered with ciliated epithelium. It does not look at all alive an adult fluke but does show resemblance to some free-living flat worm.

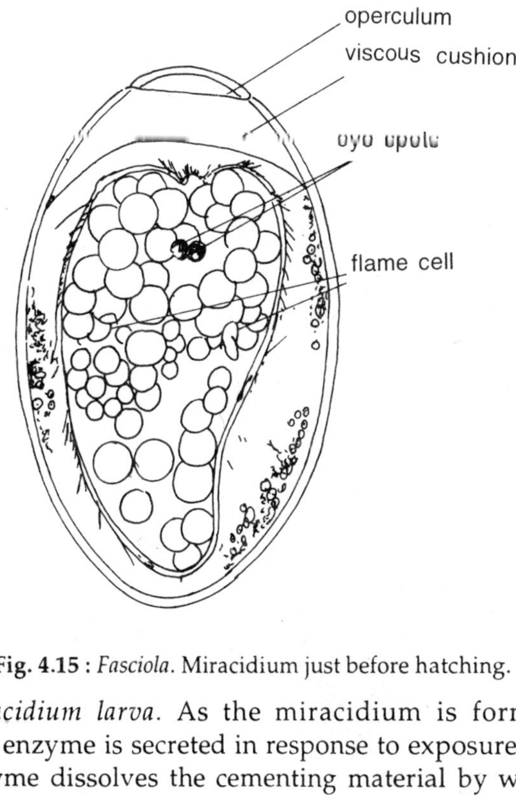

Fig. 4.15 : *Fasciola*. Miracidium just before hatching.

Miracidium larva. As the miracidium is formed the hatching enzyme is secreted in response to exposure to light. The enzyme dissolves the cementing material by which the operculum is attached, thus releasing the operculum. Expansion of the granular cushion, accompanied by exosmosis of salts and other materials from within the egg pushes off the operculum. The hatching occurs is water and free-swimming miracidium penetrates the intermediate host.

Structure of miracidium. It is an elongated, about 0.07 m m long creature with a broad anterior end and a narrow posterior end. Anteriorly there is a *conical* or *apical papilla* or terebratorium. It is cliated except in the region of apical papilla. However, anterior cilia are slightly longer.

The ectoderm is formed by 21 hexagonal plate like cells which are arranged in 5 rows. The I row consists of 6 small plates. The II row contains again 6 plates, III row has 3 plate cells, 4 plates in IV row and 2 plates in V row. The arrangement of the ciliated plates is quire constant and is of considerable value is determining phylogenetic relationship. In *F. hepatica* the cells of first row arranged are two dorsal,

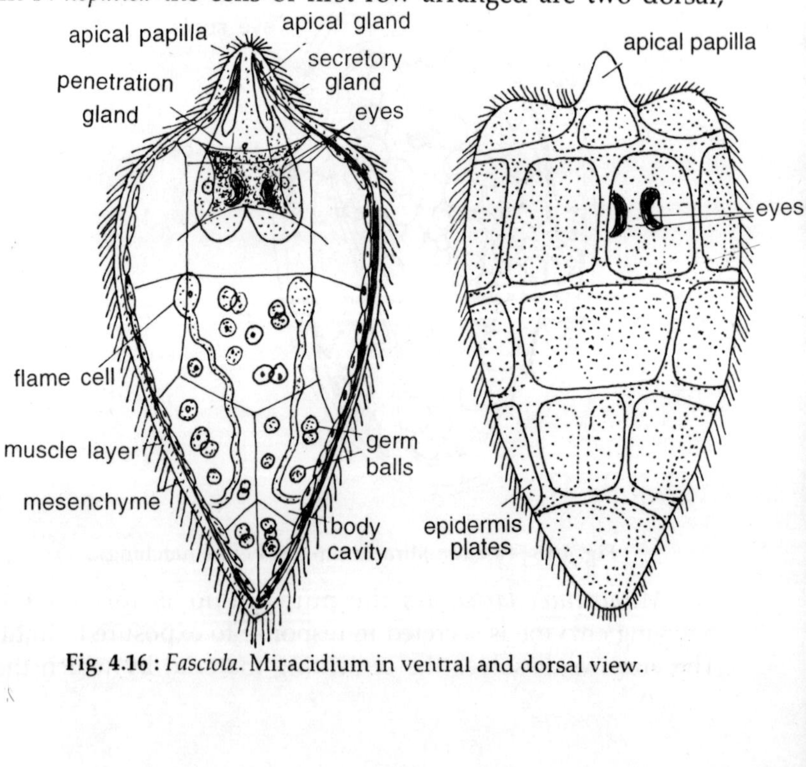

Fig. 4.16 : *Fasciola.* Miracidium in ventral and dorsal view.

two ventral and two lateral in position. The cells of second row are three dorsal and three ventral. In the third row the cells are one dorsal and two ventro-lateral. The fourth row has two right and two left cells and the fifth row has one left and one right cells.

The ectoderm is followed by musculature consisting of outer circular and inner longitudinal muscles fibres. The musculature is followed by a layer cells forming the sub-epithelium. The ectoderm musculature and sub-epithelium forms the body wall of larva.

At the anterior tip of the miracidium is found a mobile apical papilla. A pair of cephalic *glands* or *penetration glands,* located in the mesenchyme, empty to the outside near the apical papilla. The secretion of these glands aids in dissolving the host's tissue during penetration process. Besides a sac-like multinucleated *apical gland* is located between the two cephelic glands. Previously it was supposed to be the gut of larva.

The sensory organs of miracidium includes two eyespots. Eyes contain pigments and are situated at the surface of the second row of epidermal plate over the brain. The eyes may be primary receptors involved in the process of hatching. 17 kept in dark few miracidia hatch, but illumination rapidly induces hatching. The *brain* of the miracidium is a large cephalic ganglion lying in the parenchyma behind the apical region. It is X-shaped. From this nerve centre, fibres are oriented in all directions of innervate the various body tissue.

There is a pair of flame-cells or protonephridia, which are excretory organs and their ducts open to the outside between IV and V rows of epidermal plate cells.

During the differentiation of the miracidium within the egg shell, certain germinal cells become trapped in the parenchyma. These develop into germ balls by increasing in number and size and becoming enveloped, eventually giving rise to the next larval generation. Such germ balls are usually located in the posterior region of the body.

Miracidium swims about in water or moisture film where cattle usually pass out their faces. It dies in 30 hours unless it is able to reach intermediate host, which is always an amphibious snail of the genus *Limnaea.* Intermediate host in case of *Fasciola indica* is *Limnaea* acuminatisata while in the case of *Fasciola hepatica* intermediate host is *Limnea truncatula.* When larva enters the tissue of the intermediate host with the help of apical papilla, it reaches soft parts, such as mantle, pulmonary sac and hepato-pancreas. Penetration into the intermediate host is helped by the secretion of the penetrant glands or other glands. On reaching the intermediate host, ciliated epidermal cells are cast off, while eyes and brain are lost. Larva swells up and changes in shape and size, so as to become a sac-like structure, called sporocyst. Germ-balls of miracidium do not disintegrate.

Sporocyst. The sporocyst is an elongated, hollow, sac-like larva. Its body wall consists of non-cellular cuticle followed by musculature and sub-epithelium. The brood chamber is hollow space in the centre of the sporocyst within

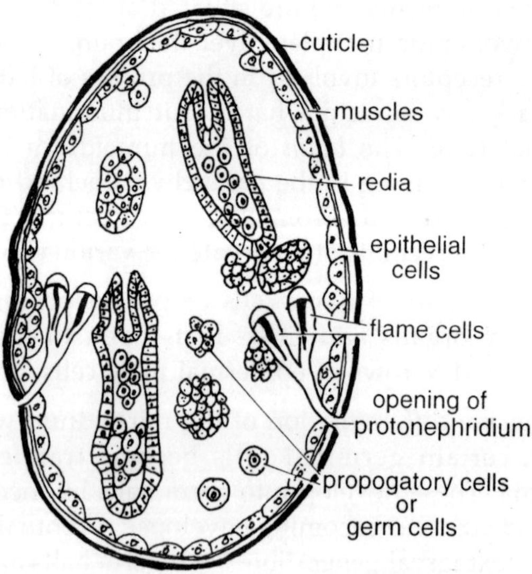

Fig. 4.17 : Fascola Sporocyst

which are found the germ balls that eventually give rise to the next generation. Sporocyst contains no alimentary canal, nervous system or reproductive systems. The protonephridia are present opening laterally by pores.

The sporocyst moves about muscular contractions in the host tissues, absorbing nourishment from them and causing them a considerable damage. Its germ balls develop into the next larval forms, the redial. Each sporocyst produces 5-8 redial. Some of the germinal cells are again set aside within the redial for the formation of the next larval forms.

Redia larva. The rediae emerge from the sporocyst by the rupture of body ball of sporocyst. Each radial is an elongated and cylindrical form, measuring 1.3-1.6 mm in length. In has a circular ridge, called collar near the interior end and a pair of projections termed *lappets* near the posterior end. The body-wall consists of the usual cuticle, musculature and subepithelium. Unlike the sporocyst, the redia possesses a mouth located at the anterior terminal of he elongated body. The mouth leads into a muscular pharynx, which in term leads a blind-sac, the gut. Numerous unicellular pharyngeal glands open into the pharynx.

Protonephridia divide further and form a much branched system. All the flame cells of one side open out through a common excretory duct. A birth pore is present, lateral to and in proximity of the mouth. Within the brood chamber are found germ balls.

Redia is a very active little creature. It moves about by the contractions of the body and lappets. It nourishes itself by sucking fluid and cells of the host tissues with its muscular pharynx. It soon migrates into the digestive glands of the snail. During summer months when the food is sufficiently available, mother redial mature in 12-18 days, the germ balls give rise to a second generation of daughter redial. Germ balls of daughter redial in the winter develop into next larval stage, the cercariae.

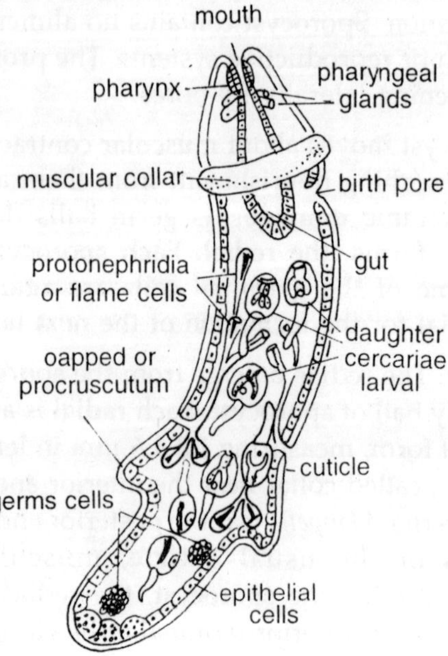

Fig. 4.18 : *Fasciola.* Redia.

Cercaria larva. Cercariae leave the mother or redial through their birth pores. From each redial 14-20 cercaria are produced. Cercaria is about 0.25-0.35 mm long and is an oval larva with a tail. The body wall consists of cuticle, musculature and subepithelium. On the anterior side, there is a mouth with a sucker, a sectorial pharynx, long oesophagus and a bifid intestine. A ventral sucker or acetabulum is present on the ventral side. A large number of dark, brown cells, called *cystogenous glands* are found beneath the cuticle. There is a pair of excretory ducts, each with several flame cells. These open near the hind end of the larva into a bladder the leads out through a pair of pair of nephridiopores. A nervous system similar to that of adult has also appeared. Rudiments of reproductive organs develop from the germ cells near the hind-end.

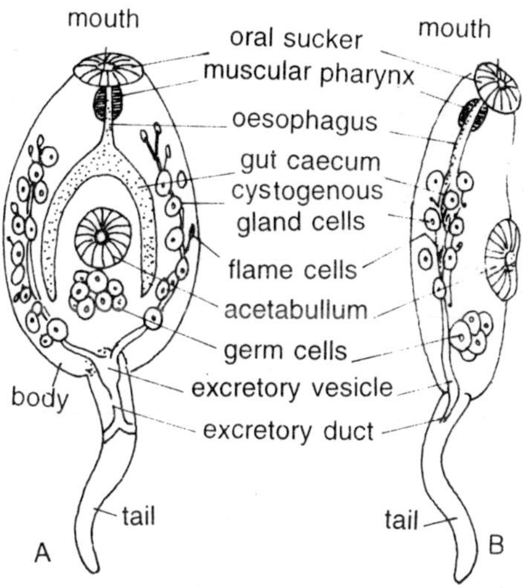

Fig. 4.19 : *Fasciola.* A = Cercaria in ventral view. B = Cercaria in lateral view.

From the digestive glands of the snail, the cercariae pass into the pulmonary sac and then escape into the surrounding water. Cercaria swims about 5 minutes to an hour in summer. Afterwards, it settles down on some green leaves of water plants. Tail is cast off and the body becomes rounded and forms around itself all thick brownish cyst wall from the cystogenous glands. The encysted cercaria is now called metacercaria.

Factors Affecting Development in Snail

The rate of development in the snail host a function of the environmental temperature. *Kendell* reported a curious effect of termperature on the development of *Fasciola*. In snails maintained at laboratory temperature, daughter redial were not produced, the first generation of redial giving rise to cercariae. On the other hand snails exposed daily a period of low temperature *i.e.*, 4-5°C, daughter redial were produced from redial.

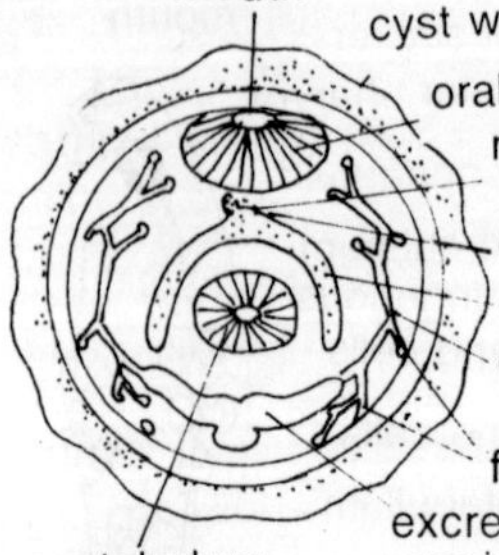

Fig. 4.20 : *Fascioda*. Metacercaria

The state of snail nutrition also effects development. The effect of food availability on development in the snail host is probably the explanation for the observation that fewer larval develop in small snails than in large ones, *Metacercaria*. It is a juvenile fluke having a rounded form with a diameter of about 0.2 mm. It resembles cercaria larva but differs from later in the absence of tail, cystogenous glands and presence of thick cyst wall. Excretory bladder opens directly through a single pore.

The metacercaria, encysted on vegetation, may live for as long as a year at low temperature, or for as short time as two or three weeks at 25°C.

Inspection to Final Host

The final excystment in the most requires an elevated temperature i.e. 38°C, low redox potential, a high carbon dioxide tension, and the presence of bile. The factors necessary for the excystment of *Fasciola* matacercariae seem to serve as stimuli to activate the worms, but the mechanism is not clear understood. When metacercaria reaches the intestine of host, the cyst is digested and young fluke is liberated. The young worm penetrate the wall of gut and passess into the peritoneal cavity. In the body cavity the young worms seem to move about rather aimlessly until the encounter the capsule of liver. On contract with the liver, the flukes burrow into it, devouring liver parenchyma as they move. It grows in size 5-6 weeks and finally settles in the bile duct. It attains sexual maturity and begins to lay eggs 11-13 weeks after ingestion by the mammalian host.

Nature of Life-history

The life-history of *Fasciola hepatica* is much complicated due to the occurrence of several larval forms in a series. Formarly, it was held that the redial and cercariae develop by parthenogenesis from the propagatory or germ cell, which were believed to the eggs (*Grobben*, 1882). The kind of asexual parthenogenetic reproduction by larval forms is known as *hetrogamy*. This view was widely accepted for sometimes and under its sway, three terms were coined. These are *parthenitae* for sporocyst and rediae larvae, *adolescariae* for cercaria larvae, and maritae for adult flukes. This view, however, is now considered erroneous and has been given up.

It has been found that the germ cells represent in the larval forms are not eggs. Instead they are diploid cells derived by the mitotic divisions of propagatory cells which is separated from the zygote at its first cleavage. Thus it can be said that they are parts of zygote. It, thus, follows that one zygote produces several larvals. Thus, this process of reproduction in sporocysts and rediae has been looked upon as *polyembryony* by *Ishii* (1934) *Chen* (1937) *Rees* (1940) and *Cart* (1944). Since all the larval resulting from a zygote are not formed simultaneously but at intervals, the phenomenon should be called delayed polyembryony.

Steenstrup (1942) and some other workers interpret polyembryony as a sexual multiplecation and hold that there is a alternation of sexual and asexual generation in life-history of *Fasciolu*. But *Hyman* is of the view that it is a continuous ontogeny involving asexual multiplication in the larval stages.

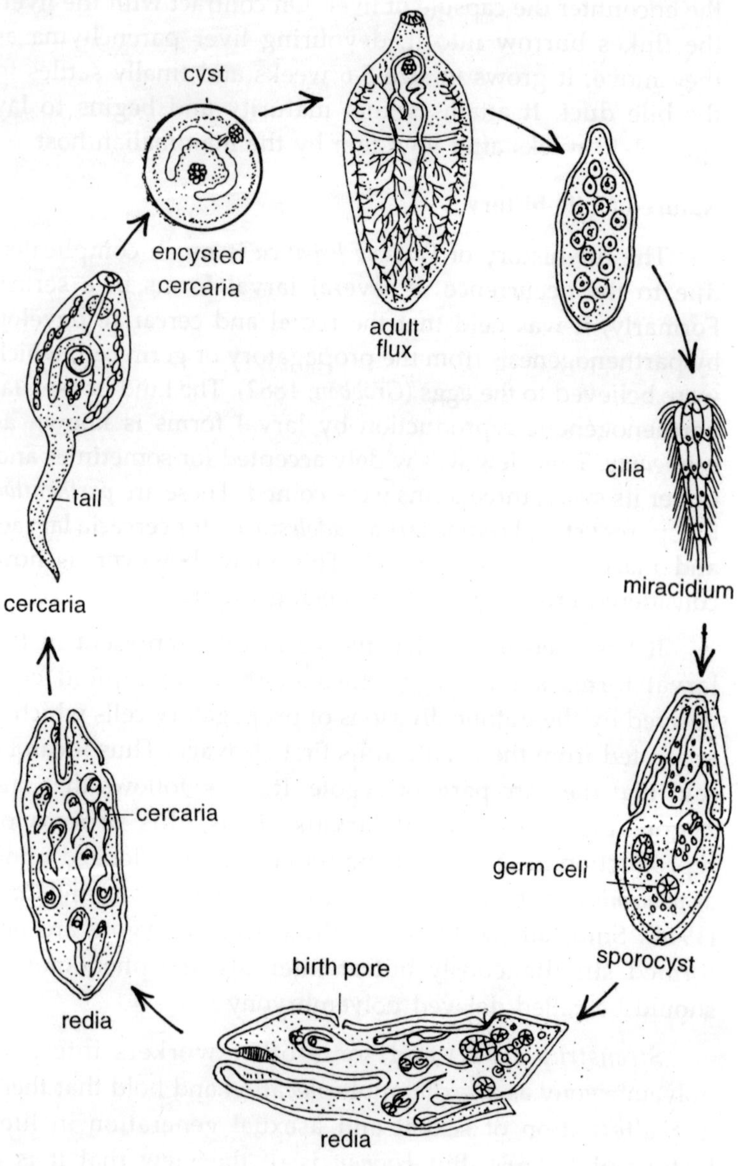

Fig. 4.21 : *Fasciola*. Life-cycle.

Pathogenesis

Pathogenesis as a direct result to the fluke's activities may be either chronic or acute. Acute *fascioliasis* or *liver rot* occurs during the preadult migration of the flukes in the parenchyma of the liver, and sometimes other organs, for about 8 weeks. The most striking changes in acute fasciolosis are seen in the liver, where there are haemorrhagic flukes. The liver becomes Swollen and in some instances, may become so swollen that the capsule ruptures. The by-products produced by worms are very toxic and *anaemia, diarrhoea, eosinophilia* etc.

Chronic fascioliasis occurs beyond 12 weeks, when the flukes have reached the bile ducts and are maturing sexually. Although milk and wool production loose are the usual results of chronic infections, deaths also occur.

The liver performs a vast array of essential functions invertebrates, and it is axiomatic that liver damage will have a variety of detrimental effects. The adult fluke causes anemia, haemorrhage and suppression in production of new red blood corpuscles. The result of liver damage is failure to again weight in young animals, weight losses in older animals, reduced milk production and lowered fleece weights in sheep. The osmotic pressure of the blood is lowered and fluid is retained in the tissue spaces.

Treatment. Carbon tetrachloride has been used in sheep at 500 mg/kg. Hexachlorethane was developed especially for use in cattles. Other drugs used are emetine hydrochloride, filcin, tetrachloroethane, etc.

Control. Control of liver flukes is effected in three major ways:

1. Snail control.
2. Treatment of infected sheep.
3. Herd management through predicting out breaks of disease.

Parasitic Adaptations in *Fasciola*

The main changes that occur in the structure and life-cycle of a parasite, with the help of which it is capable of leading a parasitic life successfully, are called the *parasitic adaptations*. Since *Fasciola* is an endoparasite in the liver of sheep, it has undergone similar modifications under the stress of the parasitic mode of life which can be studied under two heads:

I. Structural modifications.

II. Modifications in the life-history.

I. Structural Modifications

These modifications are of two types.

(A) Morphological.

(B) Anatomical

(A) Morphological Modifications

1. Body is dorso ventrally flatenned leaf-like and is nearly triangular due to which the danger of injury to its body is minimized.
2. Adhesive organs or suckers are present, and *oral* and a ventral or *acetabulum*. These are muscular and help in attaching to the tissue of host.
3. External features simple and is not divisible into head, trunk and tail. Appendages are also absent.

(B) Anatomical Modifications

1. Body is devoid of cellular epidermis, thick cuticle is present to prevent the animal from the effect of chemicals from host body. Cilia absent in adult. Cuticle is provided with spinules for attachment.
2. The thickening of cuticle in the oral and ventral regions leads to the formation of two suckers.
3. The locomotory organs are lacking except in free living larval stages as adult *Fasciola* is and

endoparasitic and it need not make excursions for food and safety.

4. The digestive system is incomplete. The *mouth* is anterior and terminal and surrounded by oral sucker. The pharynx is also muscular and sectorial. Bother are adapted to suck the nutrient.
5. Intestine is highly branched so that digested food can be distributed of different parts of body. Anus is absent.
6. Circulatory and respiratory systems are absent. Anaerobic respiration takes place.
7. Excretory system is well developed. It consists of extremely branched excretory canals with terminate into flame cells.
8. Nervous system is poorly developed. Sense organs are absent
9. Reproductive organs are extensively developed. The highly branched ovary and tests are capable of producing large number of gametes. Bisexuality offers full chances of fertilization.

II. Modifications in Life-history

1. The production of eggs in a very large number, about 10 million, due to which the threat of survival of parasite is removed.
2. Eggs are shelled and provided with an operculum.
3. A secondary host is present for the dispersal of species.
4. Multiplecation takes place by *paedogenesis* during larval stage.
5. Miracidia are free swimming larval stages because they have to search new hosts. The cercariae are encysted when out of the body of secondary host which ensures protection from external effects.

Chapter—5

Schistosoma

The schistosomes are members of order Digenea, but are different from most members of order in that the sexes are separate. The terms *Schistosoma* means split body and refers to the fact that the males have a ventral groove called *gynecophoral canal* in which females lodge. Adults of the family Schistosomatidae are founding the blood vessels of vertebrates, both birds and mammals. Three species namely *Schistosoma mansoni, S. japonicum* and *S. haematobium* are found in the blood of man and cause *Bilharziasis* or *Schistosomiasis*. No other disease caused by animal parasite causes as much harm to man as *Schistosoma* does with the sole exception of *plasmodium* causing malaria. Schistosomiasis was discovered in 1779 in France. In 1851, *Bilharz* reported *S. haematobium*. *Harley* and *Cobbald* (1864) reported that some molluses are intermediate host. The molluscen hosts are species of *Tropicorbis, Biomphalaria* and *Australorbis*.

SYSTEMATIC POSITION

Phylum	-	**Platyhelminthes**
Class	-	**Trematoda**
Order	-	**Digenea**
Family	-	**Schistosomatidae**
Genus	-	***Schistosoma***

Distribution

The human schistosomes are distributed thorough some of the countries of the world at the lower end of the economic scale, or at least the parasites are found in relatively poor rural people. *S. haematobiumis* found in Africa especially in Egypt and Sudan where up to 90% of the rural masses may carry the infection. The intermediate host is the freshwater snail *Bulinus truncatus* and related species of *Bulinus*. *S. japonicum* is found in the Far East (China, Japan, Philippines, Thailand, Vietnam etc.). The mollusc host for *S. japonicum* are species of the snail genus *Oncomelania*.

S. mansoni is found in many parts of Africa (including Egypt) and Tropical America including West Indies.

Habits and Habitat

They live in the blood vessels of the bladder, abdominal cavity etc. *S. mansoni* found in the portal blood of the host, usually in the smaller branches of the inferior mesenteric vein in the region of the lower colon. *S japonicum* adults inhibit the branches of super mesenteric vein in the proximity of the small intestine. *S. haematobium* adults reach the vesical, prostatic and uterine plexuses by way of inferior haemorrhoidal veins.

External Morphology

The adult worms are beautifully adapted of life in blood vessels. The mature worms, male and female, live in *copula* (in almost permanent copulation).

Colouration. The colour of the body is greyish or pinkish. The alimentary canal presents reddish colour due to degenerating blood cells of host.

Shape and Size. The body is thin, cylindrically elongated. The male are comparatively thicker and small than females. In *S. mansoni* the male, measuring 6.4-9.9 mm in length, is broader then the female. The female measures 7.2-14 mm in length. In *S. japonicum* the males may reach upto 22 mm in length while females 26 mm in length. In *S. haematobium* the adult males may become 15 mm long and the females 20 mm in length.

Suckers. Both male and female possessing two suckers each. Both the suckers are present and are located in the anterior portion of the body. The oral sucker surrounding the mouth and the acetabulum present on ventral side lies close to the oral sucker. Acetabulum is very powerful in the males.

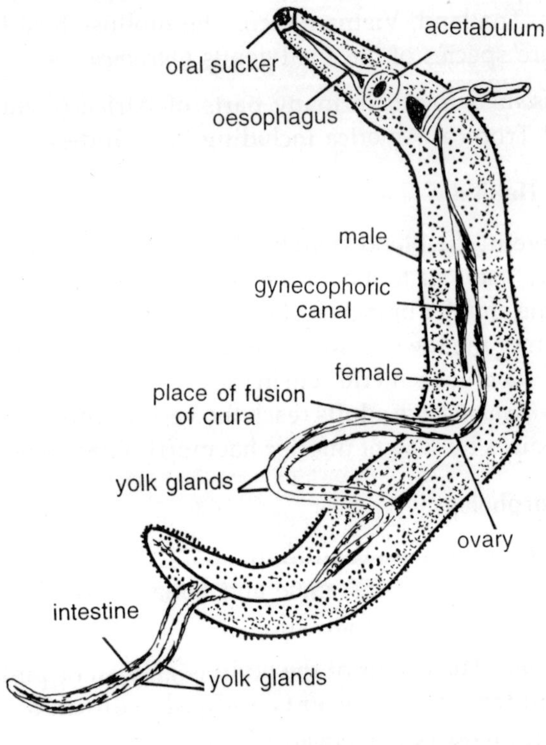

Fig. 5.1 : *Schistoma*. Male and female

Tubercles. In S. mansoni minute sensory tubercles cover the body surface of the male, and microscopic tufts of cilia are found on these protuberances. In *S. haematobium* the tubercles are very fine. In *S. japonicum* these are absent.

Gyanophoric canal. In male the two lateral borders of its body are curved (on ventral sides) to form a groove the gyanophoric canal, in which the female lodges.

Internal Structures

Digestive system. Anteriorly placed mouth surrounded by oral sucker, leading into a small oesophagus. The pharynx is absent. The oesophagus leads into a forked intestine which again fused posteriorly to form single structure for a distance. The intestine ends blindly. The oesophagus is surrounded by clusters of oesophageal glands.

Excretory system. It is represented by a bladder and a pair of collecting tubules. The bladder opens out through a single minute excretory pore lying ventrally near the acetabulum. The flame cells collect waste products from mesenchyme and discharge into collecting tubules to the bladder.

Nervous system. It is similar to other trematodes consisting of a circum oral ring with coerebral ganglia and two lateral longitudinal nerve cords.

Reproductive System

Sexes are separate. Males are broader and enclose the females in the gynecophoric canals.

1. *Male Reproductive system.* The male reproductive system consists of 4-7 *testes* which are spherical and are situated anteriorly. Each testis gives rise to a minute duct known as vas *efferens*. All the *vasa efferentia* unite to form a common spermduct or *vas deferens,* which after a short distance dilates into a structure, the *seminal vesicle* just before opens into a genital aperture situated just below the acetabulum or ventral sucker.

2. *Female reproductive system.* Female reproductive system consists of a single ovary. From the posterior portion of *ovary* arises a coiled tube knows as *oviduct*. The oviduct bends upwards and runs after a short distance to open into a ootype. The ootype in turn gives off a wide coiled tube, the

uterus. The uterus is thin and long tube running between the two branches of intestine and opens into a female Mehli's' gland. *Vitelline glands* are situated in the posterior region of the body. From these glands a vetelline duct arises that runs along the oviduct and joins the later before it opens into ootype. In *S. haematobium* seminal recepticles is absent.

Life-History

Copulation and fertilisation. The male and female adults in the definitive host move against the blood flow into small venules until they can move no further. Fertilisation occurs when spermatozoa from the male are introduced into the female. Fertilisation takes place in the oriduct. The fertilised eggs are encapsulated by the secretion of vitelline glands which also form yolk cells. In *S. haematobium* the uterus contains only 10-100 capsules while uterus of female *S. mansoni* contains only 14 capsules at a time.

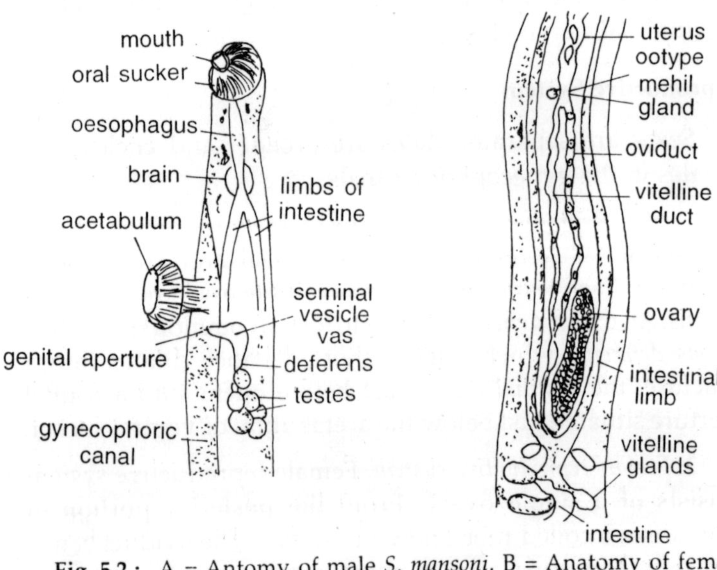

Fig. 5.2 : A = Antomy of male *S. mansoni*. B = Anatomy of female *S. haematobium*.

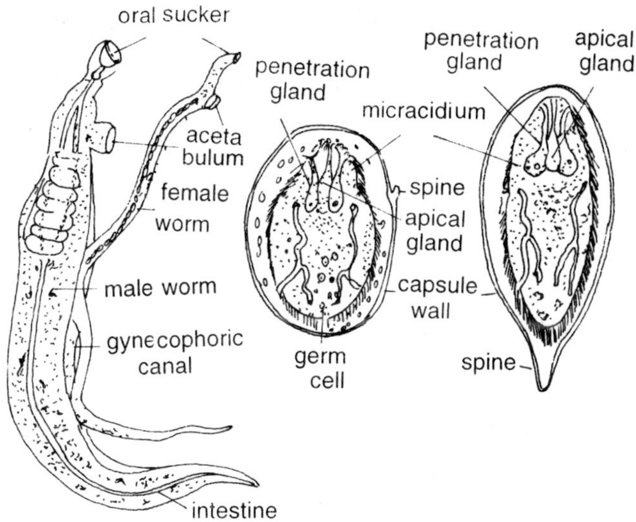

Fig. 5.3 : A = *Schistosoma japonicum*. B = Capsule of *S. japonicum*. C = Capsule of *S. haematobium*.

Egg laying. The fully formed eggs or capsules measure 114-175 m × 45-68 u in *S. mansoni*. In *S. japonicum* 70-100 u × 50-56 u and in *S. haematobium* 112-117 u × 40-70 u. The capsules are nonoperculate and bear a large posterior-lateral spine in S. mansoni. In *S. haematobium,* the spine is terminal and in *S. japonicum* the spine is small and subterminal. In *S. mansoni* such eggs are laid in the mesenteric vessels and are carried by blood to the proximity of large intestine where they gradually penetrate intestinal wall, become include in the lumen, and are passed out in feces. In *S. haematobium* the capsules are passed into the urinary bladder and are expelled in urine.

Female lays a single eggs at a time. *Pesigan et al.* (1958) estimated that a female *S. japonicum* lays about 1200 eggs per day.

Miracidium larva. Eggs do not hatch in the urinary bladder or intestine of the host. The eggs of *S. japonicum* takes 9-10 days to develop mature miracidia while passing through the tissue and will live for 10 or 12 days longer if not expelled out, but may die in the tissue. Miracidium measures about 130 u in length and 60 u in breadth. The larva contains ciliated epithelium. Two pairs of penetration glands are present, one pair anterior in position other posterior. Meracidium also possesses a primitive gut, nerve centre and two pairs of flame cells. Germ balls are usually located in the posterior portion of the body.

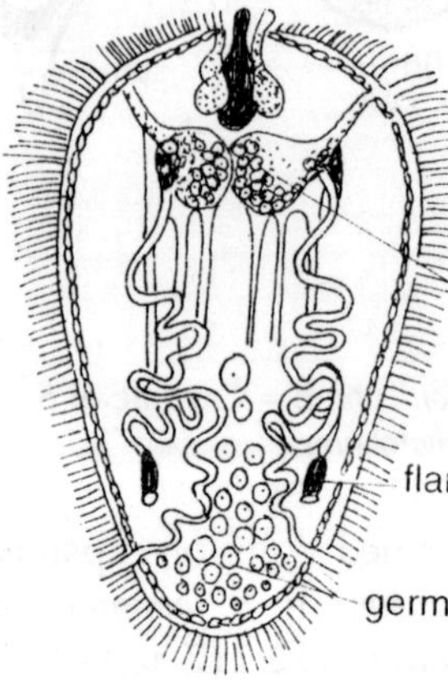

Fig. 5.4 : *Schistosoma*. Miracidium.

When the eggs come in contact with water, at 28°C the enclosed miracidia break out the shells by cracking them. The free-swimming miracidia finds and penetrates a pulmonate snail and transforms into mother sporocyst which is located in the foot region of the host. Free-swimming miracidium survive for upto 24 hours. It dies if not find its intermediate host within this single day of free life.

Sporocyst. In *S. mansoni* the sporocyst grows 1 mm in about 14 days. The sporocysts start producing daughter sporocysts which burst free of the mother sporocyst. These are differentiated from germ balls. The daughter sporocysts reach a length of 1.5 mm and produce forktailed cercaria from the germ balls. A single snail infected by a single miracidium may discharge about 3500 cercariae per day for sometimes

The cercariae are discharged during the forenoon on sunny days.

Cercaria larva. The cercariae of *S. haematobium* and *S. mansoni* have a body of about 200 M long with a tail stem slightly longer and forks about 75 M long. For about two of three days the cercariae alternately swim and rest in water. Cercaria possesses an oral sucker and ventral acetabulum are present. Mouth leads into the oesophagus which leads into a trifid intestine. Pharynx is absent. Five pairs of penetration glands are present-2 pairs of anterior oxyphilic glands and 3 pairs of posterior basophilic glands. If the cercariae fail to reach final host within two days the die.

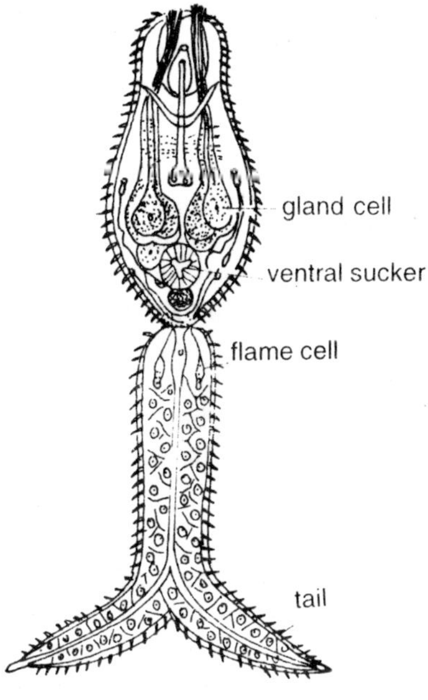

Fig. 5.5 : *Schistosoma*. Cerceria.

Infection to the Final Host

If successful the cercariae burrow through the skin, using the histolytic and hyaluronidase - bearing secretion of their penetration glands just as the miracidia do. Skin penetration requires several minutes and may or may not be accompanied by a prickling sensation and subsequent dermatitis. If ingested with water the cercariae attach themselves to the mucous membranes of the mouth or throat and similarly bore in. They leave their tail behind them. They find their way into the blood systems and are carried via the heart to the lungs. Young *S. mansoni* accumulate in the lungs on the 2nd and 3rd days. By the 16th day they appear in number in the liver, where they are well established by the 15th day. Apparently cercariae feed only on portal blood, but once in the liver they grow rapidly. Migration to the mesenteric veins begins about the 23rd day, and mating and eggs production about the 40th day. All the larval produced from one egg become adults of some sex. No female worms mature in a host, if the male worms are absent.

Pathogenicity

The injuries inflicted by schistosomes depend on the degree of infection and the prolonged period of infection and reinfection. During migration of the eggs through tissue to reach the colon or urinary bladder tissue reactions in the form of inflammation, necrosis, connective tissue encapsulation, eventually scar formation take place. During migration of the eggs through the wall of urinary bladder or colon, haemorrhage generally occurs. This is accompanied by diarrhoea.

In endemic areas, where reinfection occur, repeated penetration of colonal and sladder wall results in extensive scar formation, which prevents migration of eggs through these structures. As a results, the eggs may transported to liver and rarely to brain or spinal cord. Whenever these aberrant eggs become lodged, local inflammation, fibrosis and necrosis occur. Cirrhosis of the liver is not uncommon. Swelling of liver and spleen is of common occurrence. In the case of heavy infections, mature worms may migrate to the brain, lungs uterus, oviduct and gonads. Their presence at these sites provoke pathological changes.

Acute cases of schistosomiasis, especially by *S. japonicum*, which is highly pathogenic, often lead to death.

Treatment

Treatment of schistosomiasis has been hundred by the fact that relatively few drugs are available, their toxic effects are sometimes severe and they are active only against mature worms. Tartar emetic (potassium antimony tartrate), Sodium antimonial gluconate and Stibocaptate have been used for years. Thioxanthones, miracils, Lucanthone, hycanthone, Niridazole imidazolidin z-one, etc. are also used.

Epidemiology and Control

The basic pattern of life cycles in the blood flukes was worked out in the first decade of the 20th century. With that information came the keys to control. It was obvious that sanitary disposal of human excreta and reducing population of snails were the steps to take. As knowledge increased, additional techniques were developed, so that general methods used in attempting to control schistosomiasis are :

1. Snail control.
2. Sanitary disposal of human excreta.
3. Treatment of infected persons.
4. Treatment of elemination of reservoir host.
5. Prevention of penetration by cercariae.
6. Immunization of persons at risk of infection.
7. Prevention of infections in snails.

Each methods is directed towards the same end; to break the life cycle or interrupt the chain of infection. If all the snails could be removed from endemic area, the probability of infection would approach zero. If all definitive hosts were treated and maintained free of infection for a year, the cycle would be broken.

Table 5.1 Showing differential characteristics of *Schistosoma* species of humans

	S. haematobium	*S. japonicum*	*S. mansoni*
Adults			
Male			
Length(mm)	10-14	12-20	6-12
No. of tesets	4	7	6-9
Tegument	scattered tubercles and blund short spines	Minute spines 1	Numerous tubercules and large sharp spines.
Posterior union of intestinal caeca midbody.	Midbody	posterior to	Anterior to midbody
Female			
Length (mm)	16-20	16-28	7-17
ovary	Posterior to midbody	Midbody	Anterior to midbody
No. of eggs in uterus	20-100	50 or more	usually 1
Eggs produced by single female per day	?	1400-3500	100-500
Location in host			
Primary host	Bladder and pelvic plexuses	Both superior and inferior mesenteric veins	Veins of small and large intestine near ileocecal junction.
Secondary host	Portal and mesenteric veins	Portal vein	Haemorrhoidal and hepatic veins
Prepatant period in weeks.	10-12	5-7	7-8
Eggs			
Length(u)	112-170	70-100	114-175
Mean		82	
Width(u)	40-70	50-65	45-68
mean		60	
Colour	Golden brown	Golden brown	Golden brown
Spines	Terminal	Subterminal knob	Subterminal spine
passed in	urine	faeces	faeces

Chapter—6

Cytology of Schistosomes

The previous chapter presented the structural organization of schistosomes, showing that the surface of the parasite facing its host was always a syncytium in either the mammalian host or the snail. Here we extend the discussion into the functional properties of the synctium, particularly its surface membranes. The structural, molecular, and functional properties of the schistosome surface are critical in explaining the parasite's long survival in the human host, because the surface is a significant potential target of host defence. Unlike some protozoa that hide within host cells, schistosomes are exposed to the circulating components of the immune system. To survive, the parasite surface presumably possesses mechanisms for evading immune attack. We wish to explain how the parasite can be attacked and killed by some cells, such as eosinophils, but evade that attack of other cells. The chapter will focus on *Schistosome mansoni* because most of the experimental work has been done on this species. Presumably, the paradigm constructed for *S. mansoni* will be applicable to the other species of schistosomes that infect humans. We will consider initially knowledge of the structure and composition of the surface membranes and then review functional studies performed on the membranes themselves. The chapter will conclude with a discussion of the interactions between various human blood cells and the parasite.

Structure of the Parasite Surface

During its life span in its mammalian host, the parasite is covered by a synctium called the tegument. In all stages, the tegument is connected to the cell bodies that are located deep within the organism below the large circumferential bands of muscle that are responsible for such gross movements of the whole parasite as contraction, elongation, and turning. The cell bodies contain the basic machinery for protein synthesis, namely, nucleus, endoplasmic reticulum, and Golgi apparatus. The surface of the tegument has a varied topology consisting largely of protruding spines, which contain crystalline actin, and invaginating pits. The distribution of the spines and pits is altered during larval development. During the first 4 days after transformation from cercariae to schistosomula, the spines are lost over much of the body, there is more pitting at the surface, and the surface area increases by 30 per cent. In adults, different regions of the parasite have characteristic patterns of folding, pitting, or spines. Despite these large variations of structure, the ultrastructure of the surface membranes that face the host tissues or bloodstream remains relatively similar from the completion of cercarial transformation through adulthood.

Transmission electron microscopy reveals that the parasite surface consists of adjacent lipid bilayers that appear as two apposed membranes by transmission electron microscopy. These membranes are anisotropic, however, as shown by freeze-fracture technique. In newly transformed larvae, the inner of the tegumental membranes, that is, the one that is closed to the cytoplasm of the tegument, contains many intramembrane particles and resembles the plasma membrane of most mammalian cells. Intramembrane particles are associated with membrane proteins, and membranes devoid of these particles usually contain little or no protein. The outer membrane which faces the host in vivo or culture medium in vitro has few intramembrane particles and resembles a simple protein-poor lipid bilayer. Interestingly, the outer membranes of worms isolated from the lungs of mice about a week after infection contain intramembrane particles, whereas outer membranes of worms cultured in vitro for the same period do not. The parasite may acquire these particles or proteins from its host. However, the host may induce the parasite to synthesize intramembrane particles in its outer membrane. Electron microscopy has also suggested that the outer membrane contains cholesterol, since the membrane structure is perturbed by digitonin and the polyene antibiotic filipin, both of which characteristically alter cholesterol-containing membranes. These structural observations suggest that the schistosomula are covered by two different membranes: an inner one that probably can function in ion transport much like mammalian cell plasma membranes and an outer one that is simply a lipid bilayer composed of sterol and phospholipid. The schistosome surface the bears a strong resemblance to gram-negative bacteria, which also have a double membrane system.

The Surface Membranes

When the plasma membrane of most mammalian cells is exposed to a molecule that binds specifically to a receptor on the membrane, for example, low-density lipoproteins or transferrin, or to reagents that bind to specific chemical groups on the membrane, for example, lectins or polycations, the bound molecules are usually internalized by the process of endocytosis and transported to internal compartments within the cell, such as phagosomes or lysosomes. Some mammalian cells, notably neoplastic cells, also can shed bound ligands into the culture medium. Schistosomes bind antibodies and complement at various stages of their life cycle in the mammalian host but most extensively shortly after transformation. However, these multivalent ligands bound to the surface are not endocytosed by the parasite. Furthermore, low-density lipoproteins, which also bind to the tegument, and fluid-phase markers such as horseradish peroxidase are also not endocytosed by the tegument. Thus, the tegumental membranes appear incapable of endocytosis and are very different from typical normal mammalian cell plasma membranes.

Although complement and antibody bind and are not endocytosed, they do not remain on the parasite surface but are slowly lost. This loss has been explored by applying the lectin concanavalin A (Con A) to schistosomula. The lectin binds to a unique population of molecules that are present for 24-48 hours after transformation but are not present on cercariae or schistosomula older than 48 hours. Thus, the Con A - binding molecules are lost from the surface independent of the binding of ligand, and the loss therefore is physiological and not induced by Con A. Furthermore, the kinetics of the loss of Con A binding is the same as that of the loss of bound rat and human antibody and bound complement, suggesting that the mechanism of loss is not confined to this lectin; a more general mechanism of membrane turnover is probably at work. The kinetics of the loss of these three molecules-antibody, complement, and Con A-follows that of a decay curve and indicates that the loss of bound molecules is random event. In addition, Con A can be bound to the parasite and recovered in the culture medium over time. The recovered Con A is the same molecular weight as the Con A originally bound. These studies suggest the parasite is sloughing or shedding the bound ligand and attached membrane component into the culture medium.

In mammalian cells, multivalent ligands such as Con A or antibody bound to the plasma membrane at 0°C are capped when the temperature is raised, so the ligand becomes localized to a region of the cell surface. The studies with Con A also show that there is no capping of the bound ligand on the parasite surface, suggesting that the contractile proteins of the tegument do not functionally interact with the outer membrane. As suggested by the kinetics of loss of Con A that follows a decay curve, bound Con A is randomly distributed on the parasite surface. However, after the addition of hemocyanin that is rich in mannose and binds to the Con A, clustering of the hemocyanin and hence the lectin is observed. This clustering occurs after the parasites have been killed with azide, demonstrating that contractile proteins are irrelevant in cluster formation. Instead, the hamocyanin itself induces aggregation of the Con A in the membrane. These experiments also demonstrate movement in the plane of the membrane by the moieties to which Con A is bound.

The loss of bound Con A and its recovery in the culture medium suggest that the outer membrane or portions of it are being sloughed into the medium. This phypothesis can be tested by labelling the parasite surface and examining the culture medium over time to see if labeled membrane components appear there. Several general labeling schemes are available, particularly iodination and labeling oxidized sugars on glycoporteins and glycolipids with tritium. The sugar-labeling techniques have the advantage of being two-step procedures, permitting the nonspecific binding of the radiolalbel to be determined in the absence of oxidation. Further, the oxidation step can be performed with an enzyme, such as galactose oxidase, which limits the specific labeling to the parasite surface, since the parasite does not endocytose. Experiments of this type as well as subsequent experiments employing monoclonal antibodies show that many surface glycoproteins and glycolipids are being shed into the culture medium with a half-time of approximately 12-14 hours. Given that the surface area of a schistosomulum is 20,000 square micrometers (μm^2), this figure represents losses of approximately 800 μm^2/hour, an astronomical rate by mammalian cell standards. Furthermore, the rate of synthesis must be double that figure, since the surface area is increasing by 100 per cent per day. Studies of this type are usually not carried out on adult schistosomes, largely because such studies would be complicated by the presence in adults of a functioning gut, which can ingest sloughed proteins.

Although the surface membrane proteins and lipids are clearly being sloughed into the culture medium, some experiment do not fit this model and there are serious methodological limitations in all experiments of this type. In particular, studies with monoclonal antibodies have suggested that some proteins may not be sloughed and may persist for long periods of time on the surface. More generally, the interpretation of the turnover experiments is clouded by the fact that schistosomes are covered by two membranes and that the distribution of labeled or antigenic membrane components within the double-membrane system is usually unknown. In particular, it is unclear whether only components of the outer membrane are labeled or those of the inner membrane are also labeled. Definitive experimentation in schistosomes is complicated by difficulties in obtaining large numbers of organisms and ultimately in determining the composition and distribution of the components of the two membranes.

Chemistry of Surface Membranes

Many laboratories are now actively investigating the composition of the surface membranes, but a definitive statement cannot yet be made about which molecules constitute the two membranes. Investigations have employed isolation of the membranes, surface labeling, selective extraction, and analysis of biosynthetically labeled materials shed into the medium. Taken together, they yield a general picture. The outer membrane is primarily composed of phospholipids. (Note that phospholipids are made up of a glycerol backbone attached to two fatty acids, called acyl chains, and phosphate. The phosphate is linked to another molecule, such as choline or serine, called the head group.) The major phospholipids in the outer membrane are phosphatidylcholine and phosphatidylethanolamine, with small amounts of phosphatidylserine, phosphatidylinositol, sphingomyelin, and lysophosphatidylcholine. Cholesterol is also present as well as glycolipids, some of which are antigenic. There are a variety of antigenic and nonatigenic proteins, most of which are glycoproteins as well. The inner membrane probably contains the various enzymes known to be associated with the tegumental membranes, namely acetycholinesterase, Na-Ca ATPase, and alkaline phosphates as well as phospholipids and cholesterol.

Schistosomes do not synthesize cholesterol but instead use host cholesterol; they do not modify host acyl chain length during phospholipid biosynthesis but perform a limited number of head-group modifications. Presumably, the lipid components are assembled into membranes in the cell bodies and transported to the tegument in multilamellar bodies. There they fuse with the base of the pits and insert the new membrane into the parasite surface.

INTERACTIONS WITH CELLS

Schistosomes dwell in the bloodstream from about 4 days after they penetrate through the skin until they die years later. Being in the circulation exposes them to cells and plasma components that have the potential to destroy the parasites. On the other hand, one of the ways the parasite can protect itself from immune attack is to acquire host antigens from serum or blood cells themselves. In this way, the parasite mimics the host, is not recognized as foreign, and in a sense becomes the wolf in sheep's clothing. The attack of cells and parasite defense can be studies in vitro by mixing purified leukocytes, schistosomula, and infected patient's serum that contains antibodies against the parasite. Let us consider four blood cells: eosinophils, neutrophils, monocytes, and erythrocytes.

Eosinophils adhere to the parasite surface and degranulate within minutes. The parasites die about 18 hours later when most of the surface is covered by material discharged from the granules. Since a single eosinophil discharge over an area of about 80 square micrometers (μm^2) and since about 1800 μm^2 of the parasite are covered by discharge material after 12 hours, several hundred eosinophils have to discharge onto each parasite. The successful attack of the eosinophils should be regarded as a mass onslaught directed broadly against the entire surface of the parasite that is carried out over the course of a half a day or more.

Eosinophil degranulation is mediated through antibody bound to the parasite surface. The cells bind to the worm by means of an Fc receptor in the plasma membrane binding to the Fc portion of the IgG on the worm. The granules move through the cytoplasm to the zone where the plasma membrane is bound to the antibody. The granule membrane then fuses with the plasma membrane, thereby releasing the granule contents at the site of eosinophil-parasite contact.

During the attack, eosinophils die and lyse on the parasite surface. Cells can be seen with electron lucent cytoplasm and granules that are still intact and undischarged. Presumably the cells died before they can exocytose all of their granules. Furthermore, fragments of the plasma membranes of eosinophils are seen covering degranulated material on the parasite's surface, perhaps assisting in the attack by confining exocytosed material to the target and preventing diffusion of material from the surface. Whether the eosinophils are killed by the parasite or by material released by their own or other eosinophil granules remains to be determined.

Unlike eosinophils, neutrophils do not kill schistosomula in the presence of antischistosomal antibodies alone, but they will kill if complement is also present. Although neutrophils adhere to the worm in the presence of antibody alone, they do not degranulate, which probably explains their failure to kill the worms. The lack of degranulation is in turn explained by experiments examining how neutrophils process multivalent ligands such as antibodies and lectins on the parasite surface. Fluorochrome-conjugated antischistosomal antibodies or lectins promote cell-parasite adherence and can be observed simultaneously by fluorescence microscopy. Immediately after labeling, the ligands are evenly distributed on the parasite. Soon after neutrophils are added, dark areas appear on the parasite and fluorochrome can be identified in the lysosomes of cells. Neutrophils endocytose antibody or lectin from the parasite surface. Thus, unlike the interaction of the Fc portion of the antibody with the Fc receptor on the eosinophil, which promotes fusion of the granule membrane with the bound membrane at the cell surface, the neutrophil membrane receptor ligand complex is internalized and fuses with the granule membrane in the cell. Parasite antigens are also endocytosed along with the antidody, as demonstrated by autoradiography. Removal of both antigen and antibody from the parasite by neutrophils reduces the density of recognition signals available to eosinophils.

About 10 per cent of adherent neutrophils fuse their plasma membranes to the outer of the two tegumental membranes of the worm when worms are first coated with antibody or lectin. A hybrid membrane is derived from the plasma membrane of the neutrophil and the outer tegumental membrane of the parasite is formed as a result. No antibody or lectin is detectable by electron microscopic autoradiography in the fusions, suggesting that the ligand is removed by endocytosis before fusion occurs. These fusions are probably the largest (up to 10 μm) to be described between biological membranes. They may be a mechanism for acquiring host membrane components, particularly integral membrane proteins, such as histocompatibility antigens, which are thought to defend the parasite against being recognized as foreign by the host immune system. The fusions are not reversible and ultimately result in lysis of the fused neutrophil. After this lysis, the fused membrane appears to mix with the normal worm outer membrane.

The third major effector cells in human blood is the monocytes. Human monocytes are not as effective as eosinophils in killing schistosomes. Monocytes that have been cultured for several days are called monocyte-derived macrophages because of a certain resemblance to tissue macrophages. Quiescent monocyte-derived macrophages do not effectively kill tumor cells or schistosomula that have been prein-cubated with antibody. However, after exposure to interferon-γ (a mediator that activates the host cells), the monocyte-derived macrophages acquire the capacity to kill tumor cells; their ability to kill schistosomula is inconsistent.

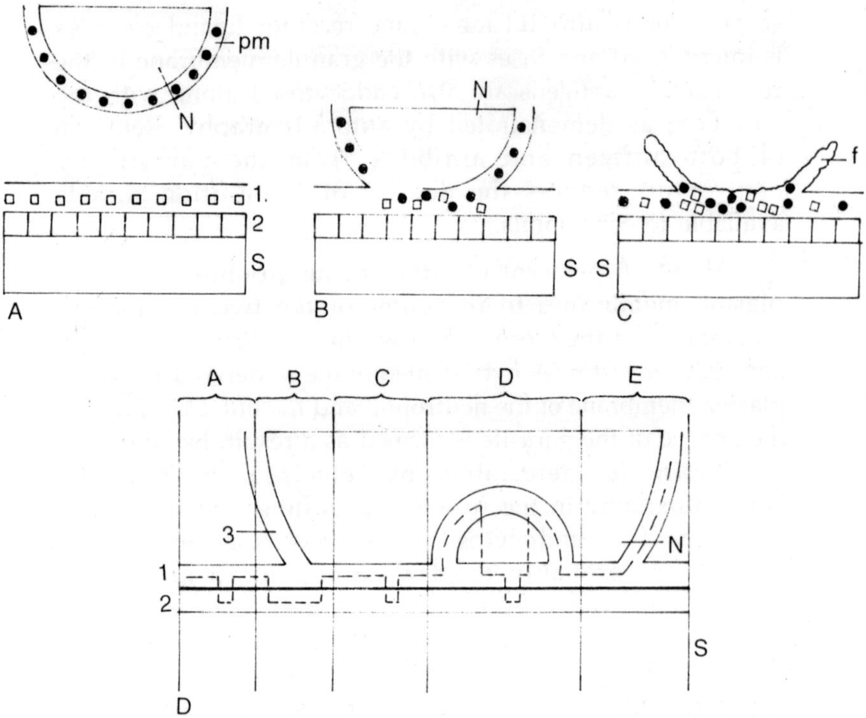

Fig. 6.1 : A hypothetical model of the transfer of membrane components from the plasma membrane (pm) of a neutrophil (N) to the tegumental membrane of a schistosomulum (S). In panels A-C, the cell approaches the worm, and the membrane components are seen as open or closed circles in the outer tegumental membrane (2) or cell plasma membrane, respectively. In B, the neutrophil has fused with the worm, and a hybrid membrane that contains components of both original membranes has formed. The worm inner membrane (1) and the unfused neutrophil membrane are both intact. In C, the neutrophil has lysed, and a fragment (f) of the neutrophil membrane remains attached to the worm by the fusion. The neutrophil membrane components have moved from the hybrid membrane into the unfused portions of the outer membrane. Panel D corresponds to the freeze-fracture micrograph. Host molecules acquired in this manner may protect the parasite from recognition as foreign by the immune system.

Human monocyte-derived macrophages fuse with the outer tegumental membrane. Murine macrophages, in contrast to human monocyte-derived macrophages, consistently kill schistosomes and apparently are major host defense effector cells in murine schistosomiasis infections.

Parasite interactions with erythrocytes are potentially important, particularly since adult worms can acquire human ABH blood group antigens in vitro. After coculture of schistosomula and erythrocytes, an electron-dense plaque is seen between the parasite outer membrane and the plasma membrane of the attached cell. Membrane proteins and glycolipids are not transferred to the parasites from adherent cells. However, carbocyanine dyes, which are anchored in the red-cell membrane by acyl chains, are transferred rapidly in 30 minutes to 3 hours. These studies also demonstrated that erythrocytes lyse on the parasite surface and that the lysed membrane fragments remain attached to the parasite. This mode of host antigen acquisition and the resultant masking of parasite antigens differs from the process of host antigen acquisition in which the antigens are presumably inserted into or absorbed onto the outer tegumental membrane. Additional modes of host antigen acquisition could exist. Since adult worms ingest red blood cells, they also might be able to transport intact glycolipid antigens of erythrocytes from the their gut to the tegumental surface. Furthermore, glycolipid antigens can be acquired from serum.

Each cell type examined-eosinophils, neutrophils, monocytes, and erythrocytes—is lysed on the surface of the parasite. With the first three cell types, studying the mechanism of lysis is difficult because the cells themselves have cytolytic molecules. It therefore is not clear whether lysed effector cells represent a form of suicide or homocide, with the parasites as the murderer. Erythrocytes, on the other hand, do not have such autocytolytic capacities. Erythrocyte membranes bound to parasites have been studied by fluorescence photobleaching recovery (FPR) methods to further address the role of the parasite in host

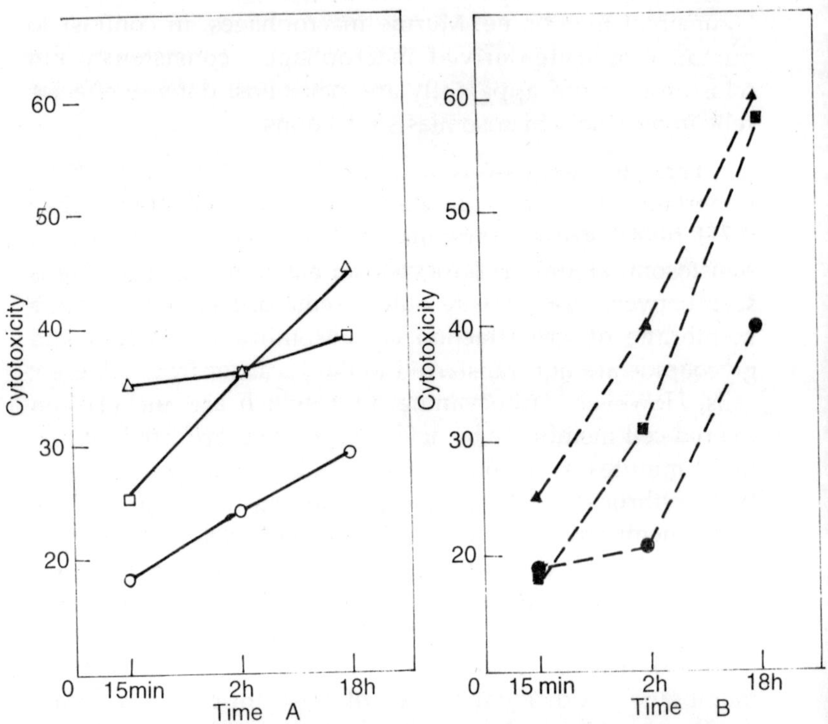

Fig. 6.2 : Demonstration of the cytotoxic effect of schistosomula on human monocyte-derived macrophages. The ordinate shows percentage of specific killing of the cells in the presence of schistosomula after total killing was corrected for spontaneous death of the cells alone. At each point for each experiment the cell death in cultures without schistosomes was subtracted from the cell death in cultures with schistosomes. The abscissa indicates the incubation time. The left panel (A) shows three experiments performed in the absence of antischistosome antibodies; the right panel (B), three experiments performed in the presence of antibodies. Each point is the mean of 12 counts, the standard deviation for each point is less than 5 per cent. (Δ, O, and □ each represent an experiment done on a different day; ▲, ●, and ■, experiments performed on the same day as the corresponding open marker.) The recovery of monocyte-derived macrophages

cell lysis. In this technique, a fluorochrome is inserted into the red cell membrane. For example, either glycophorin or band 3 can be selectively fluorescinated or the phospholipid analogue, fluorescein phosphatidylethanolamine (Fl-PE), can be partitioned into the membrane. After the labeled erythrocyte adheres to the worm, a laser is used to bleach the fluorochrome in a small spot on the surface of the cell. If the molecules bearing the flurochromes are able to diffuse in the plane of the membrane, then the fluorescence will reappear at the bleached spot. Two measurements of recovery can be obtained: the mobile fraction, or percentage of labeled molecules that are able to move, and the diffusion coefficient, or rate of movement of the fluorochrome into the bleached area.

In such experiments, both glycophorin and band 3 are totally immobilized in lysed red cell membranes adherent to the worm (e.g., they do not move into the bleached spot). (The phospholipid probe, Fl-PE, is partially immobilized (50 per cent mobile fraction). These results obtained with red-cell membranes adherent to worms contrast with what is observed in studies of intact red-cell suspensions, in which the mobile fraction of glycophorin and band 3 is about 60 per cent and that of Pl-PE is 100 per cent. Red-cell membrane ghosts produced by hypotonic lysis exhibit enhanced membrane movement, with mobile fraction of the proteins nearly 100 per cent. Relative immobilization of the proteins observed on the parasite - associated erythrocyte membrane can be reproduced by cross-linking the protein in intact erythrocytes with lectins, polylysine, or diamide. Under these conditions there is no decrease in Fl-PE mobile fraction, however.

The only treatment of labeled erythrocytes that caused red cells lysis and alterations of the FPR measurements of all three test probes in a manner similar to those found in

---

in the absence of schistosomula compared to the starting number of cells was 100 per cent at 15 minutes and 82±4 (S.E.) per cent at 2 and 18 hours.

red cells adherent to the worm is the addition of exogenous lysophosphatidylcholine (lyso-PC). Interestingly, lyso-PC is also produced by worms metabolically labelled with palmitate or choline. Furthermore, the parasite releases lyso-PC into the culture medium in quantities sufficient to account for the FPR measurements. Together, these observations and the results of photobleaching experiments strongly suggest that the parasite is capable of lying erythrocytes attached to its surface by the generation and release of lyso-PC. Perhaps the lysis of effector cells (eosinophils, neutrophils, and macrophages) is also due to the action of lyso-PC.

Coda

The schistosome appears to operate according to three general principles: (1) I'll take anything I can get; (2) If you want it back, you can have it; (3) If you get too close, I'll bite. The parasite outer membrane appears to be deceptively simple, containing mainly phospholipids and cholesterol derived from the host as well as adsorbed host proteins and lipids. The true parasite molecules, proteins and antigenic glycolipids, are probably only a small fraction of the outer membrane. The outer membrane can be shed rapidly because it is energetically cheap to make. It is also poorly antigenic because so much of its is derived from the host.

The presence of lyso-PC in the outer membrane or the ability to generate this molecule explains many of the membrane properties. Lyso-PC is a detergent very similar to sodium dodecyl sulfate. When its critical micelle concentration is reached in the membrane, proteins are solubilized in lyso- micelles or mixed micelles formed from lyso-PC and the other phospholipids and cholesterol. This property of lyso-PC explains the shedding of both membrane proteins and ligands from the parasite. Endocytosis by neutrophils of ligands such as antibody and lectins and the membrane components to which they are bound can be seen as an extension of this process. Lyso-PC is a fusigen, which accounts for the fusion of neutrophils and monocytes with the outer membrane. Finally, as suggested by its name, lyso-PC is a powerful lytic agent and is most likely responsible for the lysis of various cells on the parasite surface. The success of eosinophils in attacking the parasite probably owes to the high levels of lysophospholipase on their surface and in their granules. This enzyme cleaves lyso-PC into a fatty acid and glyceryl phosphorylcholine, thereby inactivating it.

Many important questions about the function and molecular architecture of the tegumental membranes remain unanswered. In particular, it is not clear whether the outer membrane has high steady-stage concentrations of lyso-PC or generates it enzymatically in a controlled manner to in response to stimuli. It is also not clear how the inner membrane is separated from the outer membrane, how the membranes maintain their identity, and how the inner membrane functions as a true biological membrane when covered by an apparently intact separate lipid bilayer. Finally, the biogenesis of these membranes needs further study to determine if the models based on morphological data-namely, assembly and transport in multilamellar bodies—are correct.

Chapter—7

Human Lung Fluke

A number of mammals, including humans, are infected by ling flukes of the genus *Paragonimus;* Yokogawa (1981) listed over 30 species of flukes, at least 7 of which are implicated in human disease.

Name of Organism And Disease Caused. *Paragonimus westermani* Braun, 1899, is the most important species infecting humans; it causes paragonimiasis, pulmonary distomiasis, and endemic hemoptysis (expectoration of bloody sputum).

Hosts And Location. Hosts are carnivores including tigers, lions, leopards, feral cats, pigs, dogs, monkeys, and humans. Adult flukes are found in lungs, but may also be found in almost anybody organ or tissue; 30 to 60% of extrapulmonary paragonimiasis were either cerebral or spinal cord infections.

Distribution And Importance. The organism is distributed in China, Japan, Taiwan, Korea, the Soviet Union, Philippines, Thailand, Indonesia, India, and Sri Lanka. In Korea alone, between 1.0 and 1.5 million people are infected.

Morphology. Adult flukes are 7 to 12 mm long with a thick, spiny tegument; a small acetabulum is located just anterior to the middle of the body. Often one must remove some tegument and vitellaria to discern the reproductive organ structure. Two posterior lobed testes are present; the

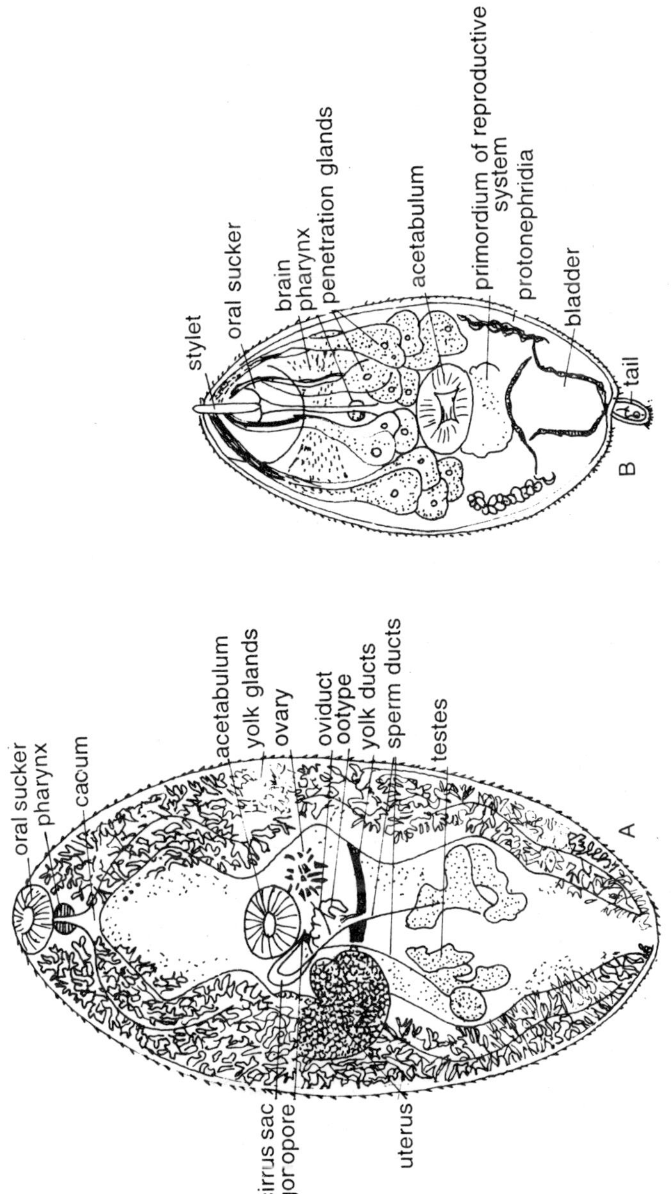

Fig. 7.1 : A, *Paragonimus westermani*, adult. B, Cercaria.

ovary is also branched and is located slightly posterior to the acetabulum. There is no cirrus or cirrus pouch. The genital pore is posterior to the acetabulum. Eggs are opercdulate, irregularly barre-shaped, and are 80 to 120 μm by 48 to 60 μm in size.

Eggs, which averages 87 × 50 μm, are brownish in color and sometimes are present in such numbers that they give the sputum a brown or rust-colored tinge. These operculated eggs are coughed up from the lungs and expectorated, or swallowed and leave the body in the feces. After two to

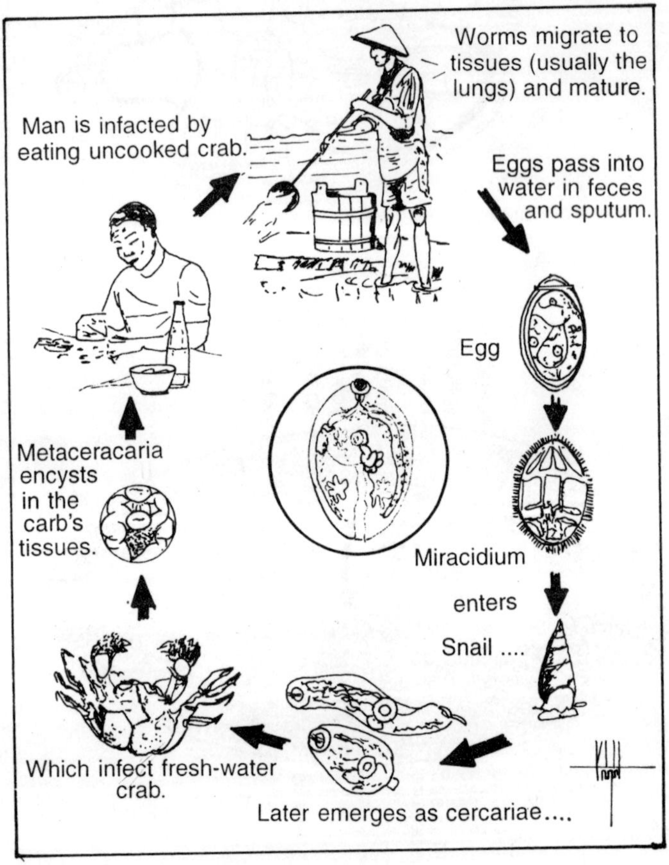

Fig. 7.2 : Life cycle of *Paragonimus westermani.*

several weeks of development in moist soil or water, the eggs hatch, and the miracidia swim about until a suitable snail appears. See Figure 7.2 for the life cycle.

Hatching is initiated by a sudden increase in activity of the larva inside the egg. The animal may, at first, lunge at the operculum, but is soon flattens its anterior end against this structure. A series of violent contractions then occurs, and the posterior portion of the body is extended until the operculum suddenly files back as if on a hinge. Water passes though the exposed vitelline membrane and causes it to swell and to extend through the opercular opening. A bubble, which may become larger than the egg itself, is produced and forms a transparent sac into which the miracidium finally forces itself. The larva swims rapidly about inside this sac until the membrane is ruptured and the ciliated worm swims away.

Mature miracidia average 0.08 × 0.04 mm in size and are covered with 4 rows of ciliated ectodermal plates. *Semisulcospira, Brotia,* and several other genera of snails are common first intermediate hosts. These snails are usually found in swift mountain streams, often away from towns or villages. Infection of the snails probably depends primarily on animal reservoir hosts rather than on human being. The incidence in snails is surprisingly low. Out of hundreds of snails examined in endemic areas in the Philippines, we have found only 2% to be infected. Age immunity apparently protects some snails. Miracidia shed their ciliated plates during the act of penetrating the snail. The sporocyst generation develops free in the lymphatic system, usually along the esophagus, stomach, or intestine. First generation rediae emerge from the sporocyst about 4 weeks after infection. Second generation rediae appear in the digestive gland 63 to 75 days after infection.

Microcercous cercariae emerge from the snail. They are ellipsoidal and posses a short tail with a spiny tip. The body is covered with small spines, and a prominent stylet lies in the dorsal side of the oral sucker. Two types of penetration glands are present, in addition to two irregular rows of mucoid glands on each side of the body. Cercariae of *Paragonimus westermani* in the laboratory do not readily escape from snails.

Cercariae enter a crayfish or a crab. Crabs belonging to the genera *Potomon, Eriocheir, Sesarma,* and *Parathelphusa* are the usual second intermediate hosts. Crabs probably ingest the cercariae. Metacercariae (0.34 to 0.48 mm) in the crustaceans may be found in the gills, heart, muscles of the legs and body, liver, or reproductive organs. Infected crayfish usually belong to the genera *Cambaroides* or *Cambarus.* The time between infection of the crustacean and the appearance of mature metacercariae on its tissues may be several months.

When a new host eats uncooked or poorly cooked crayfish or crabs, the metacercariae excyst in the intestine within an hour of ingestion. The minute worms are pinkish in color and possess a large, elongate, sac-like bladder. They penetrate the intestinal wall and usually go through the diaphragm to the pleural cavity. They may wander around the abdominal cavity for 20 days or more and may enter various organs in which they may reach sexual maturity. Among many organs that may be affected are the brain, intestine and other abdominal viscera, spermatic cord, epitomise, scrotum, vaginal wall, lumbar and gluteal muscles, spinal canal, subcutaneous tissues of the thorax and abdominal wall, and the entremities. Even after penetrating the diaphragm, the worms may never reach the lungs. A heavy infection of the liver frequently is observed in experimental animals.

In animals, usually 2 worms exist in a lung cyst, but this phenomenon rarely occurs in man. It has bee observe, in experimental animals, that worms probably locate one another by chemical attractants. Eggs appear in the sputum 2½ to 3 months after a man eats infected crustacea.

Crabs are often prepared for cooking on chopping blocks, and the juice left on the block may contain metacercariae that adhere either to foods subsequently prepared on the same block or to the fingers of the cook. The juice of fresh crabs is used in various preparations in Korea and the Philippines. Crabs are often "pickled" or salted or dipped in wine or vinegar before eating. These practices rarely kill all metacercariae. The encysted parasites can live for several days in a dead crab, one or 2 days in diluted wine, or 3 weeks in an ice chest at 10°C in 10% formalin.

Uncooked paratenic hosts, such as pigs and wild boar have been the source of infection in man.

Symptoms of infection include coughing, profuse expectoration, occasional appearance of blood in sputum, pain in the chest, brown sputum and muscular weakness. Pathologic features involve bleeding spots where worms have penetrated tissues, leukocyte infiltration, tearing of muscles, scar tissue, host allergic reaction, and fibrous cyst formation around the worms. Cysts may contain blood-tinged material, purulent chocolate-colored fluid, worm eggs, living worms, dead worms, or no worms. Hemoglobin taken from the worms has been reported to be different from that in host red blood cells. Apparently *Paragonimus* employs the heme moiety of the host directly in the formation of its own pigments. Migration of the worms to other tissues or organs may be troublesome. Cerebral involvement can be serious. Eggs distributed widely throughout the host cause inflammation, and secondary bacterial infection may occur.

Diagnosis. Diagnosis should include microscopic examination of sputum and feces for large operculate eggs. Sputum should be examined following centrifugation and sedimentation in 1 to 2% NaOH. Charcot-Leyden crystals in sputum help to confirm the diagnosis; these crystals are associated with monocytes. Serodiagnostic tests have been developed to identify *Paragonimus* infections, and are especially useful in diagnosing extrapulmonary paragonimiasis and in evaluating chemotherapy. Intradermal tests distinguish between paragonimiasis and tuberculosis, but not among various *Paragonimus* spp. Nor can they identify a clinical cure. The most reliable serodiagnostic tools available are complement -fixation and precipitation tests.

Host-Parasite Interactions. There are three characteristic lung lesions in paragonimiasis. First are foreign body granulomata around eggs. This host response is similar to that around schistosome eggs in liver, bladder, and intestines. Granulomata typically have fibroblasts, lymphocytes, and eosinophils encircling the eggs.

The second lesions are cysts or distoma cysts in which both worms and eggs are encapsulated by fibrous granulation tissue of fibroblasts, lymphocytes, eosinophils, and mononuclear cells. Calcification may occur in chronic paragonimiasis. Apparently, two worms are necessary for formation of a cyst, but surgery usually yields cysts with only one worm. Two worms appear to be present initially, but of the two dies during chronic infection.

The third type of lesion is abscess cavities caused by necrosis of lung tissue and the subsequent liquefication of this tissue, and often of the worms and eggs as well.

Treatment. Bithionol is the drug of choice for both pulmonary and subcutaneous paragonimiasis. Because 10 to 20 individual oral doses of the drug must be given on alternate days, other drugs have been sought. Nwokolo and Volkmer (1977) reported 73 to 90% cure rates with a single dose of menichlopholan for pulmonary infections of *P. uterobilateralis* in Nigeria; it is also used in treating fascioliasis and clonorchiasis.

Epidemiology and Control. Adequate cooking of the crustacean intermediate hosts kills metacercariae and prevents infection. Unfortunately, cultural preferences sometimes interfere with satisfactory control. In Japan crabs are usually eaten well cooked, but a common practice is to prepare vegetables using the same chopping block and cutlery immediately after preparation of the crab. Thus, metacercariae are transferred to other foods or to the mouth. In China "drunken crab" is a delicacy. Crabs are immersed in wine for about 12 hr. before being eaten, resulting in infection rates approaching 100%. The raw or under-cooked

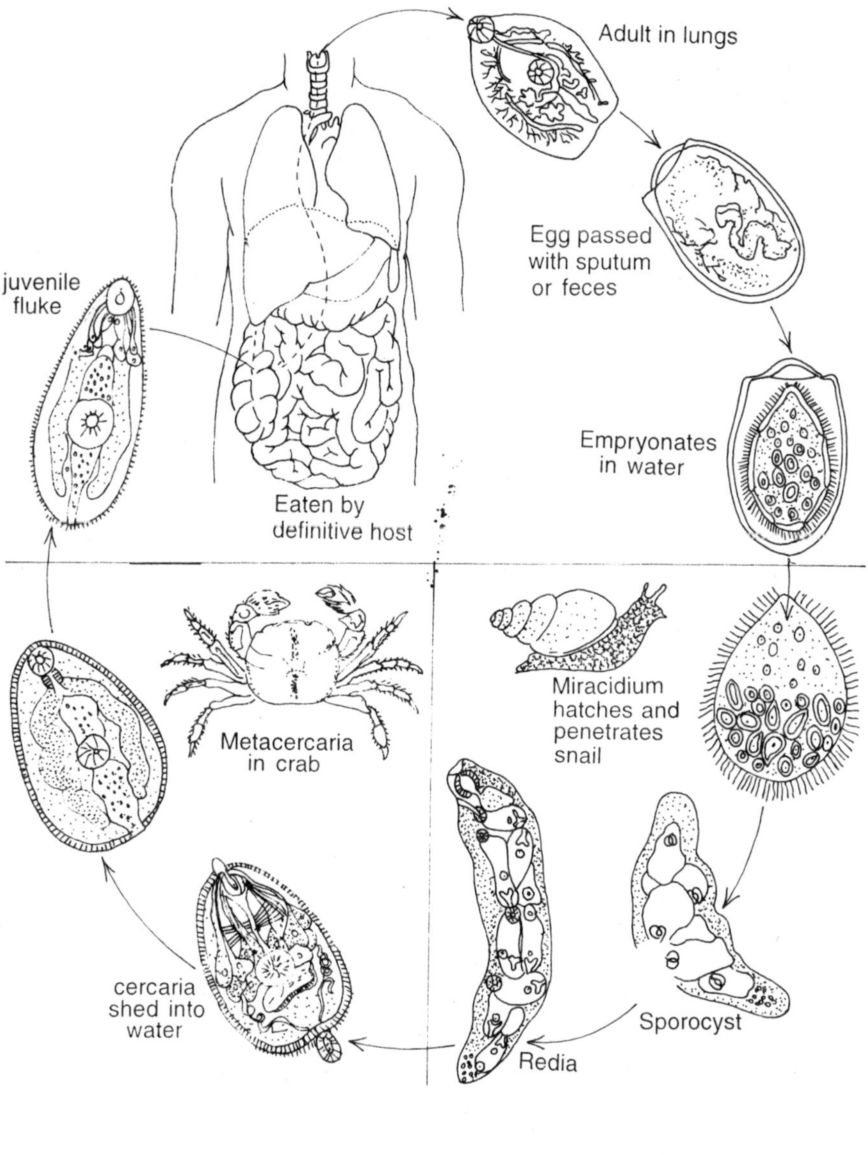

Fig. 7.3 : Diagram of the life cycle of *Paragonimus westermani.*

crab or crayfish eaten in many other parts of the world leads to paragonimiasis.

Another sources of human infection was reported by Miyazaki and Habe (1976). In Japan 136 human cases of paragonimiasis were acquired when people ate raw pork and wild boar which were paratenic hosts for *Paragonimus westermani;* metacercariae were located in the somatic muscles of these hosts.

Control of paragonimiasis through eliminating the snail host has not been successful because their habitat is rapidly flowing water. Elimination of crustacean second intermediate hosts is impractical because they are both a valuable source of protein and a culinary delicacy. Adequate cooking of crabs and crayfish, adequate cleanliness of food preparation utensils, and drug treatment for adult worms offer the most promise for control. Also, other meat from possible transport hosts must be well cooked.

Other Species Of Paragonimus. Knowledge of the taxonomy and biology of most species of *Paragonimus* is still fragmentary. *P. miyazakii* in Japan, *P. africanus* in the Cameroon and Congo, and *P. uterobilateralis* in Nigeria are also etiological agents of pulmonary paragonimiasis in humans; subcutaneous paragonimiasis is caused by *P. skrijabini* in China and *P. heterotremus* in China, Thailand, and Laos.

In the Western Hemisphere human pulmonary paragonimiasis was previously attributed to *P. westermani.* Only *P. preuvianus (= P. mexicanus, P. ecuadorensis),* found in Ecuador and Peru, is now thought to cause human paragonimiasis. The status of other Western Hemisphere *Paragonimus* species is currently uncertain.

P. kellicotti is a lung fluke wide distributed in the eastern United States and Canada. Definitive hosts include cat, dog, pig, goat, mink, muskrat, bobcat, raccoon, opossum, skunk, red fox, coyote, and weasel. *P. kellicotti* can be distinguished from *P. westermani* and other species by its ovarian shape and the thickness of the inner metacercarial wall. Stromberg and Dubey (1978) described its life cycle in cats in detail; *P. kellicotti* dies not infect humans.

Chapter—8

Opisthorchis Snensis

Opisthorchis sinensis is commonly known as 'Chinese liver-fluke'. It was formarly known as *clonorchis sinensis* and was detected in 1857 from a Chinese Hospital patient whose liver, on post-mortem, was found to be enlarged and turgid and from whose distended bile duct many small trematode worms were recovered. Since than its epidemiology, therapy, morphology, prevention and life-cycle have been studied in details.

SYSTEMATIC POSITION

Phylum	-	**Platyhelminthes**
Class	-	**Trematoda**
Order	-	**Digenea**
Family	-	**Opisthorchidae**
Genus	-	***Opisthorchis***
Species	-	***sinensis***

Distribution

Opisthorchis is common in Japan, Korea, China, Hong Kong and Indo-China. This distribution coincides with the distribution of the snail first intermediate host, *Parafossanulus manchouricus* and closely related hydrorbid snails (*Malek*, 1980). Nearly 100% of people in certain villages of China (*Komiya*, 1966) and over 65% of population Hong Kong are infected.

Host and Location

Humans are definitive host for *Opisthorchis.* Most fresh-eating mammals and fish-eating birds can be infected, but doges and cats are probably the most important reservoir hosts.

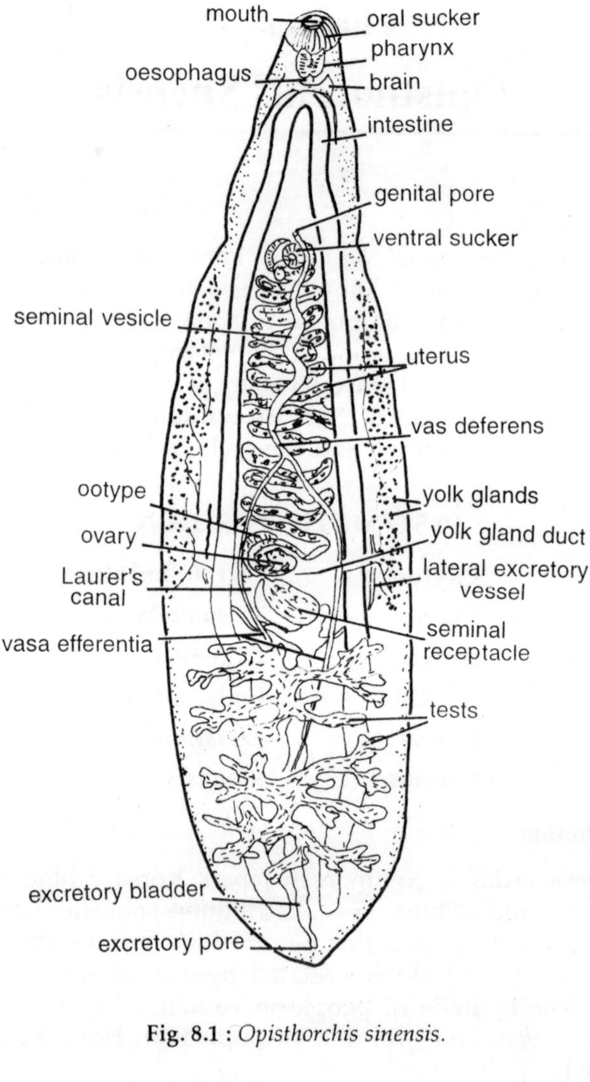

Fig. 8.1 : *Opisthorchis sinensis.*

Morphology

Opisthorchis is a smaller and narrower than *Fasciola* and not commonly related to it. Adult flukes are 10-25 mm in length and 2-4 mm in width. The body remains covered by rough cuticle. There are two suckers an oral and a ventral or acetabulum. The acetabulum is slightly smaller than the oral sucker and is located near the end of the anterior third of the body. Genital pore opens ventrally just infront of the acetabulum. A excretory pores lies at the posterior end.

Integument and Muscular System

A thin, minutely wrinkled cuticula which is devoid of spines overlies the hypodermis and the peripheral muscular layers of the body.

Parenchyma. The space between the organ systems is the interior of the body is filled with a mesenchyme called parenchyma. This tissue consists of irregularly shaped cells between which is a matrix of fluid and intertwining fibres.

Digestive System

A transversely oral *mouth* which is situated subterminally opens into buccal cavity. It is surrounded by well-defined oral sucker occupying most of the space of the extremity. Posterior to the oral sucker is a conspicuous ovoid *Pharynx*. The sucker and pharynx which are provided with radiating muscles fibres, serve for attachment, abrasion of the bile ducts and aspiration of loosened food particles. Pharynx opens into a short *oesophagus* approximately equal to the size of pharynx. Oesophagus joins two mildly distended gut caeca. The buccal cavity, pharynx and oesophagus are lined with thin cuticula while the gut caeca, which reach nearly to the posterior end of the body, are lined with a vacuolated columnar epithelium. The epithelial cells secrete enzymes. The food consists of blood corpuscles and a mucin-like exudate produced by laceration of the inflamed mucosa of bite duct.

Respiratory System

The respiratory organs are absent and the animal is anaerobic because of low oxygen contents in its surroundings. Energy in obtained by anaerobic fermentation, the carbon dioxide, a by -product, is diffused out through general body surface in the surrounding medium.

Excretory System

A clear, irregular *excretory bladder* is constricted posteriorly to form efferent canal which opens to the outside by a posterior *excretory pore*. The bladder proper is approximately has the same diameter as that of a gut caecum. It follows anteriorly between the main bodies of the rests to the level of the seminal receptacle.

Two lateral *excretory vassels* join the bladder slightly behind its blind anterior end. These vessels curve laterally and forward and follow an extracaecal course to the level of the acetabulum, where they branch into anterior and posterior recurving tributary vessels. The continued branching of these major vessels results in increasing finer vessels, the terminal units of which are the flame cells.

Nervous System

A bilobed brain with medial commissure and anterior and posterior nerves is located in the dorsal aesophaged region.

Reproductive System

Opisthorchis is hermaphrodite trematode possesses well-defined reproductive organs and a common genital pore.

Male reproductive System

Testes are posterior, branches and lying one behind the other. From the main body of each testis arises a thin duct the *vas efferens*. This duct proceeds forward to a point slightly anterior to the ovary where it joins its mate from the other testis to form a vas deferens vas deferens enlarges into a sperm-filled seminal vesicle. A narrow ejaculatory duct connects the vesicle to the genital atrium which opens at the ventral genital pore. A cirrus sac, prostate and cirrus all are absent.

Female Reproductive System

A singe branched ovary is present in the median line slightly anterior to an oblong sperm-filled seminal recepticle. The *seminal recepticle* lies just ahead of the anterior testis. Anteriorly the seminal recepticle gives off a short duct that joins Laurer's canal and which terminally joins a short *oviduct*. *Laurer's canal* curves posteriorly around the seminal recepticle and also runs abruptly dorsally to open by a small aperture, the *opening of laurer's canal.* This opening lies between the anterior testis and seminal recepticle near the middle. The oviduct, after receiving the seminal recepticle, proceeds forward a short distance and receives a *common vitelline duct*. Vitelline glands are small and numerous located in the middle third of the body. Two lateral vitelline duct arises from the middle of transverse vitelline duct. The oviduct receiving all these ducts (e.g. median vetilline duct, Laurer's canal and seminal recepticle) then penetrates an *ootype* which surrounded by *Mehli's* gland. The Mehlis' glands are loosely organized and composed of small acinus shaped cells. A very conspicuous uterus arises from the ootype anteriorly and proceeds an on irregularly coiled duct to the genital atrium. The uterus become larger in diameter anteriorly because of the massing of eggs.

Fertilization

Mutual cross-fertilization occurs after copulation between two flukes. The ova liberated from ovary are fertilized by the sperms from the seminal recepticle. In the ootype the fertilized egg yolk are enclosed in a thin walled, oval, operculate shell which is often provided with a comma-shaped posterior appendage. The egg gradually acquires a darker brown colour and proceeds with cleavage until a fully formed miracidium is formed. The embryonated capsules, pas to out side through the genital pore and are carried by the bite to the host's intestine, they are finally eleminated with the faces.

Life-History

Man is the primary host besides snail and fish are two intermediate hosts.

Miracidium and Sporocyst

As stated above the eggs when laid contain fully matured miracidia. Each ovoid miracidium bears long, motile cilia on an outer ectodermal sheath. The apical region bearing no cilia. In the apical region there is no apical spine and an apical gland opening. On the dorsal side there is a single excretory pore. An apical histolytic gland and a large, ventral, sausage-shaped secretory gland with a narrow duct opening besides the spine. About 25 germ balls are present, the fills the miracidial body space. Two dorsally placed flame bulbs and their ducts comprise the excretory system. Nervous system is situated ventral to the apical gland.

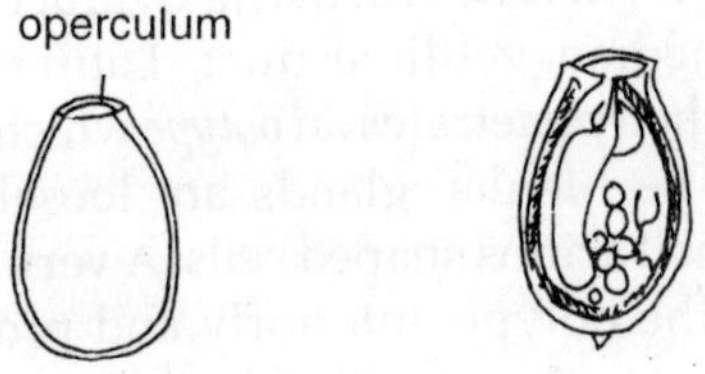

Fig. 8.2 : *Opisthorchis.* Egg and egg containing miracidium.

Miracidium do not hatch free in nature, rather they are released from the eggs only after ingestion by appropriate snail host. The miracidium than actively penetrates the gut wall into the peri-intestinal blood sinus. Now it becomes a thin walled *sporocyst*. Sporocyst produces the redial generation which migrates to the hipatic blood sinuses.

Redia. A mature redial is elongated in appearance. It anterior end is rounded whereas posterior end is some what pointed. The larva internally lined with sub-epithelium. A birth pore is absent. An anterior pharynx and a brownish sacculate gut extending about half-way posteriorly are present. Several germs balls are present which develop synchronously into cercariae. About 6-8 cercariae are liberated by the ruptuse of redial wall.

Cercaria. The long-tailed, oculated cercaria emerges from the snail host into the water and swims vigrously in the search of another host *i.e.*, a fish. The tail is provide with muscle -fibres and bears thin, lateral, transparent elytra. Two bean-shaped eye-spots are located near the anterior end of body. An oral-sucker, pharynx and acetabulum are present but oesophagus and gut caeca are not formed yet. Seven pairs of unicellular penetration glands occupy. The middle of the body. Their ducts open near the mouth. Dorsal and ventral peripheral mesenchymal cells are packed with cystogenous glands. These glands help in the formation of cyst of metacercaria. Excretory system comprises a triangular excretory bladder with anterio-lateral collecting ducts and a bifurcated posterior caudal excretory tubule.

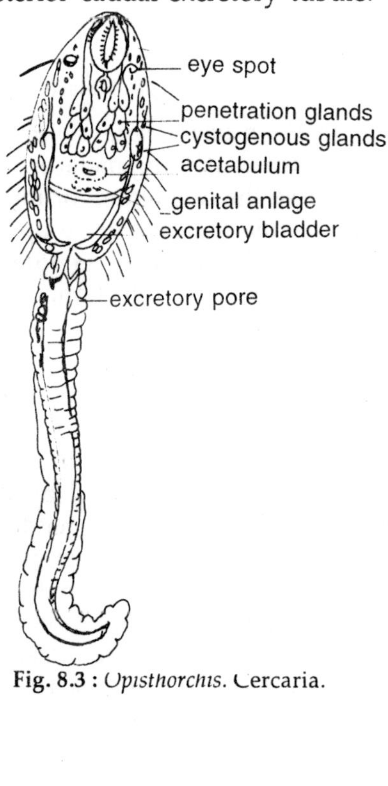

Fig. 8.3 : *Opisthorchis*. Cercaria.

Metacercaria. On coming into contact with the fish, member of Cyprinidae family, the cercaria attaches to the surface, discards its tail, and creeps under a scale into the fish. A cyst wall is formed by the material secreted by cystogenous glands. The cyst is thin, tough, and spherical. Organ formation continues e.g., the gut caeca etc.

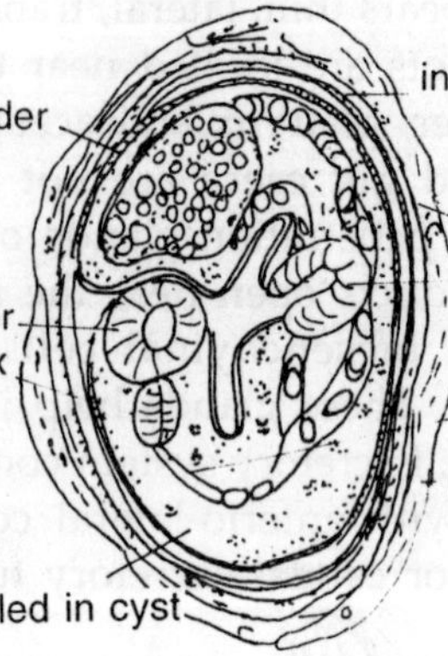

Fig. 8.4 : *Opisthorchis.* Metacercaria

Inspection to the primary host. Primary host i.e. man ingests uncooked fish containing metacercaria. Metacercarial cyst walls dissolve in the duodenum or intestine (*Gibson* and *Sun,* 1971) and juvenile flukes migrate up the common bile-duct into the small bile-duct of liver (*Sun et al,* 1968). Egg laying beings 3-4 weeks after infection.

Effects on the Host

Symptoms are classified as mild, modrate or severe, depending upon the burden of worm, length of infection and health of host. Light infections may be asymptomatic. Heavy infections are characterized by chills and fever, followed by liver enlargement and tenderness, sometimes with splenomegaly (Spleen enlargement). Chronic heavy infections cause cholecystitis (inflammation of gall bladder) and hepatitis (inflammation of liver). Modrate infection causes diarrhoea, edema and some hepatomegaly (enlargement of liver). The disease it termed *opisthorchiasis* or *clonorchiasis.*

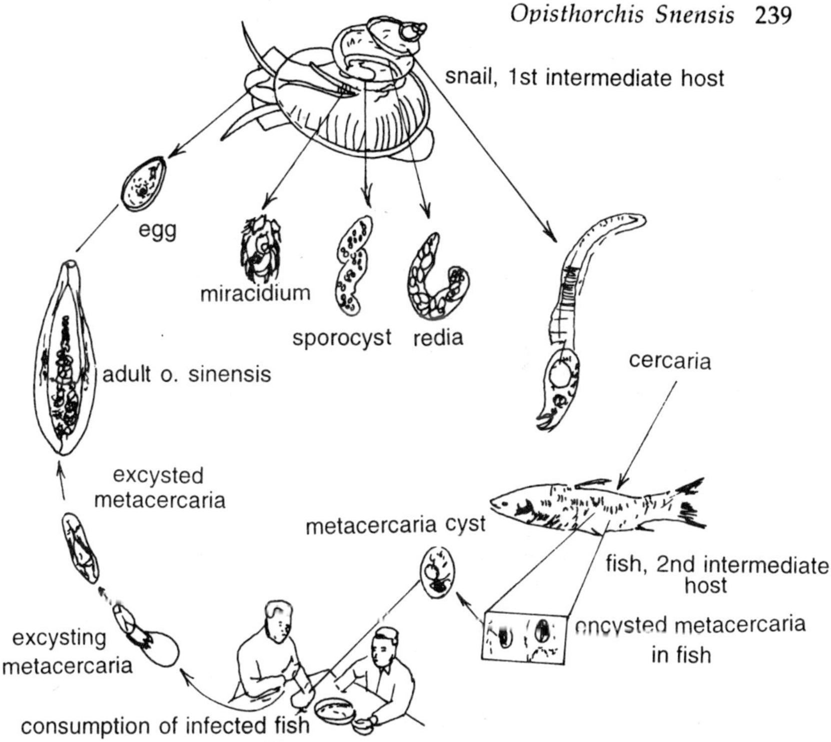

Fig. 8.5 : *Opisthorchis.* Life cycle.

Treatment

Hexachloroparaxylol, Hetol and Paraziquantal are used. The first two appear to be useful.

Control

All types of fishes, used as food, should be properly cooked to avoid infection.

Molluscicides to control the snail (first intermediate host) are generally not used because they are harmful to important fresh water fishes (*Malek,* 1980). Animal and human faces are important fertilisers for fish ponds in the Orient. Faeces must be properly composted. In additions, snail populations have been significantly reduced by removal of water plants when snails lay their eggs, thereby permitting fish more easily to feed on larval snails.

Chapter—9

Taenia Solium

Taenia solium belong to class Cestoda. The members of this class are commonly known as `Tapeworms'. All are endoparasites in the alimentary canal and associated duct of various vertebrate hosts. During their life cycle one or two or more intermediate hosts, vertebrate and invertebrate, are required in which the tapeworms undergo a phase of development. The cestodes differ, however, from the members of other two classes (Turbellaira and Trematoda) in the complete absence of a digestive tract.

The present text mainly related to *Taenia solium,* the pork-tapeworm of man

SYSTEMATIC POSITION

Phylum	-	**Platyhelminthes**
Class	-	**Cestoda**
Subclass	-	**Eucestoda**
Order	-	**Cyclophyllidea**
Genus	-	***Taenia***
Species	-	***solium***

Habits and Habitat

Taena solium or pork tapeworm is commonly found in those parts of the world where pork is eaten as food without proper cooking. It is found as an endoparasite in the gut of vertebrate hosts. One host is man and another is pig. Formerly, it was very common intestinal parasite in Europeans, especially is some places, of Europe. It is rare in many part of the world viz., India, Philippines, North America. It is not at all found in Jewish and Muslim countries as in those countries eating of the flesh of pig is considered as an evil deed.

External Morphology

Shape, size and colour. The body is elongate and flatenned dorsoventrally like a ribbon or tape. The body is narrow anteriorly and gradually broadens laterally towards posterior end.

The size of adult worm varies from 3-5 metres but a few are recovered to attain a length of about 8 meters.

The body is usually opaque white in colour but creamish, yellowish or greyish colouration is also common.

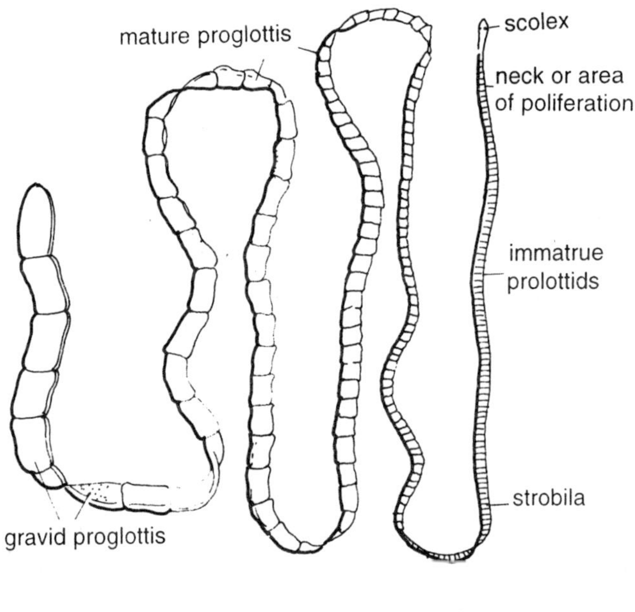

Fig .9.1 : *Taenio solium.* Entire

Divisions of the Body

The body of *Taenia* can be divided into three regions.

1. Scolex
2. Neck
3. Strobila

1. *Scolex*. The scolex is the holdfast organ, which is generally considered the anterior end. There are two schools of though concerning the interpretation of the structure of tapeworms. Some consider a single scolex and connected proglottids to be a single animal. Others consider each proglottid to be an individual-hence, a single tapeworm is a colony of many individuals. Some helminthologists argue that the scolical end of the cestode is actually the posterior end. However, *Hyman* (1951) and *Chang* (1964) are not convinced with the argue that scolex is the posterior end. Furthermore, the more recent literature indicates the scolocal end should be considered the anterior end.

Scolex is a pear-shaped structure, but is four-sides at the posterior part. It possesses a small terminal cone called the *rosetellum*. Around the base of the terminal cone or rostellum is a double row of curved, pointed, chitinous

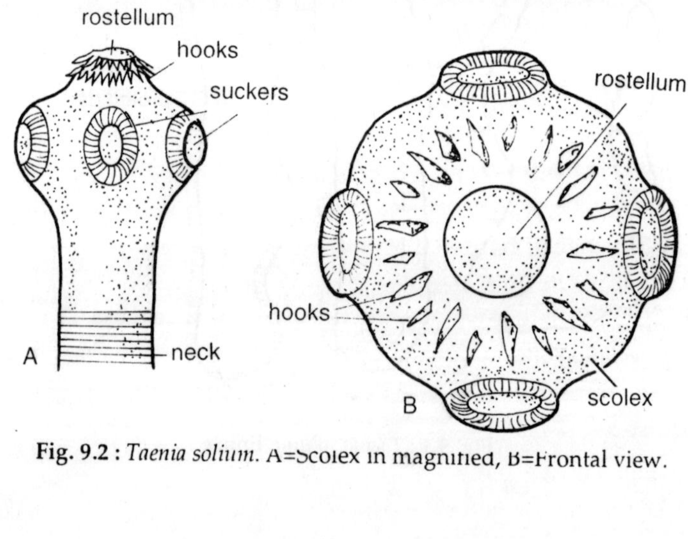

Fig. 9.2 : *Taenia solium*. A=Scolex in magnified, B=Frontal view.

hooks, which are usually 28 number but are sometimes between 22 to 32. Hooks are of two types, long and small. They are arranged alternately. Long hooks are approximately of 180u and short ones are of 130 dimension.

Each hook consists of a *base* or *guard* by which it is fixed, a blunt projection or handle directed towards the apex and a conical outwardly directed *blade.* Rostellum is capable of slight protrusion and retraction. Dmuring protrusion hooks are directed forwards but when rostellum is retracted, hooks are directed backwards.

The mouth is absence in scolex. The term "head" frequently used for the scolex, is, thus, inappropriate, because the head is essentially that part of the body where organs for *perceiving* and catching food are aggregated. It measures about 0.6-1 mm in diameter.

2. *Neck.* It is situated immediately posterior to the scolex. This is an unsegmented, poorly differentiated area that is generally narrower than the scolex and the strobila proper, and is the continuously differentiating or budding zone or area of proliferation that gives rise to immature proglottids.

3. *Strobila.* The strobila, which conatitutes the main mulk of body and is made of a chain of proglottids. There are about 800-1000 such *proglottids.* Each proglottid bears a complete set of genitalia. This linear repetition of genital organs is termed as *proglottisation.* Since the proglottids are budded off from the neck region in an orderly succession, the youngest proglottids are towards the neck and the oldest proglottids are posterior most.

Depending upon the development of genital organs, the proglottids are distinguished into three types:

(i) *Immature proglottids* - These are young, undifferentiated proglottids neck to the next region. The reproductive organs are either absent or in stages of development. Their width is more than the length. They comprises about 200 anterior proglottids.

(ii) *Mature or reproductive proglottids* - The mature proglottids are sexually mature consisting of a male and a female set of reproductive organs. They are usually as long as broad and are founding the middle of the body. They comprise about 450 proglottids *Taenia* is protandrous, about 100 or 150 anterior mature proglottids contain only male reproductive organs and remaining 300-350 contain both, male and female reproductive, organs.

A mature proglottid is a complete reproductive unit in its own right and may undergo self or cross fertilization.

(iii) *Gravid or ripe proglottid* - These are found in the posterior part of the body. These are characterised by the possession of a large, highly branched uterus filed with developing embryos. The rest of the genital organs have already disappeared from them after performing their function. They are longer than broad. The posterior most proglottids are cut off from the body and are expelled out of houst body along with its faeces. This process of separation of proglottids is known as *apolysis*.

Body-wall

The body wall of cestodes is made up of several layers but lacks a cellular epidermis. It consists of (i) cuticle (ii) Basement membrane and (iii) Musculature.

(i) *Cuticle.* The outer most layer of the body surface is called the cuticle or tegument. This layer is more or less transparent, noncellular, and without visible pores. The cuticle consists of two layers. (a) outermost hair-like or fringe -like *comidial layer* and (b) a thick *homogenous* layer. The cuticle is secreted by modified mesenchymel cells. These cells lie beneath the musculature and have long processes extanding up to the cuticle.

When studied under electron microscope, minute projections known as *microtriches* are present on the cuticular surface. *Rothyman* (1963) suggested that the distal portion of each microthrix, which is solid, is concerned with two functions (a) it may serve as a means of resisting the intestinal current, since the body surface always in contact with the microvilli of the straited border of the cells lining of small intestine of host (b) it may serve to agitate the micro habitat in the vicinity as the worm moves, thus string up the intestinal fluids so that mutrient materials as well as waste products are always in a state of flux. The proximal position of microthrix is medullated and could very well serve as sites of absorption. In addition to microtriches, ultramicroscopic pore canals that communicate with the parenchyma are present in cuticle. These may be the sites of absorption also. Mitochondria and endoplasmic canals are also reported in the cuticle. Thus, it is apparent that this outer body layer is a dynamic structure rather than a mere covering.

(ii) *Basement membrane.* Immediately beneath the cuticle is the basement ment membrane. This noncellular connective tissue layer contains vacuoles and granules. The protoplasmine layer beneath the basement layer is thick.

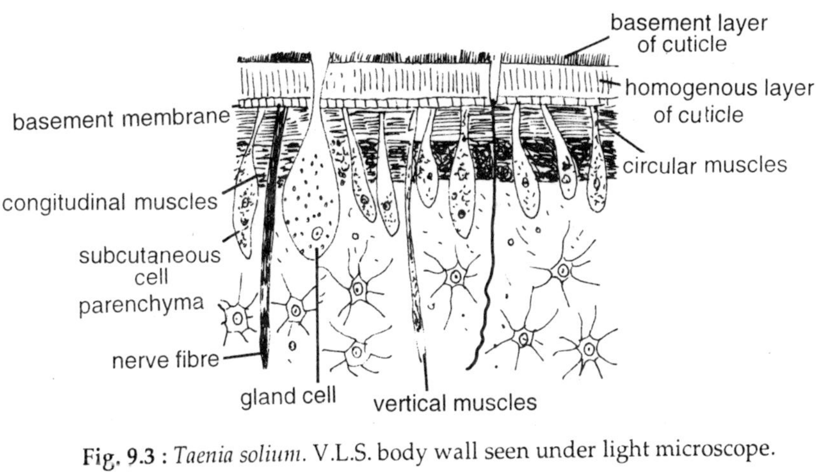

Fig. 9.3 : *Taenia solium.* V.L.S. body wall seen under light microscope.

(iii) *Musculature.* Immediately beneath the basement membrane is a layer of *circular muscles* followed by *longitudinal muscles* and vertical or *dorsoventral museles.*

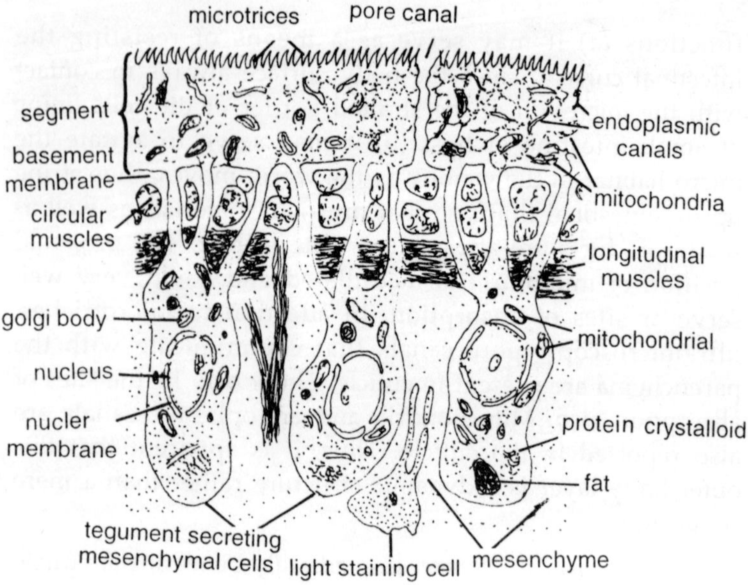

Fig. 9.4 : *Taenia solium.* V.L.S. Body wall seen under electron microscope.

Mesenchyma or Parenchyma

The space enclosed by the body wall, except for the occupied by the reproductive organs, osmoreguletroy structures, muscles fibres and nervous tissue, is filled with a spongy type of tissue, the parenchyma. Certain syncytially arranged cells form a network. In living animals, the space between the cells are filled with fluid. Numerous rounded calcarious bodies composed of concentric layers of calcium carbonate are present in mesenchyma secreted by *lime cells.*

Except at the two sides in each segment the parenchyma is divided into (i) outer *cortical* past and (i) inner *meduallary* part by circular muscle fibres of the mesenctymal musculature.

The parenchymal cells, and to some extent the spaces, serve as sites for the storage of glycogen.

Nutrition

In *Taenia* there is no alimentary canal and as it lives in the intestine of host, there is totally absence of alimentary canal in all the stages of its life-cycle.

Its chief source of nourishment is obtained from the intestine of host. There are amino acids, glucose, glyccrol etc.

Chandler thought that tape-worm gets its sugar supply from digested food-material of a host but *proteins* and *vitamins* are made available to the animal from the living tissue of the host by the secretion of the digestive enzymes, which effect the cells.

Respiratory System

In *Taenia,* the respiratory organs are absent so that respiration takes place though general body surface. Like other endoparasites, *Taenia* lives in the intestine of man where the oxygen content is very low or absent. So mode or respiration is anaerobic in which the energy is obtained by the break-down of glycogen into fally acids and carbon dioxide. The carbon dioxide diffused out in surrounding medium. If oxygen is available, it respires aerobically.

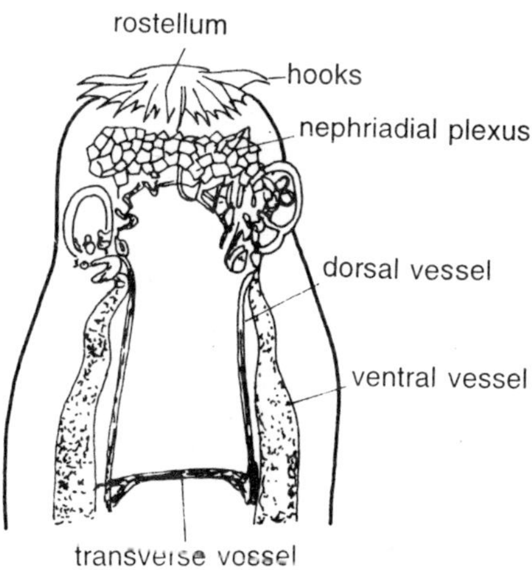

Fig. 9.5. *Taenia solium.* Nephridial plexus.

Excretory System

Excretory system, also known as osmoregulatory system, is same as the protonephric type found in tremetodes.

For main collecting canals traverse the entire length of the strobila. The two canals are ventrolaterally located, and the two dorsal canal are dorsolateral. All these canal are situated in the peripharynageal zone of the medullary parenchyma. A single transverse canal connects the two ventral canals at the posterior end of each proglottid. The ventral vessel carry the water away from the scolex, the dorsal vessels toward it. Within the scolex, the four longitudinal canals may be joined by a network of canals forming *nephridial plexus*.

Along the length of ventral canals, a series of secondary tubules arise that may in give rise to tertiary tubules. At the free end of terminal tubules are flame cells. The flame cells are generally arranged in a group of four. A flame cell is of irregular shape, with granular cytoplasm and nucleus. The `flame' is actually a group of cilia that arise from a concave basal plate located nearly the cell nucleus. The cilia are enveloped in the funnel-shaped enlargement of the free end

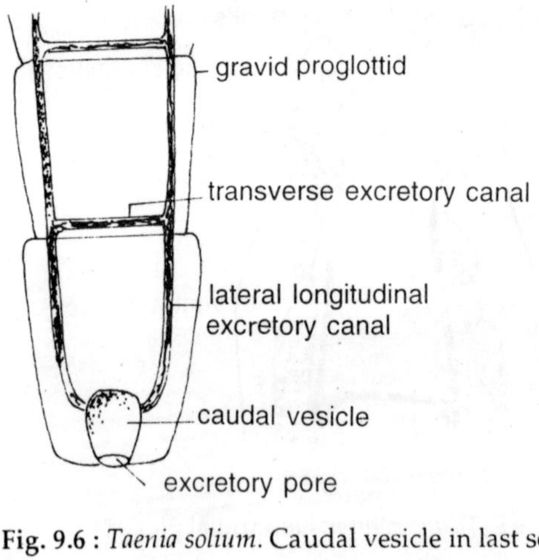

Fig. 9.6 : *Taenia solium.* Caudal vesicle in last segment.

of the tubule waste products collected though the flame cells is passed down the tubules into the main canals.

In the young worm, there is an *excretory vesicle* in the posterior most broglottid, into which the ventral canals open. However, as the neck region produces more proglottids, the older proglottids are pushed further back, and eventually the proglottid containing excretory vesicle is detached by the process of apolisis. Thus, in most older animal the posterior end of the tubes open independently to the exterior.

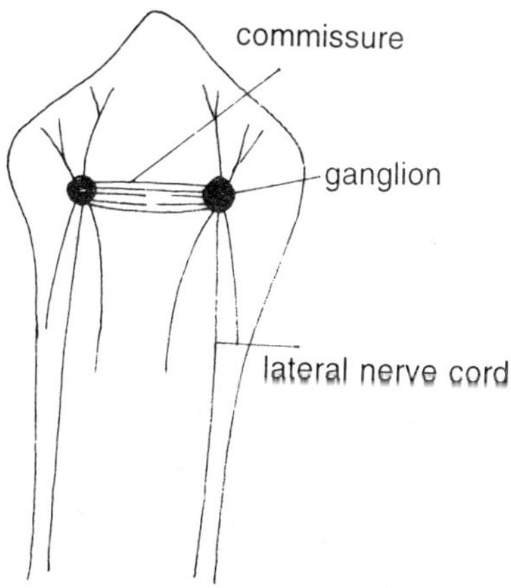

Fig 9.7 : *Taenia solium*. Nervous system.

Nervous System

A thick band in the scolex with two ganglia constitute the brain or central nervous system. From the brain few nerves are given out anteriorly which innervate the scolex. Three pairs of longer nerves are given out towards the posterior. Out of these only one pair, the *lateral nerve, cords* extend through the strobila passing just outside the excretory canal in each proglottid.

Sense organs are neither present nor they are necessary. Free nerve-endings are found in the body wall and detached proglottids for some time show independent muscular movement.

Reproductive System

Taemia is a *hermaphrodite.* Reproductive organs are usually found in the proglottids after 200th partition. Some anterior proglottids are functionally male. Middle proglottids are male as well as female. Proglottids at the posterior end are filled with enlarged *uterus.*

Structure of the proglottids in the middle region is as follows:

I. The male *reproductive system* consist of the following organs:

1. Testes
2. Efferent ducts
3. Vas deferens.
4. Cirrus or penis.
5. Cirrus sac.

1. *Testes.* They are numerous small, rounded bodies situated in the medullary past of the parenchyma. There number is about 15-200 and are distributed throughout close to the dorsal surface of the proglottid. Some workers are of the opening that the numerous rounded follicle, in fact, constitute a single and highly sub-divided testis. However, the number of testes is of texanomic value in tapeworms.

2. *Efferent ducts.* From each testis arises a fine ductule or capillary known as *vas efferens.* All the *vasa efferentia* unite to form a *vas deferens.*

3. *Vas deferens.* It is a long and tube, arising from the mid-plant of the proglottid. It runs transversely to the left of right to open into cirrus. In some cestodes the vas deference enlarge to form the seminal vesicle for the storage of sperms. But in *Taenia* no such enlargement is seen.

4. *Cirrus or penis.* The distal end of vas deferens in modified into a protusible *cirrus.* The cirrus is armed with bristles, hooks or spines. It is eversible through a male genital pore into a cup like *genital atrium* of the genital papilla.

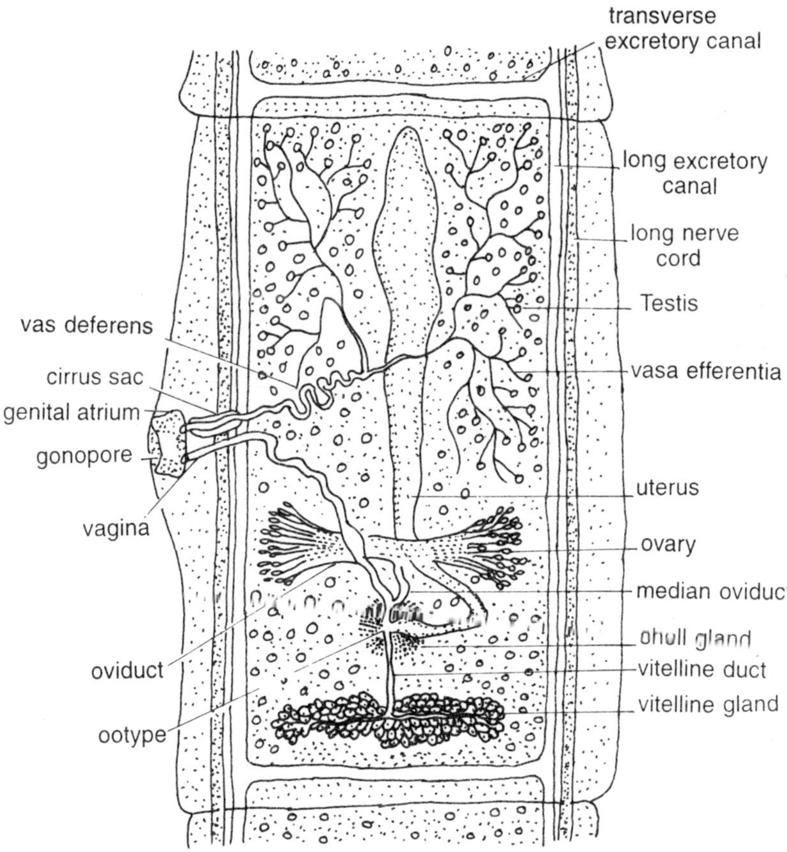

Fig. 9.8 : *Taenia solum.* Mature proglottid showing reproductive system.

5. *Cirrus sac.* Cirrus is surrounded by a kind of muscular sac known as cirrus sac. Its function is to erect penis during copulation.

II. *The female reproductive system* comprises:

1. Ovary
2. Oviduct
3. Ootype

4. Uterus
5. Vagina
6. Vitelline gland
7. Mehli's gland

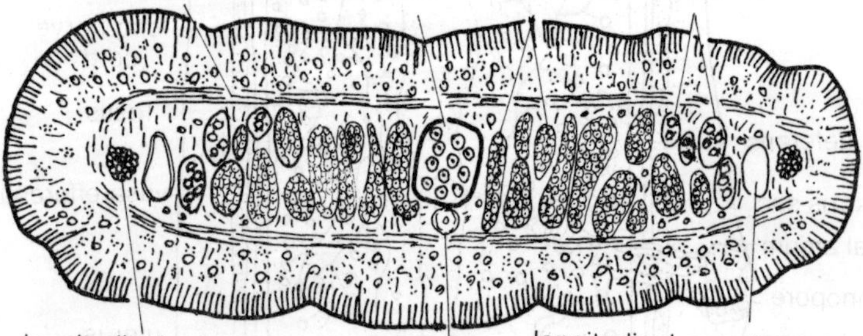

Fig. 9.9 : *Taenia solium*. Diagrammatic T.S. of a mature proglottid.

1. *Ovary*. A single bilobed ovary lies ventrally in the posterior past of the proglottid. Each lobe of ovary is dorsoventrally flattened and consists of a number of radially arranged follicles. The right and left lobes are connected medially by a tranaverse bridge, the *ovarian isthmus.*

2. *Oviduct*- The oviduct arises from the middle of the bridge. Its is short but wide duct and joins the ootype distally.

3. *Ootype* - It is a small rounded chamber developed at the junction of oviduct with the vitelline duct. It is surrounded by numerous unicellular *Mehlis glands,* the cells of which form an amorphous complex.

4. *Uterus* - A blind sac-like or tube-like uterus arises from the ootype and runs forward in the segment. In gravid proglottids the uterus enlarges and gets branched to occupy the whole of the proglottid. It remains filled with the fertilized ova or developing embryos.

5. *Vagina* - The narrow tubular vagina arises from the receptaculum seminis and runs obliquely outwards to open into the genital chamber by female reproductive opening.

The seminal receptacle or receptaculum seminis is a small wide tube connected one hand with the ootype by a narrow *spermatic duct* and on the other with vagina.

6. *Vitelline gland* - The vitelline gland is a compact, elliptical mass of numerous follicles situated posterior to the ovaries. A single short median duct, *median vitelline duct* arise from the gland and opens into the ootype. The secretion of the vitelline duct is rich in yolk and forms a covering of yolk and shall around the fertilized egg.

7. *Mehlis glands* - these are numerous minute unicellular glands situated around the ootype and emptying into it. The function of the secretion of Mechli's gland is to ease the passage of shelled eggs into the uterus.

LIFE CYCLE

Fertilization - Fertilization occurs by introducing the cirrus into the vagina of the same proglottid and sperms are stored in the receptaculum seminis, which later on fuse with eggs in uterus. This in *Taenia* self-fertilization takes place.

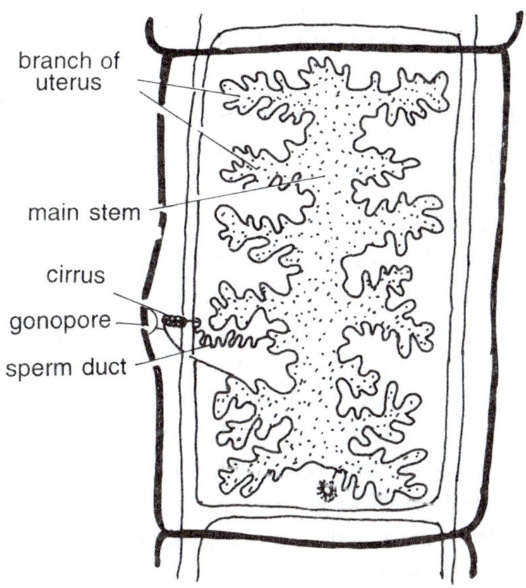

Fig. 9.10 : *Taenia solium*. A gravid proglottid

Sometimes, the anterior mature proglottids, devoid of female genitalia, can enter into copulation only with posterior mature proglottids with fully developed female genital organs. Fertilization, following copulation between two different proglottids, is sometimes termed *cross-fertilization* to distinguish in from that occurring between gametes of the same proglottid (self-fertilization).

Thus the eggs are fertilized inthe oviduct and get surrounded with yolk and egg-shall in the ootype. These materials are provided by vitelline gland. The capsulated eggs enter the uterus and are collected there. The uterus enlarges in size, gets branched and occupies the whole space. The eggs are small measuring 40μ (micro) in diameter.

Cleavage - The zygote are egg cell undergoes cleavage when the capsule is in the uterus. The first unequal division results in a larger megamere and a smaller *embryonic cell.* The megamere divides further and froms several similar megameres, which the embyoonic cell divides repeatedly producing two types of embryonic cells, larger mesomeres and smaller micromeres. Thus, three types of cells result from thew zygote; small micromeres, medium mesomeres and large megameres.

Onchosphere - By repeated division of micromeres a solid ball of cells, the *morula* is formed. The megameres divides, redivides and becomes filled up with yolk from the yolk cells, which surrounded the embryonic cells just like an envelope. It provides nourishment to the embryonic cells and finally disentigrates. The mesomeres form the inner thick striated membrane known as *embryophore.* Beneath embryophore lies basement membrane. The, morula, develops three pairs of chitinous hooks. These *hooks* are secreted by the differentiated cells known as *onchoblasts.* Now the six - hooked embryo is called *hexacanth.* It possesses a pair of large penetration glands and is surrounded by tow membranes, the haxacanth membrances. The hexacanth, together with all the membranes surrounding it is known as *onchosphere.* The onchosphere loses the shell.

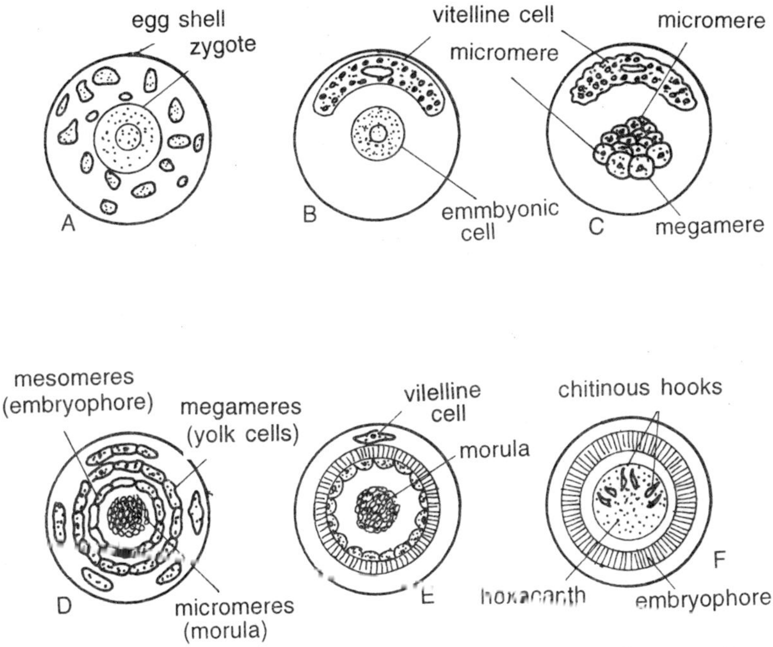

Fig. 9.11 : *Taenia solium.* A=Zygote, B,C,D,E, = Cleavage. F = Onchosphere.

These stages of development are found in the proglottids at the extreme posterior end of the body. The terminal proglottids are liberated from the body 4 or 5 at a time and are passed out quietly with the faeces of man. The proglottids show slow movements of contraction and later on perish. *Embryos* remain active for a considerable time. They are capable of further development only if they are taken in by a secondary host, which is a pig.

Infection of secondary host. The pig is the secondary host which regularly feeds of the faeces of man. Dogs, monkeys and sheeps are also known to get infection.

Inside the alimentary canal of a pig, proglottids and thousands of egg's shells, nearly 30,000 to 40,000, are digested and *hexacanths* are set free. Later on, hexacanths bore their with the help of hooks and penetration glands, through the wall of the intestine and reach intestinal blood vessles. The process requires only 10 minutes.

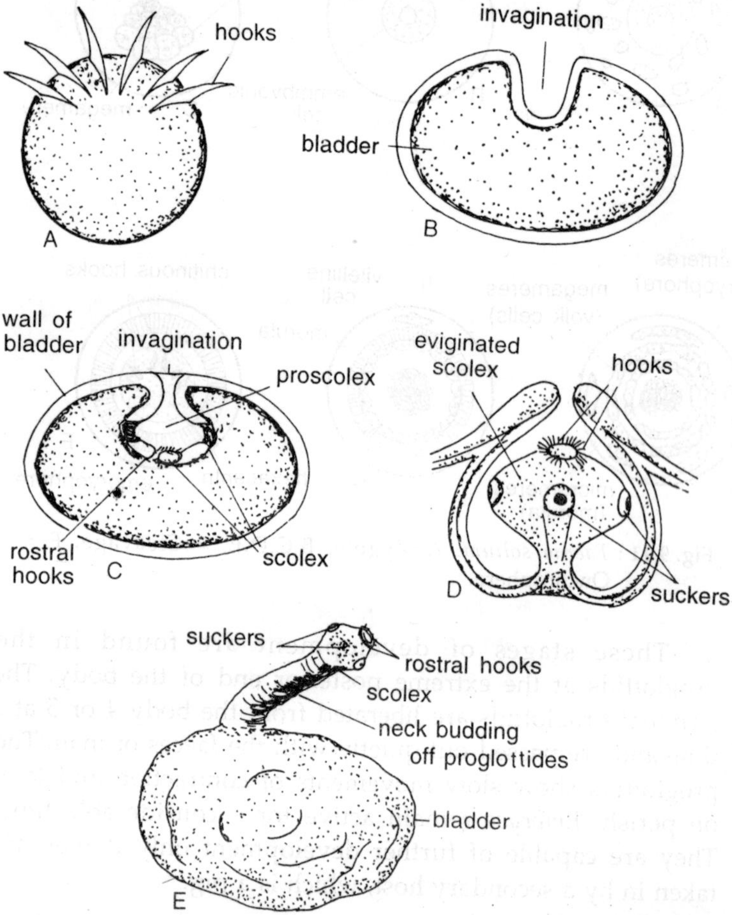

Fig. 9.12 : *Taenia solium.* A = hexaconth, B = bladder worm with invaginetion, C = bladder with proscolex, D = bladder worm with evaginated scolex, E = Cysticercus.

Migration. Travelling though the heart, these enter the muscles of various part in the body. The usual site where the hexacanths get encysted are the voluntary muscles of tongue, heart, lover and shoulder.

In the muscles, hexacanths lose their hooks, grow in size and develop a fluid-filled cavity in them. They now encyst in a cuticular covering. The encysted larval are called *cysticerci* or the *bladder worns*. They are oval somewhat whitish in colour and measuring 6 mm × 10 mm. Their wall consists of cuticle and mesenchyme layer. The fluid within them is mainly the blood plasme of the host.

The side of the bladder opposite to the originally occupied by the hooks thickness and invaginates into the cavity.

The invagination, which looks like a hollow knob, differentiates into an inverted scolex or the *proscolex*, possession suckers, hooks and rostellum. The embryo, at this stage, is called *bladderworm*. In *T. solium* the bladderworm is of *cysticercus* type which is characterised by a large vesicle and one scolex. That is why the baldderworm is also referred to as cycticercus. It appears to have a wall of cellulose and is, for this reason, sometimes called *cysticercus cellulosae*. Formation of cysticerci is completed in about 10 weeks in the pig. The cysticercus with proscolex is infective and waits for its ingestion by man.

The cysticered remain alive and infective in muscles for 5-6 years before the die, and are than calcified by the host. There may be 500-3000 cysticerci in one kilogram of infective pork (pig's flesh). The infected pork has brownish spots and is called the *measly pork*.

Infection to primary host. If the bladderworm or cysticercus is eaten with the port by man, the bladder is digested and the young scolex attaches itself to the wall of intestine by hooks and suckers. A region of proliferation of neck is formed, which develops progiottids of the adult tapeworm. While anchoring itself to the wall of small intestine in matures in two or three months.

Pathogenesis

Both adults as well as cysticercus stages are harmful:

1. *Adult worms* - The scolex of adult worm remains embedded in the mucous lining of alimentary canal for the

attachment. The injury of musoca may lead to bacterial infection. There may be abdominal discomfort, diarrhoca, frequent hunger pangs, nervousness, insomnia, nausea and epileptic etc. These symptoms are due to the presence of adult worms are collectively known as *Taeniasis.*

2. *Cysticercus larvae* - Sometimes, man may accidentally ingest or proglottids with contaminated food. Serious effects may result from cycticerosis in humans. Although they may be found in almost any organ, cysticerci are commonly located in striated muscles, liver, eye and the central nervous system, where they most often are in the meninges and ventricles of brain (*Marquez-Monter,* 1971). In an aberrant from of *T. solium,* apparently found only in the central nervous system, a cysticercus has daughter cysticerci budding from the bladder wall and forming an invasive, proliferating parasite. This type of metacestode is referred to as a *racemose cysticercus.* No viable scolices have ben found in recemose cysticerci (*Beaver* and *Rolon,* 1981).

Living cysticerci usually elicit little inflammatory response. They damage the host by forming space-occupying leisons, that is, their growth puts pressure on the adjacent tissue and interfers with its function or causes necrosis due to reduced blood circulation. In the brain ventricle the organisms cause *hydrocephalus* because the cerebrospinal fluid cannot be properly severe inflammatory reactions.

Treatment

Niclosamide is the drug of choice for intestinal infections, it is highly effective and well tolerated. Carbon tetrachloride and oleoresin are also used as an anthelmintic. For removal to tapeworms, does of 0.4 mm each of atabrin and acranil or 1 gm of hexylresocrcinol with 2 gm of acacia may be given. Cysticercosis is best treated in human by surgical removal of cysts, when possible. If surgery is not possible meberidazol can be administered.

Epidemiology and Control

Intestinal infections in humans are acquired by eating inadequately cooked pork. Pigs are infected by eating gravid proglottids passed with human faeces. Since a single proglottid may have hundreds of eggs, a pig may obtain a heavy infection throughout its body by eating only a few proglottids.

Following are the principal means of control:

1. Adequate cooking of food.
2. Freezing pork after slaughter and before it is marketed.
3. Sanitary disposal of human waste.
4. Meat inspection.
5. Prompt treatment of persona found to be infected.
6. An educational program aimed at changing habits of personal hygiene so that person do not reinfect themselves or others they have contact with.

PARASITIC ADAPTATION IN *TAENIA*

Taenia is an endoparasite in the intestine of vertebrate hosts. It has undergone a number of structural modifications of adjust in its specific environment. The modifications due to parasitic mode of life can be discussed under following heading:

I. Strucural

II. Physiological

III. Modifications in life-cycle.

I. Structural modifications can be divided into:

(A) Change in external form

(B) Change in internal form

(A) Change in External Form

1. Body is dorso-ventrally flatenned and ribbon like
2. Neck possess the power of forming new proglottids.

3. Scolex or the anterior end is modified into a hold fast for attachment for this purposes scolex hears suckers and spines.

(B) Change in Internal Organs

1. Body wall in devoid of cellular epidermis and remains invested by a thick and resistant cuticle.
2. Alimentary canal is absent because the animal lives in intestine where digested food materials always available.
3. Excretory system consists of protonephridia and which helps in osmoregulation.
4. Respiration anacrobically.
5. Sense organs are absent except tangoreceptors.
6. Highly developed reproductive system.

II. Physiological Adaptations

1. It can resists host's digestive enzymes through some unknown mechanism, probably by the secretion of anti-enzymes.
2. It maintains an equilibrium with the osmotic pressure of host.
3. It stimulates the mucous membrane of the intestine of host to secrete a large amount of mucous that cover for further protection against host's digestive enzymes.
4. Power of anacrobic respiration enables it to lie in the oxygen free intestinal contents.
5. Long flatenned body provides large surface area for absorption of digested food of the host that is why the mouth and alimentary canal are absent.
6. Long flatenned body provides large surface area for absorption of digested food of the host that is why the mouth and alimentary canal are absent.

III. Modifications in Life-cycle

1. To cover the risk of finding individuals of different sexes in one host the organism and moreover, each segment is bixexual and self-fertilization occurs.
2. Each proglottid produces 30-40 thousand eggs.
3. A secondary host i.e. pig is introduced to ensure its dispersal from one primary host to another.

Chapter—10
Echinococcus Granulosus

Echinococcus granulosus or the `hydatid worm', primarily parasetizes members of the canine family. The larval stage, the hydatid, is normally a parasite of sheep, but almost any mammal will serve as a host. It is potentially dangerous to human beings, in whom it may give rise to a fatal metastasizing, cancer-like tumor.

SYSTEMATIC POSITION

Phylum	-	**Platyhelminthes**
Class	-	**Cestoda**
Sub-class	-	**Eucestoda**
Order	-	**Cyclophyllidea**
Genus	-	***Echinococcus***
Species	-	***granulosus***

History and Habitat

The hydatid disease, which is applied for any tumor or swelling of a cystic feature caused by *Echinococcus*. *Redi* (1684) and *Tyson* (1691) who had reported actual animal nature of hydatid. But is was first discovered by *Goeze* (1782). Next *Hartmann* (1695) who found the adult in the intestine of dog.

Thus *Echinococcus* is generally found in the small intestine of dogs, wolfs, foxes and other carnivorous mammals. But larvae of this parasite develop into hybrid cysts mainly inhuman being and domestic individuals.

Distribution

Echinococcus is cosmopolitan in distribution, especially in sheep and cattle-raising areas. It has been reported from Iceland, South Australia, Tasmania, New Zealand, southern south America, southern and northern Africa, southern and eastern Europe, Siberia, Mangolia, northern China, Japan and the near East. In India, it has been reported from Punjab.

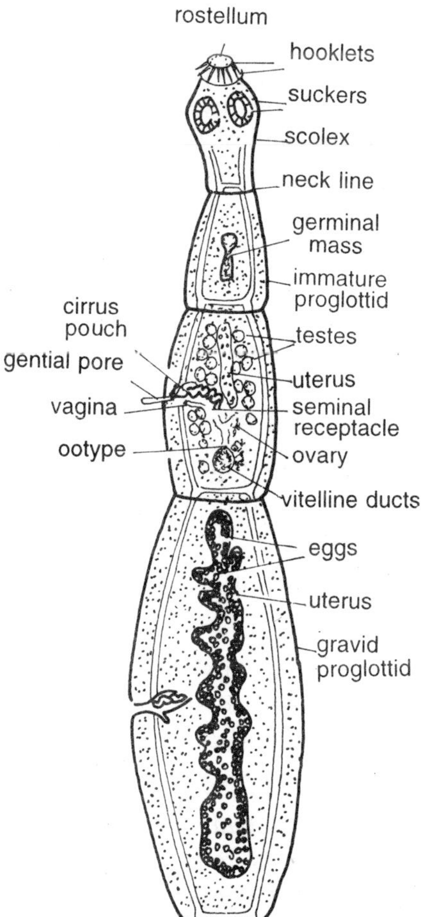

Fig. 10.1 : *Echinococcus granulosus.* Entire.

External Features

Shape and Size. The adult of *E. granulosus* is typically cyclophyllidean (*Taenia*), but it differs from it in size. It measures from 2-8 mm. in length.

Division of body. The body is divisible into (i) Scolex (ii) Neck (iii) Strobila.

(i) *Scolex.* The scolex is anterior most part of body measuring 500 u in diameter. The scolex bears four cup-like muscular structures, the *suckers* for the attachment of the intestine of host. Anteriorly situated, rectracile *rostellum* is armed with a double row of 20-80 hooks. *Smyth* (1963) has reported the presence of a group of cells lying just beneath the anterior tip of the rostellum. These cells not only contain droplets but actually secrete small viscid droplets to the exterior. The function of this secretion is still unknown probably the secretion maybe important from the immunological standpoint, acting as an antigenic agent.

(ii) *Neck.* The narrow part of the scolex, behind the suckers, forms the neck. It is the area of strobilization.

(iii) *Strobila.* The strobila consists of three or four proglottids.

(a) *First proglottid.* First proglottid is indistinet. It follows the neck and is slightly wider than the neck. It consists of a genital rudiment only in the form of dark patch of cells.

(b) *Second proglottid.* This proglottid is well distinet. It is much larger than the first proglottid. It contains a full compliment of taenioid - like male and female reproductive ogans.

(c) *Third proglottid.* It is gravid proglottid and is comparatively bigger in length as well as width than the other proglottids. It contains a uterus filled with a large number of eggs.

Internal Structure

Body wall. It is made up on the same pattern as in *Taenia.*

Digestive System. The digestive system is absent. The predigested food of the host is taken through general body surface. The food is reserved in the form of glycogen.

Respiratory System. The respiratory organs are absent. The protoscolices of *Echinococcus* carry out aerobic fermentation of carbohydrates. The protoscolices excrete pyruvic, acetic, succinic and lactic acids, and small amount of ethyl alcohol. In the absence of oxygen there is an increased production of succinate and no pyruvate is excreted. Carbon dioxide which is formed as by product is diffused out.

Excretory System. It is well developed containing longitudinal and transverse excretory canals with flame, cells. Excretory system mainly help in regulating the amount of water in body, hence they are mainly osmo regulatory in function.

Nervous System. Well concentrated in the scolex.

Reproductive System

Echinococcus is hermaphrodite. The male and female reproductive organs are present in the second proglottid or mature proglottid.

Male Reproductive System. In the middle region of proglottid, encircling the other reproductive organs, are about 20-30 spherical structure known as *testes.* From each testis a fine duct arises known as *vas efferens.* In the postmedial portion of the proglottid the vasa efferentia join to the base of a vas deferens. Vas deferens is an extremely convoluted, sperm filled tube, which passes lateral to an enlarged, coiled, marginal *cirrus* or penis. The cirrus lies in a bulbous *cirrus pouch* or *sac*. The cirrus, during copulation is protruded through a lateral *genital atrium* and genital pore.

Female Reproductive System. Opening into the genital atrium closely behind the cirrus is a female copulatory tube, the *vagina*. The vaginal opening is guarded by a muscular sphincter. Beyond the sphincter, the vagina passes downwards and enlarges at its inner end to from a sperm filled sac, the *seminal recepticle.* On the posterior side of the proglottid there is a bilobed *ovary* with a narrow, central *isthmus* from which an oviduct arises posteriorly. The oviduct joins a duct from the seminal receptical, and this combined fertilization duct continuous posteriorly into a median compact *ootype*. The ootype is surrounded by unicellular *Mehlis' gland*. In the posterior most region there lies a bilobed *vitelline gland.* The two lobes of vitelline glands are oriented dorsoventrally from the vitelline gland a median *vitelline duct* arises that joins the ootype. From the ootype a tube-like *uterus* arises which extends to a point near the anterior border of the proglottid.

Eggs. Eggs, resembling those of taenoid, are monoperculate and measuring about 30x38 u. These are ectolecithal i.e. the zygote is surrounded by the yolk cells provided by vitelline gland and shall material. After an egg-shall is laid down the eggs pass from the ootype into the uterus. When the uterus of a gravid proglottid bursts, the eggs are liberated an passed out along with faeces of dogs. These eggs gain enterence into the intermediate host (man or other mammal) either through water or in forage. Each egg contains a fully developed *onchosphere.* Human infection develops from too intimate an association with dogs. Infection of children is commonly established when dogs are permitted to lick their months. Eggs can be ingested from contaminated fingers.

The eggs, upon being swallowed by a mammalian intermediate host, hatches in duodenum as a typical six - hooked onchosphere. The matching occur in slight alkaline medium. Activation of embryos take place in either human bile or intestinal juice of ruminants, they are not activated in bile or intestinal juice of dogs and cats. The escaping onchospheres penetrate in the intestinal wall, enter the mesenteric venules, and become lodged in the capillary beds of various organs. The hooks are now shed. After three hours of egestion, the embryo reaches its final place of development. Among moose, deer etc. they reach lungs where as in man and other mammals, they reach in the liver, lungs, brain, kidneys, spleen etc. but liver being the most common site. Implantation and growth occur here, Onchosphere now develops into *hydatid cyst*. The cysts formation is a slow process, with the host lying down an envelope of connective tissues around the parasite. In about 4-5 months, the cyst reaches approximately 10 mm. in diameter. It goes on increasing in size and eventually becomes as big as the organ or even sometimes as football.

The various types of larval cysts, formed in the life-history of *Echinococcus* are:

1. *Unilocular hydatid cyst*. The hydatid cyst is composed of three layers. The inner most layer is called *endocyst*. It is thin syncytium forming the germinal layer. The middle layer is called *ectocyst*. It is made up of cuticle which is amorphous and laminated. The outer layer is *pericyst*. It is composed of connective tissue and fibres secreted by the host.

   The cysts produces daughter cysts upon invagination of its wall. These daughter cysts project into the lumen of mother cysts. The cavities are filled with a sterile fluid known as *hydatid fluid*. Within some daguther cysts, tertiary or grand-daughter cysts, from the invagination. These are endogenous or exogenous cysts or brood capsules.

   The endogenous cysts fall free into the hydatid fluid of mother cyst white the exogenous cysts, after detaching, migrate to other part of body, where they continue to grow further . The daughter cysts have a fibrous cysts wall internally lined with endocyst.

Rounded and hollow brood capsules are also continuously budded off from the endocyst of mother cyst. These capsules remain attach for some times by means of a narrow pedicle.

From the wall of these brood capsules, minute scolices are formed. Scolex formation may occur from the invagination of walls either of daughter or grand-daughter cyst.

The anterior end is inturned to form a central canal into which the four suckers open and into the expanded base of which the hooklets project. The body of each scolex is filled mesenchymal calls aggregated to form a subcuticular layer and to form a compact cup-shaped mass below the hooklets. The syncytium of the endocyst retains its embryonic potentialities, and new brood capsules may be produces indefinitely. Hydatids thus may expand to enormous size and contains nearly 50 litres of fluid as well as thousands of scoleces. The pressure so produce may severely damage adjacent host organs.

When the infective brood capsule is ingested by the final host, each scolex quickly evaginate and attach to the duodenal mucosa. For the first four to five days in the host they undergo little growth By the 14th day the scolex has grown in size, and first proglottid has appeared. The genital anlage and a second proglottid appear by the 18th day. Growth continues and fully developed shelled embryos are formed by about the 35th day, and the life cycle is begun again.

2. *Multilocular cyst.* It is characterized by many small bladders embedded in a common enclosing membrane. These are found in cattles. It is

considered to be the result of differential reaction inside the host.

3. *Alveolar hydatid in man.* It is characterized by a malignant type of growth with jelly-filled proliferating vesicle embedded in a common dense stroma. The very thin laminated membrane does not restrict the germinal membrane that grows out into surrounding host tissues. It is commonly sterile and it undergoes degeneration and calcification.

4. *Multivesticular.* It is characterised by many adjoining and connected bladders, each having its own separate germinal membrane.

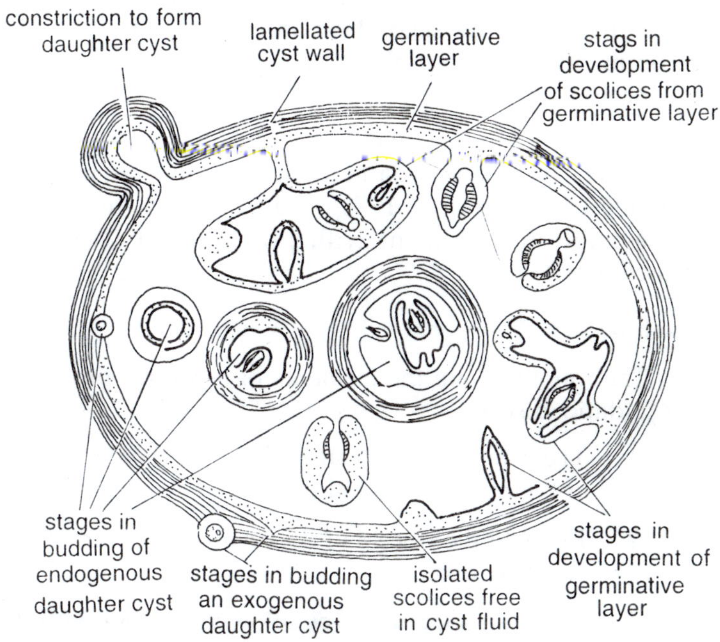

Fig. 10.2 : Echinococcus granulosus. Hydatid cyst.

Pathogenesis

Bladders or cysts are harmful when they develop rapidly. It causes inflammatory reaction in the surrounding tissue of the host result in the formation of fibrous tissue. The escaping hydatic fluid not only causing diarrhoea, abdominal pain etc. but also death of the host sometimes.

Treatment

Surgery is the recommended treatment for hydatid disease, but removal of a hydated cysts must be approached with care. An individual harboring a hydatid often develops hypersensitivity to the fluid in the cyst and if the cyst is broken during surgery and fluid leaks into the body cavity, the patient may die within minutes from anaphylatic shock. If the patient survives, protoscolices, which are liberated from breakage of cyst, may set up secondary infection. In order to check the secondary infection sometimes it is recommended to inject the cyst with alcohol or iodine to kill protoscoleces.

Mebendazol has been successfully used in some individuals who have hydatid in locations that are inoperable (*Keystone* and *Murdoch* 1978). Adult worms in dogs are suscesptible to treatment with praziquantel (*Wilson* and *Rausch*, 1980).

Control

The following are the elements of a control programme for hydatid disease:

1. Treatment of infected dogs with an effective anthelmintic or to kill them.
2. Keeping the viscera of domestic animals from being eaten by dogs.
3. Children should not be allowed to play with dogs frequently.
4. Avoiding play with dogs and hands should be thoroughly washed before eating.

Chapter—11

Reproduction

Introduction

This chapter contains a very brief account of the processes of asexual and sexual reproduction in parasites, from Protozoa to Nematoda. Limitation of space precludes a detailed description of the physiology of reproduction so that subjects such as sexual attraction, mating, spermatogenesis and oogenesis, development of the fertilized eggs, egg shell formation and hatching of the mature egg, can only be treated in the most superficial of ways, or in some cases neglected altogether.

The most reproductive biology of most parasites has not been studied in any great depth and, while the details of the morphology or reproductive systems as well known, relatively little physiological information is available.

Asexual reproduction

Asexual reproduction is achieved by the splitting or budding of a single cell or an entire organism, and therefore does not involve the fusion of two haploid gametes, with a subsequent meiosis. Asexual multiplication is a feature of many parasitic protozoa, all of the Digenea and a small number of the Cestoda. In many of these parasites there is an alternation of sexual and asexual phases during the life-cycle, sometimes occurring in the same host.

In the Protozoa there are at least five distinguishable asexual processes the occur. *Binary fission* is by far the most common from a asexual multiplication and involves the parent cell dividing into two daughter cells. The division is preceded by a nuclear mitosis; the plane of division is transverse in the Ciliophora and Sporozoa and longitudinal in the Mastigophora. *Multiple fission* the secondary division of daughter cells before they become separated from the parent cell. It is a feature of the gregarines and trypanosomes and gives rise to spheres or chain of daughter cells. A third form a asexual reproduction is *schizogony*, a form of multiple budding, in which the parent cell (the schizont) produces daughter cells (merozoites) first by undergoing one or more nuclear division, and then by dudding off many uninucleate daughters. Schizogony is typical of the Sporozoa, such as the Haemosporina, (e.g. malarial parasites). *Endodyogeny*, in which two daughter cells are budded internally within the parent cell, is only known to occur in some sporozoans, such as *Toxoplasma* and related forms. A fifth type of asexual reproduction, *single budding*, occurs in free-living but not parasitic protozoans.

A.

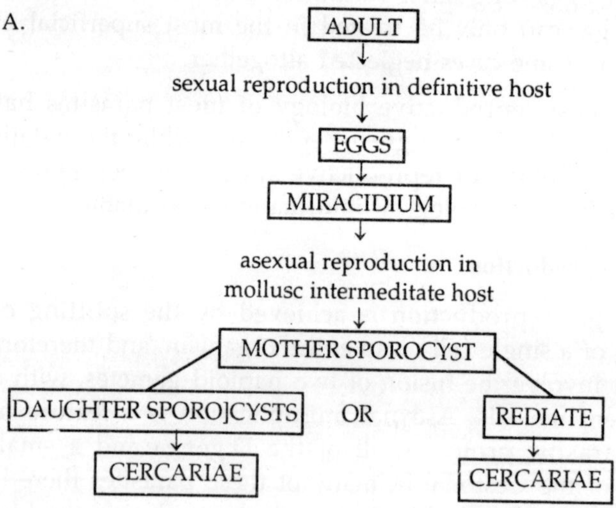

B. THE ASEXUAL STAGES OF CESTODA

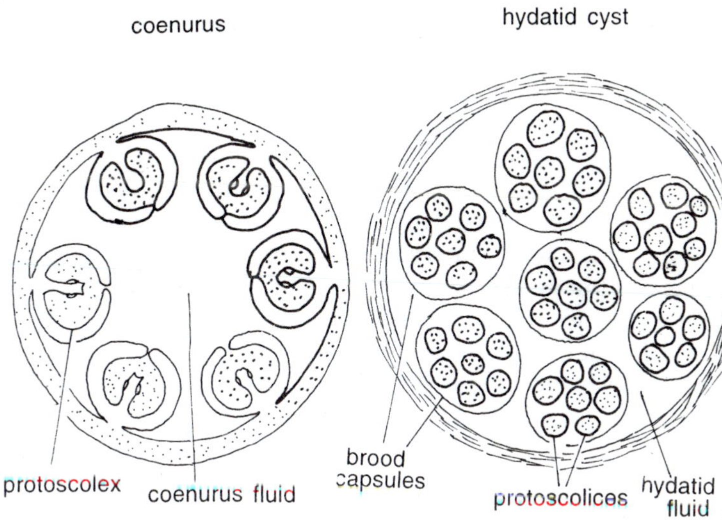

Fig. 11.1 : Asexual multiplication in helminth parasites. The Monogenea, Acanthocephala and nematodes to not have asexual stages in their life-cycle.

Asexual multiplication in the helminths occurs exclusively by *internal budding* or *polyembryony*. In the Digenea, the asexual phases are found only within the snail intermediate host, and involve the sporocyst and redia stage. During the intramolluscan phase of development of a digenean there may either be two generations of sporocysts, i.e. mother and daughters, or the initial sporocyst (arising from the penetrated miracidium) may give rise to a redia, within which are developed daughter rediae- the precise pattern depends upon the species involved. Cercariae are developed asexually within daughter sporocysts or daughter rediae. The intramolluscan phase of the digenean life-cycle is highly proliferative, since one miracidium may produce as many as several million cercariae by asexual development involving sporocysts or rediae.

The sporocyst is generally a hollow, elongated organism, with an anterior birth pore and well developed excretory system, but lacking a gut. At maturity, a mother sporocyst (derived directly from a miracidium) will be replete with growing daughter sporocysts or rediae; these emerge through the birth pore and migrate within the snail, usually to the hepatopancreas (digestive gland). Daughter sporocysts are morphologically similar to the parent sporocyst, and produce, by internal budding of germinal tissue, large numbers of cercariae, which escape through the birth pore when "mature". The redial stages are not unlike the sporocyst but possess a mouth and gut. Asexual multiplication in the Digenea involves the division of *germinal masses* within these larvae; these masses contain large numbers of germ cells, which proliferate during development. In the mother sporocyst germ balls are formed, each of which will become a daughter sporocyst—many hundreds of daughters can be produced by the division of this germinal tissues. Daughter sporocysts and rediae, likewise contain germinal tissue from which embryonic cercariae are produced. Cercarial production by individual sporocysts or rediae may continue, uninterrupted, for several months, or even perhaps years, as has been recorded in one case in a laboratory infection.

There is some argument concerning the exact nature of asexual multiplication in the Digenea. Some authors regard this form of multiplication as parthenogenesis (ameiotic or meiotic), while others consider it to be regenerative budding or polyembryony. It has even been suggested that the intramolluscan stages of digeneans reproduce sexually and that sporocysts and rediae are not larval forms at all. These vies have not been substantiated and the debate is really philosophical and adumbrated by semantic problems. Nevertheless, however one regards the germinal lineage of the intramoluscan stages of digeneans, a clear function emerges, that of enormously increasing the numbers of infective individuals. This may be an adaptation to the great wastage that must occur once cercariae are liberated into the external environment (figure 11.1 *a*).

In the cestodes, larval development involving asexual budding occurs in only a few groups (Diphyllobothriidae, Hymenolepididae and Taeniidae). The majority of tapeworms form only a single larva from each egg; of the many types of larval stage, only a small number are capable of proliferation by asexual budding. External budding occurs in two larval types, the urocystis (an budding cysticercoid) and the urocystidium (a budding strobilocercus). Internal budding is the more common pattern and is typical of the polycercus larva (*Parictotaenia paradoxa*), the coenurus larva (*Taenia multiceps*) and the hydatid larva (*Echinococcus granulosus*). Larval scoleces are budded from the cyst wall in the polycercus and coenurus, but they develop within brood capsules in the hydatid cysts, budding from the wall of the capsule (figure 11.1*b*). Asexual budding in the cestodes is a highly effective way of increasing the number of individuals during the stay in the intermediate host, e.g., hydatid cysts can grow continually, reaching a diameter of several hundred millimetres and containing several million larval scoleces, each a potential adult tapeworm. Coenurus larvae normally possess a less dramatic reproductive potential and contain only a few hundred larval scoleces.

Sexual Reproduction

Protozoa

Sexual reproduction in one form or another occurs in the Sporozoa, the Ciliophora, the Opalinata and the Hypermastigida (insect gut parasites). One problem of sexuality in the Protozoa concerns the differentiation between fusion of gametes and fusion of individual organisms, such as occurs in conjugation; these processes are often difficult to distinguish. Where recognizable gametes are formed, they may be morphologically identical (iso-gametes) or dissimilar (anisogametes), e.g. microgametes ('male'') and macrogametes ("female").

In the Sporozoa, an alternation of sexual and asexual phase frequently takes place, with gamogony leading to gamete production and sporogony leading to the asexual production of sporozoites (schizogony is an addition asexual phase that is found in some sporozoans). The major division of the Sporozoa into Coccidia and Gregarinida is based upon overt differences in sexual reproduction; in the gregarines gamete production occurs in both "sexes", while the coccidians form only microgametes, that resemble metazoan spermatozoa. In the Opalinata, which are typically parasites of the posterior gut of amphibians, sexual reproduction

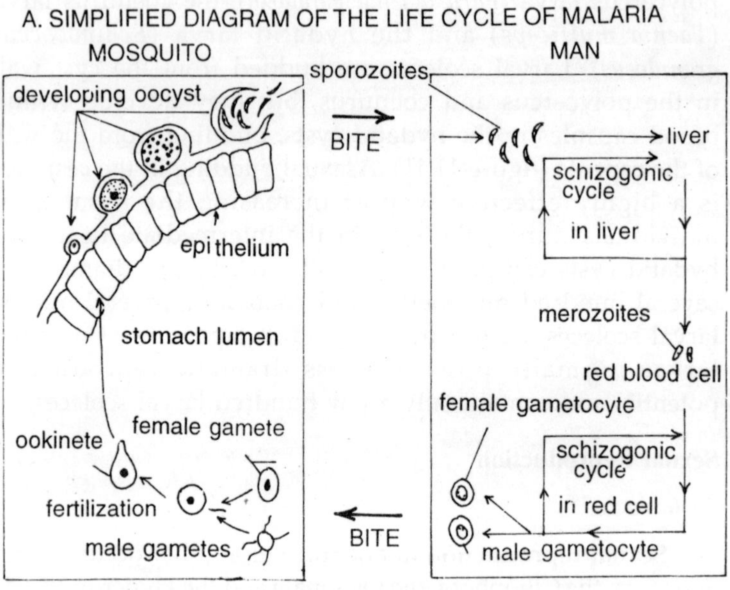

Fig. 11.2 : Patterns of sexual and asexual reproduction in the parasitic Protozoa.

A. Sexual reproduction takes place within the gut of the mosquito; the asexual phases occur in the liver and red blood cells of the bird or mammal final host and are termed schizogony.

B. Asexual schizogony precedes gamete formation, but both asexual and sexual phases take place in the gut of the same host.

C. The complete life-cycle takes place within the gut of the insect.

B. SIMPLIFIED DIAGRAM OF THE LIFE CYCLE OF A COCCIDEAN

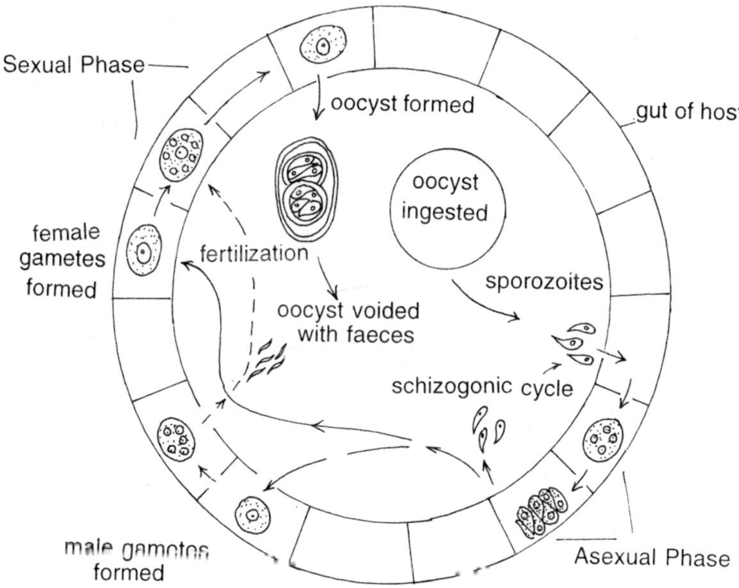

C. SIMPLIFIED DIAGRAM OF THE LIFE OF A GREGARINE

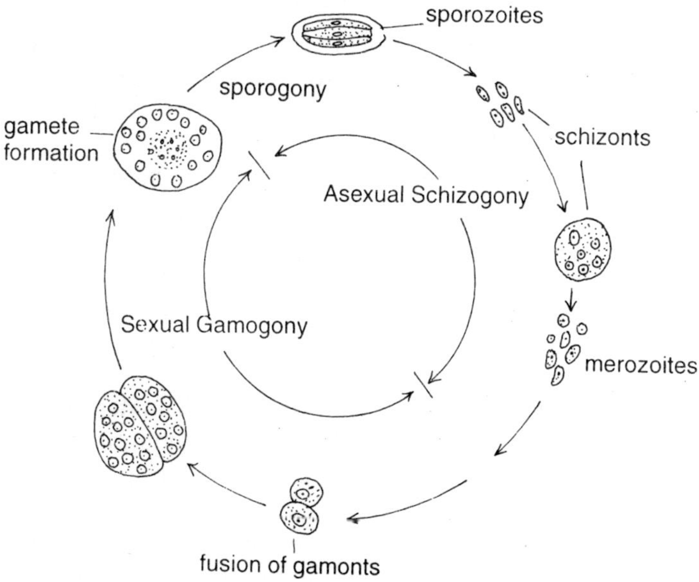

involves the fusion of anisogametes which are released from cysts produced only during the breeding season of the host. Fertilization takes place within the newly infected tadpole. In the flagellates, particularly those that inhabit the gut of insects such as termites, sexual reproduction involves the fusion of flagellated gametes. This process is under hormonal control and related to the moulting cycle of the host.

Where two different hosts are involved in the life-cycle (e.g. in the plasmodia) sexual reproduction will take place in the invertebrate host, while asexual sporogony and schizogony take place in the vertebrate. By contrast, asexual multiplication in the trypanosomatids occurs in both invertebrate and vertebrate hosts.

Monogenea

All monogeneans are hermaphrodite and there usually common genital opening for both male and female systems (figure 11.3A). Cross-fertilization, i.e. between two different individuals, is probably the normal method of reproduction, but self-fertilization may also occur.

The male system consists of testes, which vary in number and disposition, seminal duct for sperm transport, various glands and a copulatory organ ("penis" or cirrus). The female system tends to be variable in structure from group to group, but basically contains a single ovary, extensive vitelline glands, various ducts, including a vagina which may have one or several openings to the outside, and, in the Polyopisthocotylea only, a genito-intestinal duct, whose function is unknown. The ootype is a muscular extension of the oviduct, surrounded by Mehlis' gland, a mucous and serous gland. The function of Mehlis' gland is not known. It was originally believed to contribute shell material to the developing egg, but this function is now thought to reside exclusively with the vitelline glands. Monogeneans are normally oviparous, laying shelled eggs into water, but one family, the Gyrodactylidae, is viviparous; bearing one, or frequently several developing embryos, one within the other.

Monogenean eggs are operculate, with filaments at one or both ends. The eggs shells are composed to tanned sclerotin, formed in the vitellaria and hardened by the quinone tanning system. The majority of eggs are liberated into water where they are presumed to sink to hatch, though the filaments may allow them to attach to a new host and hatch there; the latter is thought to be a rare occurrence. The period of incubation of the egg varies with species, e.g. it may be as short as three days, as in *Dactylogyrus* at 22-23°C, or it may take many weeks, as in *Dictyocotyle* at 10°C. The process of hatching of the monogenean egg has not received a great deal of attention and the roles of temperature, light, pH pCO_2 , pO_2 and salinity have not been determined. Light is an important factor for activating the developing larva within the egg and this may facilitate hatching, whereby the free-swimming oncomiracidium emerges through the opening of the operculum.

Digenea

The majority of digeneans are hermaphrodite (figure 11.3B) and probably reproduce sexually by cross-fertilization. Only in two families, the Schistosomatidae and the Didymozoidae, are separate male and females found.

The male system consists of paired testes, vasa efferentia, a single vas deferens, a seminal vesicle, an ejaculatory duct, and a cirrus enclosed within a sac. Spermatogenesis is typical of the phylum Platyhelminthes and sperm are stored within the seminal vesicle until copulation takes place. The female reproductive system consists of a single ovary and oviduct, a seminal (sperm) receptacle, paired and extensive vitellaria, an ootype surrounded by Mehlis' gland, Laurer's canal, which is homologous with the monogenean vagina, and an extensive uterus. There is a common genital opening for male and female systems, usually situated in the anterior half of the body. The male system normally develops first (protandry). At copulation the cirrus of one partner is inserted either into genital pore of the other partner or into the opening of Laurer's canal.

The development of the digenean egg has been extensively studied and the process of egg shell formation is well understood. Ova are produced in the ovary and are liberated periodically into the oviduct. Simultaneously, vitelline cells-from the vitellaria—and spermatozoa—from the seminal receptacle—are released. Fertilization of the egg

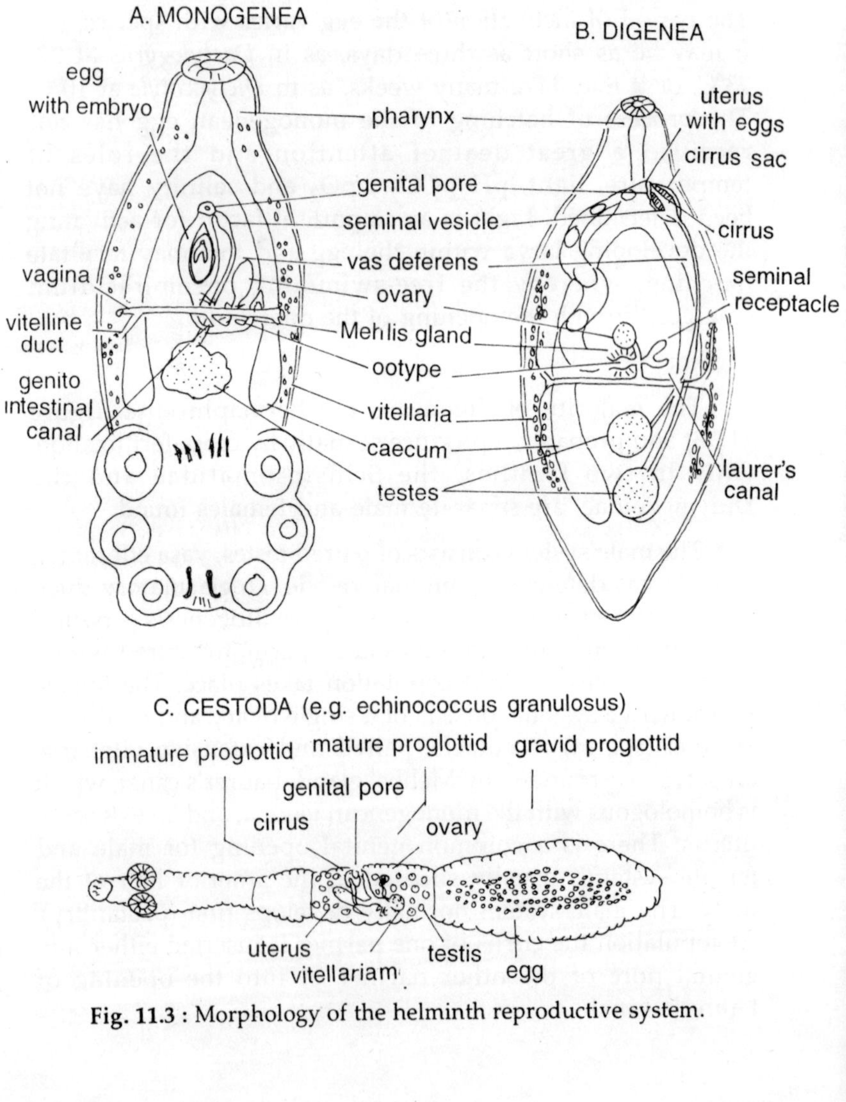

Fig. 11.3 : Morphology of the helminth reproductive system.

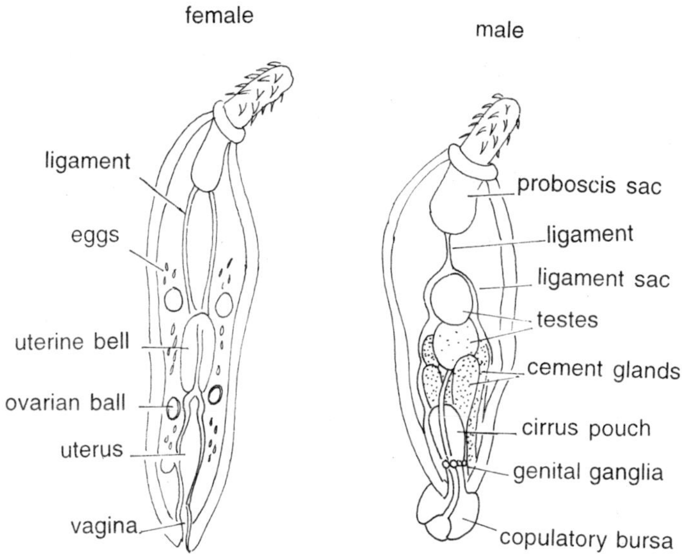
female
male
ligament
eggs
uterine bell
ovarian ball
uterus
vagina
proboscis sac
ligament
ligament sac
testes
cement glands
cirrus pouch
genital ganglia
copulatory bursa

E. NEMATODA

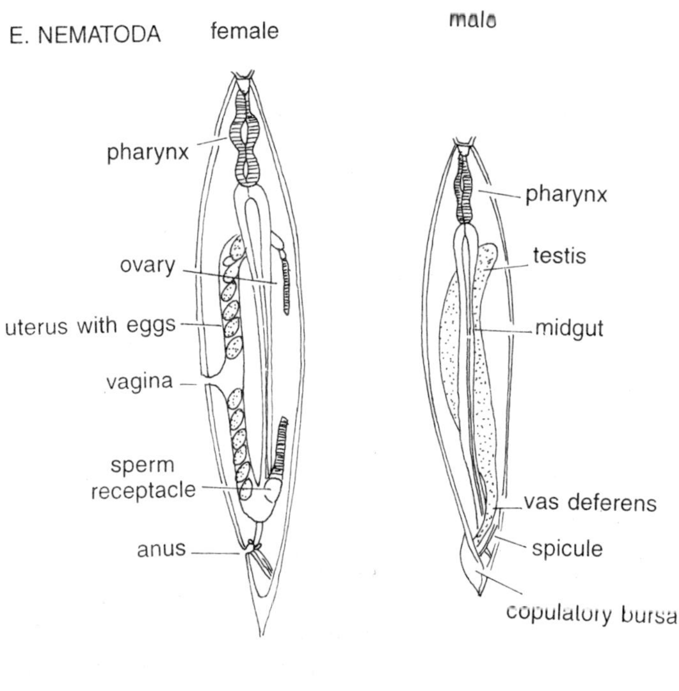
female
male
pharynx
ovary
uterus with eggs
vagina
sperm receptacle
anus
pharynx
testis
midgut
vas deferens
spicule
copulatory bursa

takes place within the ootype and then the vitelline cells release their stored, shell precursors to form a soft shell around the fertilized egg. The role of the secretions of Mehlis' gland is unknown, but they may have a nutritional function. Shell formation proceeds as the fertilized egg is transported along the uterus. In a mature digenean the uterus is typically filled with eggs whose shells are opaque and brown due to the process of tanning. Quinone tanning is regarded as the most common method of providing the egg with a resistant covering, but this form of tanning may not be universal, e.g. *Fasciola* egg contain proteins that are cross-linked by disulphide and dityrosine bridges rather than by quinones. Despite this, *Fasciola* does posses phenolases the enzymes which convert o-phenol to o-quinone during the process of quinone tanning. Thus the position remains somewhat enigmatic.

The tanned digenean egg is released into the external environment in host faeces or excreta. In most species the egg remain unembryonated until it is outside the host, but some do release eggs that contain well developed embryonic miracidia. The eggs are normally viable and if they are liberated into an aqueous environment since, despite the tanned shell, they lack the ability to resits dessication. In suitable conditions the egg develops to produce a single miracidium which is release from the egg during hatching-this depends upon the occurrence of appropriate stimuli such as light, temperature, gaseous conditions and salinity. Light appears to be a major hatching stimulus, stimulating the enclosed miracidium to become active or, perhaps, acting on light sensitive substances within the egg that facilitate the loosening of the operculum.

Cestoda

All cestodes, with the single exception of *Dioecocestus*, are hermaphrodites. Male and female systems are present in each proglottid (figure 11.3C), and the reproductive apparatus is therefore greatly replicated in an individual tapeworm. The Caryophyllaeidea, being unsegmented cestodes, contain only a single reproductive system. In the segmented worms the most mature proglottids are those that are furthest removed from the scolex and protandry marks the normal course of sexual development.

The male reproductive system is morphologically similar to its digenean counterpart, consisting of either discrete or diffuse testes, associated ducts, and a well developed cirrus within a pouch. The female reproductive system is also of the basic digenean pattern; the vagina is homologous with Laurer's canal and the morphology of the uterus is somewhat variable. Both self-fertilization and cross-fertilization occur, but little is known of the details of these processes for many species.

The cestode egg varies in its morphology according to life-cycle. In the Pseudophyllidea, the egg is operculate with a thick surrounding capsule and matures outside the tapeworm, normally in water. The cyclophyllidean egg, by contrast is, surrounded by a thin non-operculate capsule, matures within the uterus of the gravid tapeworm proglottid, and hatches in the gut of the host (either an arthropod or a vertebrate, depending upon species). The fully developed embryo within the egg is termed the *hexacanth,* since it possesses six hooks. In the Pseudophyllidea the hexacanth develops into a coracidium larva upon hatching in water. The precise stimuli associated with hatching of the cestode egg will depend upon whether the egg hatches in the external environment or within the gut of the next host in the life-cycle. Pseudophyllidean eggs may require light to stimulate the opening of the operculum, though hatching of diphyllobothriid eggs has been recorded in the absence of light. Other factors that may be involved include temperature, pO_2 and salinity, but little precise information is available. There is rather more information on the factors that contribute to hatching of the cyclophyllidean egg, a two part process involving firstly, the breakdown of the embryonic membranes, and secondly, the activation of the hexacanth larva. The hatching mechanisms of taeniid eggs are regarded as somewhat different from those of other cyclophyllideans, due perhaps to the thin embryophore of the latter in contrast with the thicker embryophore of the taeniids. *In vitro* studies on hatching of taeniid eggs show that digestive enzymes, such as pancreatin and pepsin, and bile salt bring about the breakdown of the egg membranes, while an increased ambient temperature and bile salts appear to be responsible for larva activation. The eggs of other cyclophyllideans will hatch in saline solutions and do not require enzymes or bile salts—usually mechanical disruption of the egg membranes is an important factor for hatching.

Acanthocephala

Separate sexes and considerable sexual dimorphism are features of the Acanthocephala (figure 11.3D); the female worm tends to be rather larger than the male, and is thought to be longer lived. Though the latter is a matter of some controversy.

The male reproductive system is completely enclosed within a ligament, which attaches anteriorly to the proboscis sac. The system contains two testes, a sperm duct, cement glands and a copulatory bursa. In the female worm, the reproductive system is in contact with the ligament, but is not enclosed by it. The female system consists of the ovary and ovarian balls, the uterine bell, uterus and vagina. At copulation spermatozoa gain access to the pseudocoelom, which contains the eggs, by way of the vagina, uterus and uterine bell. Copulation involves the insertion of the male bursa into the vagina of the female and after copulation the male plugs the vaginal opening with secretions that emanate from the cement glands. It seems improbable that female acanthocephalans can mate more than once. In the pseudocoelom, the spermatozoa fertilize the eggs, which are contained in the ovarian balls. The fertilized eggs are then liberated from the balls and continue their development within the pseudocoelom, entering the uterus when mature. The uterine bell acts as a sorting device allowing passage of "mature" eggs, but keeping "immature" eggs in the pseudocoelom to complete their development. Ripe eggs are released from acanthocephalans at varying rates, e.g. *Polymorphus minutus* can liberate 2000 eggs per day, while *Macracanthorhynchus hirudinaceus* released up to 260000 eggs per day; the prepatent period is from three to nine weeks depending upon species. The eggs, released into the external environment along with the faeces of the host, are surrounded by a protective covering consisting of three to four layers, which contain keratin, chitin and proteins. The shelled embryo, or acanthor, can withstand a range of environmental conditions and will only hatch after being ingested by the arthropod intermediate host. The egg, therefore, acts as a resting stage in the life-cycle which may remain quiescent for extended periods. Hatching in the arthropod gut is dependent upon pH and pCO_2, as well as on digestive enzymes, such as chitinases, to disrupt the shell membranes.

Nematoda

The great majority of nematodes parasites have separate sexes (figure 11.3E), and exhibit varying degrees of sexual dimorphism, but hermaphroditic and parthenogenetic species also occur.

The male reproductive system consists of a single testis, a seminal vesicle, a vas deferens and a cloaca, but it is though that the system was originally double. Spicules, which are hardened, extrusible, cuticular structures used to open the vulva of the female during copulation, are present in the males of most species, as either single or paired structures. They are located at the posterior extremity of the male. The female reproductive system contains paired ovaries, oviducts and uteri, which connect with a single vagina, opening via the vulva. Frequently, part of the vulva is modified to form a muscular ovijector, responsible for extruding the eggs.

In the male system, spermatogonia develop either at the distal end of the testis, or, lets commonly, along the entire length of the testis. In the female, the oogonia often develop in association with a cytoplasmic rachis, whose function is unknown. Subsequent growth and maturation of the oocytes takes place following their detachment from this rachis. Fertilization of the egg occurs within the seminal receptacle of the female. After fertilization the egg becomes surrounded by three shell layers; an inner ascaroside (lipid) layer, a thick chitinous layer and a thin outer layer. These layers derive from the egg itself, while the uterus may contribute a fourth layer, giving the egg a roughened and often sticky surface. In general, nematode eggs are remarkably resistant to chemical damage and dessication, due , in part, to the ascaroside layer which is composed of esterified glycosides. The eggs of ovoviviparous nematodes, such as *Trichinella* and the filarial worms, are, not surprisingly, less complex in their structure since the larvae hatch *in utero*.

Hatching of the eggs of the parasitic nematodes takes place either in the external environment, to release a free-living larva, or within the gut of a host animal, either intermediate of definitive host depending upon the life-cycle pattern. In *Trichostrongylus* and *Meloidogyne,* both of whose eggs hatch in the external environment, hatching is brought about by the egg shell becoming permeable to water and rupturing due to increased turgor pressure. The larvae of *Heterodera* actively cut their way to of the shell. Environmental factors such as light, temperature, pO_2 and humidity may all play a role in providing suitable conditions for hatching. Eggs that hatch within a host gut respond to the action of digestive enzymes and the prevailing physico-chemical conditions especially temperature and pCO_2. The factors that render one host suitable and another less suitable for a particular species of nematode, by affecting the hatching of ingested eggs, have not been well investigated.

Synchronization of Parasite Reproduction with Host Cycles

There are a small number of cases known in which the reproductive activities of a parasite are coordinated or synchronized to the reproductive or other cycles of its host. Such synchrony often serves to bring about the co-occurrence of infective parasites and juvenile animals of the host species. We have already noted that, amongst the Protozoa, the Opalinata, which parasitize amphibians, release their gametes only during the breeding season of the frog or toad, and fertilization occurs within the recently infected tadpole larva. A related ciliate genus, *Nyctotherus,* which also parasitizes the posterior gut of amphibians, switches from asexual to sexual reproduction during the breeding season of its host. Experiments with various hormone treatments, using both *Opalina* and *Nyctotherus,* suggest that the levels of host sex hormones directly affect the reproductive activities of these parasites. In the hypermastigid flagellates inhabiting the arthropod gut, sexual reproduction is initiated by the moulting hormones of the host, e.g. in termites, the hypermastigid gut fauna is lost with each moult, so that sexual reproduction of the parasite is synchronized to the moult cycle; this is one method of reinfection where no cystic stage is produced. Reinfection is essential for the host since the flagellates produce the cellulases that digest the cellulose diet. Only one species of coccidian, *Coelotropha durchoni* in the polychaete worm *Nereis diversicolor,* shows reproductive synchrony with its host, though other sporozoans, including *Plasmodium, Leucocytozoon* and *Haemoproteus,* all synchronize to some extent with their hosts' breeding cycles.

Amongst the helminths, examples of reproductive synchrony are extremely rare and are limited to the monogenean *Polystoma integerrimum* in amphibians, and to synchrony of mitotic rate in the cestode *Diphyllobothrium dendriticum* with the circadian rhythms of the host. Many other cases will undoubtedly occurs but these are yet to be described.

Chapter—12

Ecology, Physiology and Behaviour

The Turbellaria are in general marine animals; only the Catenulida and Temnocephalida are typically fresh-water groups without evident marine relatives or forebears. The Turbellaria have obviously spread from marine into fresh-water habitats and from thence into humid terrestrial regions. There has also been a considerable amount of return invasion of brackish and even salt water by fresh-water turbellarians, notably among the Catenulida, Macrostomida, triclads, and such rhabdocoel genera as *Dalyellia, Castrada,* and *Phaenocora.* The Turbellaria are characteristic and constant members of a variety of habitats. They are typically bottom dwellers. The large forms, as the triclads and polyclads, are generally found on hard bottom, under rocks and among shells and gravel, *etc.*; the small and microscopic members are more apt to occur on sandy and mucky bottoms. A characteristic assemblage of sand-dwelling rhabdocoels and acoels has been found on north European coasts . These have certain adaptive characteristics in common: long, slender muscular body, long sensory hairs, tail appendages, eyelessness, loss of rhabdites, especial development of food-catching devices, overdevelopment of adhesive organs including girdles of adhesive papillae in a number of species, some reduction of the reproductive system, posterior displacement of the gonopores (Fig. 12.1). The species of such sand communities appear to have a very limited distribution, *i.e.*, those of the north German coast are different from those the west French

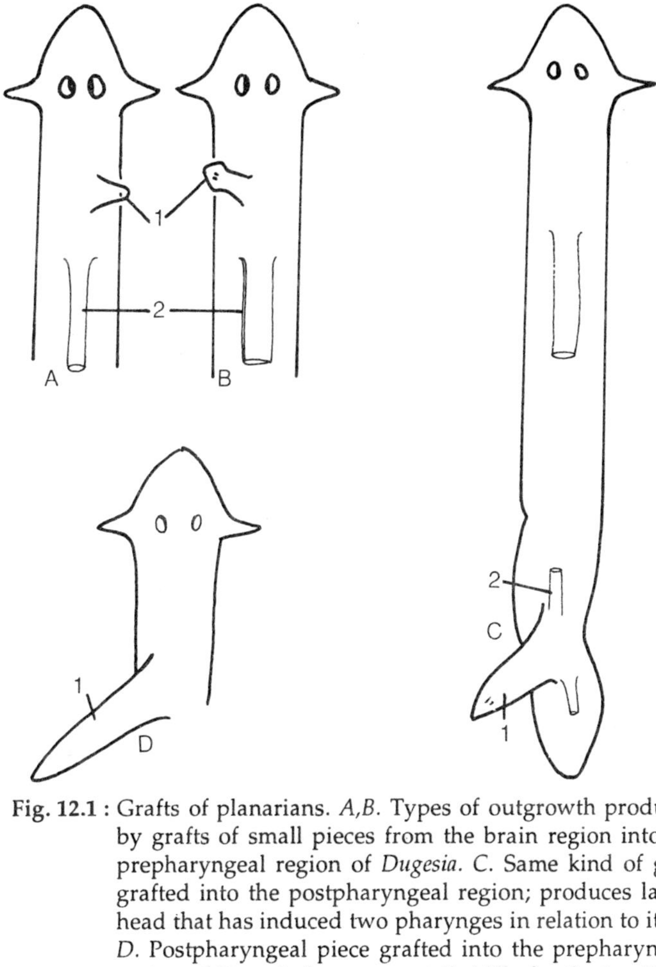

Fig. 12.1 : Grafts of planarians. *A,B*. Types of outgrowth produced by grafts of small pieces from the brain region into the prepharyngeal region of *Dugesia*. *C*. Same kind of graft grafted into the postpharyngeal region; produces larger head that has induced two pharynges in relation to itself. *D*. Postpharyngeal piece grafted into the prepharyngeal region of *Dugesia*; becomes a tail. (*All after Santos*, 1931.) 1, regenerate induced by graft; 2, pharynx.

coast, although ecologically similar. Whether such sand-dwelling turbellarian communities occur along American coasts has not been ascertained but it is highly probable. The benthonic turbellaria are mostly limited to the littoral zone of the ocean, although some occur at considerable depths; however, there are no abyssal species. Some acoels and polyclads have pelagic habits and show some adaptation

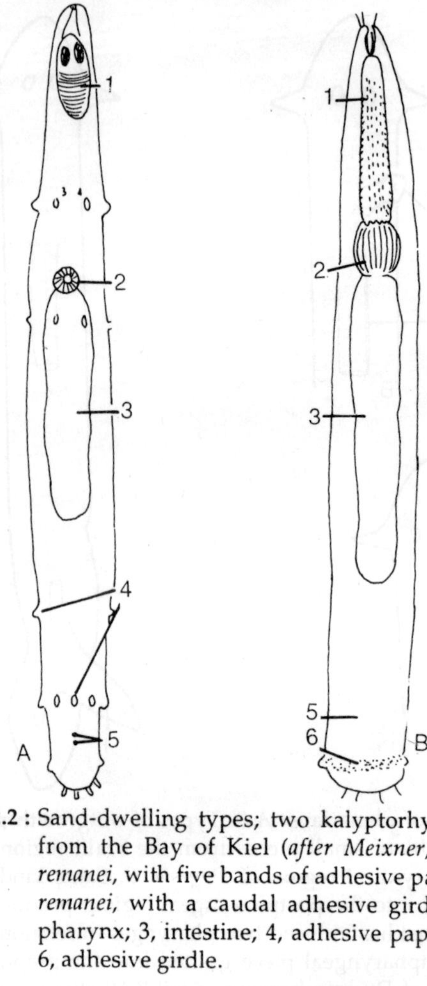

Fig. 12.2 : Sand-dwelling types; two kalyptorhynchid rhabdocoels from the Bay of Kiel (*after Meixner,* 1928). *A. Acerina remanei,* with five bands of adhesive papillae. *B. Rhinepera remanei,* with a caudal adhesive girdle. 1, proboscis; 2, pharynx; 3, intestine; 4, adhesive papillae; 5, gonopores; 6, adhesive girdle.

of pelagic life by transparency and broadly oval form; many polyclads also have pelagi larvae or juvenile stages. Acoels and polyclads are regular components of the animal association of the floating Sargassum and may have the adaptive brown-and white colour pattern characteristic of Sargassum animals. Nearly all pelagic Turbellaria are limited to tropical or subtropical waters; the chief exception is the rhabdocoel *alaurina composita* in the North Atlantic of which large number may be found in late summer in the Zuider Zee and adjacent waters (Hofker, 1930). Although the ocean is in general a very constant habitat, turbellarians that live along shores in the tidal zone are subject to wide changes of temperature and salinity. The alloeocoel *Monocelis fusca* lives in the tide pools on the Welsh Coast where the nature temperature range is - 1 to 28°C. and can recover from exposure to 44.5°C. (Rees, 1941). This species and the marine triclad *Procerodes ulvae* also great changes in salinity (*see* further below).

The fresh-water Tubellaria are chiefly inhabitants of lentic waters and often have seasonal cycles related to such habitats, especially in the case of planarians and rhabdocoels that inhabit vernal pools. Some tuberllarians are *rheophilous, i.e.,* limited to flowing water, such as springs and hill and mountain streams. Outstanding examples of stream planarians are the European *Dugesia gonocephala, Polycelis felina* (= *cornuta*) and *Crenobia alpina* which occur along streams with increasing altitude and decreasing temperature in the order named. *Crenobia alpina* requires cold temperature and running water. No alpine planarians are known for North America, but the species *Phagocata morgani,* Appalachians, and *Polycelis coronata,* western mountains, seem limited to flowing waters. Our common *Dugesia dorotocephala* appears to live only in springs or spring-fed waters, probably because of the high calcium content of such habitats. Although some tuberllarians are *stenothermous* (enduring only a limited temperature range) as *Crenobid alpina,* most fresh-water forms are *eurythermous, i.e.,* indifferent to wide changes of temperature. An extreme case in *Microstomum lineare* found in a variety of habitats ranging from alpine lakes to hot springs at 40-47°C. Rhabdocoels are characteristic parts of the fauna of a alpine lakes, but most of these are common species with a wide distribution in fresh waters. More specific are the rhabdocoels, chiefly of the genera *Phaenocora* and *Mesostoma,* found in *shallow* waters in high altitudes and latitudes, and having in common a very dark coloration. Several *Mesostomas* occur in such pools in the mountains of Tibet at elevations of over 15,000 feet, and the author described *Mesostoma arctica* from a pool on the Canadian tundra near the arctic circle and from swamps in the Wyoming Mountains at 10,500-foot altitude. A considerable number of fresh-water planarians occur only in cave or subterranean waters; they are generally white, eyeless, and provide with adhesive organs. The most notable cave planarians are the North American ones, of the family Kenkiidae. The turbellarian fauna of tropical fresh waters appears to be very scanty, consisting chiefly of a few widely distributed or cosmopolitan species that tolerate a broad range of conditions.

The terrestrial Turbellaria, comprising the land triclads, and a few rhabdocoels and alloeocoels, being devoid of protective armor, require high humidity and most seem unable to mature their sex organs except in warm temperatures. Hence they are mostly limited to tropical and subtropical jungles with thigh rainfall where they live by day surrounded by mucus under stones, fallen logs, and leaf accumulations, in earth burrows, or between the leaf bases of tropical plants, coming forth at night to hunt food. *Bipalium,* subjected to desiccation, lost water according to the simple laws of evaporation and could recover if the water loss did not exceed 45 per cent of its weight (Kawaguti, 1932). However, mucous secretion plus aggregation into the most humid spot should furnish some protection against drying.

b. Relations with other Organisms: Symbiosis, Commensalism, Parasitism, Parasitic Infections, Enemies.- Most acoels and some rhabdocoels and alloeocoels harbor symbiotic chlorellae and xanthellae; these occur in the mesenchyme chiefly just inside the subepidermal musculature and appears to be of the same nature and function as in other animals. They are absent in the eggs and newly hatched young, and it is known that they are acquired during juvenile stages by ingestion with food. They pass into the gastrodermis whence they are ingested by free mesenchymal cells that convey them into the mesenchyme (von Haffner, 1925). Much study has been devoted to the chlorellae of the acoel genus *Convoluta*, especially *Convoluta roscoffensis*, which occurs on the channel coast of France in such vast numbers as to colour large areas green. Keeble and Gamble (1907) have shown that the green cells in this worm are flagellates of the chlamydomonad group and that they occur abundantly in the habitat and on the eggs masses of *Convoluta*. These authors state that this acoel feeds while young but after attaining sexual maturity subsists entirely by digesting its chlorellae. A number of fresh-water rhabdocoels of the families Dalyelliidae and Typhloplanidae contain green cells that can be cultivated apart from the worms and appear to be green algae of the genus *Chlorella* or close to this genus. It is notable that the *Dalyellia* species harboring these symbionts are much larger (2 to 5 mm.) than other *Dalyellias*. Sojourn is darkness almost but not entirely eliminates the green cells of *Dalyellia*. *Dalyellia* digests senile and degenerating green cells but dies not damage healthy ones. The green and brown symbionts of Turbellaria contain starch, fats, and other lipids and undoubtedly possess some nutritive value for their hosts. Their chief advantages are, however, that they utilize in their metabolism waste products of their hosts (it will be recalled that the Acoela have no excretory system) and that they give off oxygen enabling the host to endure stagnant conditions. Thus the acoel *Amphiscolops langerhansi* will survive enclosure in small airtight bottles of sea water for 2 weeks or more in the light but succumbs in the dark under otherwise similar conditions (Welsh, 1936); in the light, oxygen is added to the water by the photosynthetic activity of the brown cells of the worm.

As already noted, all groups of Turbellaria, except the fresh-water and terrestrial triclads, contain members that inhabit other animals in a relation that appears to be commensal rather than parasitic. Here may be recalled the acoels that live in the interior of echinoderms; the rhabdocoel families Graffillidae and Umagillidae, entocommensals of mollusks, echinoderms, and some other invertebrates; the wholly ectocommensal group Temnocephalida; the marine triclad ectocommensals of *Limulus* and of skates; and the polyclad habit of dwelling in shells of live snails or in dead snail shells occupied by hermit crabs. The epizoic or ectocommensal forms are characterized by loss of pigment, cilia (especially dorsally), rhabdites, and often eyes, and especially by the strong development of adhesive areas and organs. The entocommensals, on the other hand, usually lack adhesive organs and are generally completely ciliated although they also tend to lose rhabdites and eyes; their most outstanding feature is the enlarged female system, exemplified by the branching yolk glands, elongated ovaries, *etc.* Such adaptation for increased egg production is a common feature of genuine parasites and no doubt indicates a tendency toward true parasitism among the entocommensal Turbellaria. However, probably the only genuine entoparasites among the Turbellaria are the Fecampiidae and *Oekiocolax*.

The Turbellaria are themselves hosts to commensals and parasites. Euglenoid flagellates of the general *Astasia* and *Euglena* occur in catenulids. A trypanosome-like flagellate, *Cryptobia (= Trypanoplasma) dendrocoeli*, inhabits the digestive tract and copulatory bursa of *Dendrocoelum lacteum*. Holotrichous ciliates of the mostly free-living genera *Holophyra* and *Ophyroglena* may occur in *Stenostomum* and planarians. Members of the parasitic ciliate group Astomata are also represented, as *Hoplitophyra uncinata* with an anterior hook in the intestinal branches of marine triclads and *Siebolidiellina* (= *Haptophrya*) *planariarum* with an adhesive disk in fresh-water planarians. *Coleps* and *Trichodina* species often epizoic on fresh-water planarians. The most common endoparasites of planarians are gregarines, chiefly of the genus *Lankesteria*. Encysted stages of larval trematodes are sometimes found in Turbellaria but probably represent accidental penetrations of wrong hosts. Land triclads seem much subject to nematode infestations.

The Turbellaria are seldom eaten by other animals as their surface secretions appear to be distasteful if not actually toxic. Planarians, if taken into the mouth by fish and salamanders, are usually quickly ejected with signs of distaste. Bolen (1938) saw dragonfly and damsel fly nymphs eat planarians. The worst enemies are other turbellarians for the group is highly cannibalistic, promptly attacking weak or injured members or new fission products; the larger species may also eat the smaller ones.

c. Duration of Life, Cycles, Senescence, Rejuvenescence.- The length of life of turbellarians is known chiefly from laboratory cultures. The smaller forms, except those capable of fission, probably live but for a short time, a few weeks or months, usually dying in summer or autumn after laying dormant eggs. Some, however, may survive a year or more by encysting during unfavorable seasons. Species capable of fission can presumably under favorable conditions propagate indefinitely by the asexual method. Thus Nuttycombe and Waters (1938) cultivated *Stenostomum tenuicauda* for 11 years during which there were over 1000 asexual generations without the appearance of sexuality. They also had male specimens of *S. grande* that reproduced continuously by fission for 4 years. Sekera (1926) kept a *Microstomum* colony going for 3 years. Similar facts are known for asexually reproducing species of fresh-water planarians. Child (1914) maintained *Phagocata velata* in asexual culture for nearly 3 years during which there were 13 cycles of fragmentation, encystment, and hatching of small worms from the cysts. Individuals of the larger turbellarians survive in laboratory cultures 2 or 3 years and may have several periods of sexual propagation alternating with periods of fission or lack of any reproduction.

Something has been said in the preceding pages concerning sexual reproduction in relation to season. This matter has been studied in fresh-water planarians by several workers. In general sexual reproduction is intermittent, alternating with periods of fission or of lack of any reproduction; the periods of capsules lying correspond in nature with external factors, chiefly temperature. In laboratory cultures, sexual reproduction is also intemittent but does to recur with the regularity noted in nature. Kenk,

Vandel, and others believe there is an inherent cyclic reproductive rhythm in these planarians, but the author found that the same species used by Kenk (*Dugesia tigrina*) can, when fully grown, be caused at will to develop a copulatory apparatus and lay viable capsules by exposure to low temperature followed by rise of temperature. In some individuals even 2 or 3 days of exposure to lowered temperature is sufficient to bring on sexuality. Some time must elapse before another sexual period can be induced in the same individual by temperature manipulation. This time is much less than a year, mostly a few months, so that the alleged inherent rhythm is probably only an expression of such a "refractory" period following a period of reproductive activity.

After continued laboratory culture or after sexual reproduction in nature, turbellarians often evince signs of senescence such as loss of pigment, sluggish behaviour, unwillingness to feed, and disintegrative processes in the eyes, head, or pharynx. Death by general disintegration frequently follows such changes. In species with high regenerative powers, however, rejuvenescence may be brought abut experimentally or occurs naturally. Thus fission products or animals emerging from cysts are physiologically like young animals. Through starvation, adult worms may be reduced to the size of very young worms and so far as can be tested have actually become young again. Adult worms may be cut into small pieces, and these after regeneration are also indistinguishable from young worms. The already much mentioned *Phygocata velata* has in temporary pools a natural cycle of senescence and rejuvenescence; the worms grow to mature size, show the above symptoms of senescence, then fragment into small bits, each of which inside a cyst regenerates to a tiny worm that in all respects is identical with young worms hatched from capsules. Apparently this cycle is repeated in nature indefinitely with no loss of vitality. By underfeeding this species so that it does not grow, senescence and fragmentation and prevented, and the worms remain in a state of perpetual youth.

d. Osmotic Relations.- As already noted, both fresh-and salt-water turbellarians may invade brackish habitats. In a fresh-salt-water gradient as the entrance of a river into the sea, there is generally a barren zone, poor in animals, lying between the limits of increasing salinity endurable by the fresh-water animals and of decreasing salinity endurable by the marine forms. To endure changes of salinity animals must have some mechanism for regulating loss or entrance of water and salts. Generally fresh - water forms invading brackish watei show reduction in size and greater tissue density as a result of water loss, and the reverse changes occur in marine forms adapting to lowered salt content. The best-studied case is that of the marine triclad *Procerodes vlvae* living on the English Coast at the mouth of streams. This animal is subject in nature to changes lasting several hours from the fresh stream water (salt content, 0.03 g. per liter) to sea water (salt content, 35 g, per liter). It can live indefinitely in all dilutions of sea water to 5 per cent; it swells and loses salts in the greater dilutions but not at all in proportion to the decreased osmotic pressure. In pure fresh water (*i.e.,* without salts), it swells to nearly twice its volume, loses salts, and survives only 1 to 3 days. The worm can recover after loss of 85 to 90 per cent of its salts. In the natural stream water, swelling and salt loss are much less and survival is better than in pure water. Osmotic pressure appeared not to be an important factor in the effect of the various media tried. It was found that calcium, in which the stream where *Procerodes* lives is rich, is the most effective factor in reducing swelling and salt loss in diluted media. Hence it appears that the presence of calcium in fresh water enables marine animals to invade such waters. The cause of the protective action of calcium is not clear, although probably calcium makes the epidermis resistant to the passage of water and salts. Beadle (1934) studies sections of *Procerodes ulvae* that had been exposed for varying periods to diluted sea water and found that water, passing through the epidermis, at first accumulates in the mesenchyme but later is transferred to the gastrodermis where is accumulates in numerous vacuoles.

Through this vacuole formation swelling is reduced, the mesenchyme is restored to normal, and the animal regains more normal appearance and behavior. In time the epidermis develops resistance to further water intake, although the gastrodermal vacuoles remain as long as the worms are kept in diluted media; *i.e.*, there is not discharge of excess water through the mouth.

A number of other littoral turbellarians live under conditions in which they are exposed daily to large changes in environmental salinity. Thus *Monocelis fusca* inhabits tidal pools subject to evaporation and hence concentration of the sea water. The salinity of these pools was found to vary from 26 to 64 parts per thousand; the alloeocoel endured in tests a range from 5.6 to 76.8 parts per thousand but began to show sluggish behavior at 80 parts per thousand. It can, however, form a mucous cyst within which it can survive exposure to salinities of 298 parts per thousand, or over eight times the salinity of sea water (Rees, 1941).

Fresh-water planarians when exposed to solutions hypertonic to natural waters lose water and weight but not in proportion to the increased osmotic pressure, and they show some ability to resist water loss. *Dendorcoelum lacteum* was studied by Gresens (1928) in stream opening into the Danish Wiek where it lives in a salt gradient from pure stream water (salt content, 0.01 to 0.04 per cent) to the Wiek water (0.05 per cent salt). Individuals from the more brackish water proved more resistant to high salinities than stream individuals. Worms could be acclimated by gradual changes so as to live normally and reproduce in 1.5 per cent salinity. Low temperature is an important factor in enabling planarians to resist increased salinity.

Some data are available on the salt content of the turbellarian body. According to the work of Pantin (1931), the salt content of *Procerodes vlvae* is about half that of sea water, or about 1.7 per cent. In *Stylochus frontalis*, the salt content is a little over 2 per cent. Hence the osmotic pressure of the body fluids of marine turbellarians is considerably less than that of sea water and the epidermis must be able to resist water loss or salt intake or some to other regulating mechanism must be present. Dilution of the sea water cause increased oxygen consumption in marine invertebrates, and this represents energy consumed in resisting water intake. The osmotic pressure of fresh-water turbellarians is greater than that of fresh water, and they require some mechanism to resist water intake and salt loss or to eliminate excess water. Murray (1927), seeking a medium suitable for growing planarian cells in tissue culture, found that the balanced salt solutions (*e.g.*, Locke's, salt content, 0.9 per cent) in use for vertebrate tissues are far too concentrated for fresh-water planarians and require one-tenth to one-twelfth or more dilution. This indicates that the somotic pressure of the species used (*Dugesia dorotocephala*) is equal to about a 0.08 per cent solution of sodium chloride. This species also requires a higher ratio of calcium to potassium than is customary in salt solutions made up for vertebrate tissues.

The water content of the polyclad *Stylochus frontalis* is 76.2 per cent; that of the land triclad *Bipalium* averages 81.5 per cent, grading from 80.7 per cent in the anterior third to 82.9 per cent in the posterior third (Kawaguti, 1932); and that of a fresh-water planarian, *Dugesia dorotocephala,* is 78 per cent.

e. Hydrogen-ion concentration.- Turbellarians are in general indifferent to the pH of their environment, enduring without evident effect wide changes in this factor. *Dendrocoelum lacteum* will live indefinitely in a range from pH 4.2 to 9.5; greater acidities or alkalinities are fatal. The toxicity of acids for *Dugesia dorotocephala* depends on the acid used, being less for mineral acids such as hydrochloric and nitric, which kill in 2 to 3 hours at pH 3.0 to 3.4, than for organic acids, especially butyric, which is fatal in the same time at pH 5.0. Similarly alkalinity caused by ammonium hydroxide is more toxic (lethal at pH 9.0) than that caused by sodium hydroxide (lethal at 9.2). In alkalinities and acidities less than those mentioned, *D. dorotocephala* survives indefinitely, although it is to be noted that planarians are able to reduce the alkalinity and acidity of the medium and it is very difficult to maintain a constant pH in solutions in which they are immersed. In general, alkalies accelerate and acids depress the rate of oxygen consumption of planarians, but the acid effect results, not from the pH as such, but from the carbon dioxide set free when natural water is acidified.

f. Locomotion.- The smaller Turbellaria, including juvenile stages of the larger forms, probably move entirely by ciliary action. Although they probably can swim, few of them habitually do so, but for the most part they glide along the substratum or the underside of the surface film. The ciliary waves pass from anterior to posterior, thus driving the animal forward, and appear to be independent of the nervous system. Forms exceeding 2 to 3 mm. in length mostly crawl along a substratum with a characteristic gliding movement. Although the ventral cilia probably play some role in this gliding, if the worm is not too large, this type of locomotion is caused largely or wholly by muscular waves that pass form the anterior end backward either across the whole width (monotaxic) or alternately along the two sides (ditaxic). Triclads have mostly monotaxic, polyclads ditaxic waves. The waves are mostly invisible to the naked eye in planarians, more noticeable in polyclads. The relative roles played by cilia and by muscular waves in locomotion can be determined by treating worms with lithium chloride which paralyzes the muscles. The first chemical does not stop locomotion in triclads and polyclads, whereas the second does. A number of polyclads swim by means of body undulations; these also may be monotaxic or ditaxic depending on the body shape.

Most turbellarians are capable of a more hasty type of locomotion which they employ when stimulated. This consists of large evident muscular contractions and expansions accompanied in those forms with adhesive organs or areas by alternate attachments of these, so that leech-like form of progress results. Usually such an effort lasts but a short time and the usual gliding is soon resumed.

Mucous secretion probably plays an important role in gliding, even in aquatic turbellarians, and is obviously indispensable to terrestrial forms. The slime trail lubricates and smooths out the substratum on which progression occurs. The land triclads and some other Turbellaria also use slime threads in dropping from one level to another or in bridging gaps in a substratum.

The role of the nervous system in locomotion has been investigated by several workers. In general, turbellarians from which the cerebral ganglia have been removed tend to cease locomotion and to remain quiescent. They can, however, be stimulated to locomotion by mechanical or chemical means and are able to execute the usual creeping. Levetzow (1936) found that even quite small postcerebral pieces of *Thysanozoon* and *Eurylepta* are able to crawl. Such experiments indicate that the more simple muscular waves are independent of the brain and are mediated by the general nerve plexi. Most polyclads, however, are unable to swim following decerebration, and some cannot even execute the regular ditaxic crawling. Therefore, the more complicated types of locomotion depends on initiation from the brain, also on transmission along the main nerve cords, since cuts through these inhibit swimming waves posterior to the cuts.

g. Food and Feeding Behavior.- The Turbellaria area as a class carnivorous. A few acoels and rhabdocoels are said to eat diatoms and other algae, but the vast majority of turbellarians eat other animals of suitable size or ingest bits of flesh of intact, damaged, or dead larger animals as fish and mollusks. Favorite items of food of the smaller species are rotifiers, copepods, caldocerans, nematodes, annelid worms, etc., and the larger species feed on earthworms, snails, insect larvae, isopods, amphipods, other crustaceans, fish, oysters, and so on. Although the Acoela lack a digestive cavity, they are on the whole strictly carnivorous, ingesting copepods and similar small animals whole and digesting them in their interior cells. Aquatic and land triclads can attack and subdue intact animals of considerable size such as snails, earthworms, and insect larvae, and polyclads often devour bit by bit live sessile, helpless animals such as oysters, barnacles, ascidians, and the like. Turbellarians often gather, often in numbers, on animal corpses if these are not too decayed. Insects falling into the water constitute a regular food supply for pond turbellarians. Although most turbllarians will eat almost any kind of animal flesh, some appear limited to particular items. Thus *Stylochus frontalis* (= *inimicus*) of the southern Atlantic coast of the United States seems to eat nothing but oysters, slipping between the shell valves and devouring the oyster bit by bit despite shell partitions erected against it, and *Pseudoceros crozieri* apparently feeds exclusively on ascidians. In laboratory cultures, turbellarians can be fed on chopped or crushed earthworms of other annelids, meal worms and other insect, clams, snails, isopods, amphipods, clotted blood, or, most conveniently, bits of beef liver, although some species will not eat the last. Some species require live prey such as *Daphnia*. In general, turbellarians from lentic waters are most easily kept in the laboratory. Polyclads are usually difficult to maintain under laboratory conditions. Turbellarians may often be captured in numbers by baiting suitable habitats with pieces of beef or dead fish; at short intervals the bait is examined and any worms present shaken off into jars.

The presence of intact prey (not emitting juices) is detected either by direct contact with chemoreceptors or by the disturbance created in the water by the prey. The tubellarian then quickly grips the prey with the anterior end, especially the anterior adhesive organ when this is present, and wraps its body about the prey, holding the latter firmly against the substratum. Adhesive margins, disks, etc., are used by the worm to maintain a firm hold on the substratum while subduing the preying it helpless. Crustaceans and baby fish living in the same habitat with planarians may become coated with the slime left by planarians so as to be unable to move and are then attacked. Some turbellarians use the penis stylet to damage prey.

Food emitting juices is sensed in still water at a distance varying with species from 1 or 2 to 10 to 15 cm. or more. A turbellarian, previously gliding aimlessly about, when coming within the radius of diffusing juices exhibits a characteristic behviour. The worm pauses, swings the raised head about, then turns toward the food and goes to it in a direct line. Planarians living in running water detect food juices coming downstream at considerable distances, and if such places are baited with meat or fish the worms can be seen emerging in hordes from under stones and leaves and travelling in processions upstream to the bait. The European stream planarian *Crenobia alpina* is in fact more or less incapable of directing itself toward to foot in the absence of a current and even if it finds the food may often fails to feed in still water. Having reached the food, planarians test it with the anterior margin, then grip it with the head and move the body along until the pharynx can be brought into play. The protruded pharynx is inserted into the food, and ingestion begins. During feeding the worms keeps some part of its body firmly attached to the substratum.

The extirpation experiments of Koehler (1932) and Müller (1936) have shown that the lateral sensory organs of the head-the ciliated pits of *Stenostomum,* the ciliated grooves of *Phaenocora* and *Bothromesostoma,* the auricles including the auricular sense organs of planarians, or the corresponding lateral angles of the head in planarians that lack definite auricles—are the chief or only sensory areas involved in chemical perception of dissolved juices. *Stenostomum* after extirpation of the ciliated pits cannot detect the presence of food or find it, and planarians deprived of their auricles or corresponding regions show great loss in their food reactions. However, fresh-water planarians appear to have some chemo-receptors over the entire body and fatter such operations may in time find food. The testing of the suitability of food after finding it is vested chiefly in the anterior margin, which also functions to release the feeding reaction of the pharynx. After extirpation of this margin, planarians find the food as readily as before but fail to test it or grip it and often glide away without ingestion. The tip of the pharynx is, however, sensitive to food, and the pharynx may protrude on contact with food in the absence of anterior sensory regions. The pharynx tip can discriminate between normal food and the same food mixed with deleterious chemicals; but, after removal of the both the anterior margin and the pharynx tip, the pharynx loses this discriminatory ability and may protrude into such treated food. The presence of fluid nutrients such as bouillon or blood in the water may induce repeated pharynx protrusions of intact planarians. According to some workers, headless pieces containing the pharynx or isolated pharynges will not feed when placed in contact with food, but according to other such reactions do occur in some species.

According to the available experiment, then, ciliated pits, auricles, auricular sense organs, and corresponding head areas have sensory functions similar to smell in higher animals, and the anterior margin and pharynx tip function like taste.

After the prey has been subdued and firmly gripped, ingestion ensues. Turbellaria with simple and doliiform types of pharynx generally swallow their prey whole by means of muscular waves in the distended but not very protrusible pharynx. The rosulate pharynx is not distensible, and hence the contents of the food are sucked in. The triclads also do not swallow their food whole, but the highly protrusible plicate pharynx is inserted into the food or into the prey and the contents sucked in bit by bit through the peristaltic action of the pharynx. In polyclads the food is usually taken whole into the interior of the voluminous ruffled pharynx. Even those polyclads that have a cylindrical pharynx like that of triclads appear to swallow their prey whole, apparently limiting themselves to small animals. Whether the secretions of the pharyngeal glands have a softening and digestive action on the food cannot be answered categorically. These secretions possibly have such action is some species, but in general they function only to lubricated the food and facilitate its ingestion. Even in those cases where food is help in the pharynx and is seen to undergo digestion, it does not necessarily follows that the enzymes involved are of pharyngeal origin.

Extirpation of the brain (without disturbances of the main chemo-receoptors of the head) may greatly retard or altogether abolish the food-finding and food-ingesting reactions. The result, however, varies with species. The most marked affects of brain removal appear in *Crenobid alpina* which will not ingest food after this operation (Koehler, 1932), but other species may show little alteration of food reactions after decerebration.

Whether the surface and pharyngeal secretions of turbellarians have a toxic and paralyzing as well as entangling effect on prey is a question often raised but never satisfactorily answered. In reviewing this question, Arndt and Manteufel (1925) note that large land planarians may produce an unpleasant astringent effect on the human tongue and that a South American land planarian, *Polycladus gayi*, is believed by the natives to be fatal when eaten by horses and cattle. These authors found that water extracts of ground fresh-water planarians were toxic or even fatal when injected into the coelom of small mammals. Later Arndt (1925) located this toxin in the surface secretons of planarians and found that it exercised its effect by stopping the heart. Mattes (1932) observed that the cercariae of flukes are damages and killed by the surface slime of planarians. However, students of planarians, including the author, in general fail to see any toxic or paralysing action of planarian slime on the usualy prey, such as crustaceans and insect larvae.

The nutritional requirements of palanarians are not definitely known, although attempts in this direction have been made by Wulzen, Bahrs, and others. Vertebrate live, kidney, thymus, adrenal, and gray matter of the brain support excellent growth, whereas little or on growth is obtained with egg white, egg yolk, lean beef and other muscle, thyroid, pancreas, gelatin, beef fat, and the white matter of the brain.

Tissues from animals fed on deficient diets also have an adverse effect on planarian growth and may cause a pathological condition in which the worms shrived and turn back. Raw egg white alone or in high proportions in the food is very deleterious, causing disintegrating protrusions on the worms. Refrigeration at 0ºC. or heating decreases the growth-promoting power of vertebrate tissues. Greenberg and Schmidt failed to find any favourable effect on planarian growth of adding vitamins and auxins to inadequate foods, although yeast addition was effective. Of all vertebrate tissues, liver has about the best growth-promoting power for planarians, and it contains and ether-soluble heat-labile growth factor of unknown nature.

h. Digestion, Assimilation, and Food Storage.—It was discovered by Metschnikoff in 1866 for *Convoluta* and later verified for rhabdocoels and triclads (Metschnikoff, 1878) that digestion in the Turbellaria is largely intracellular. As already indicated, it is improbable that the pharyngeal secretions have any digestive action, at least in planarians. Wetblad (1922) failed to find any digestive action on protein or fat of extracts of 50 pharynges of *Dendrocoelum lacteum* and 100 pharynges of *Polycelis nigra.* It is also improbable that any digestion occurs in the lumen of the intestine in planarians Willier, Hyman, and Rifenburgh, 1925). The granular clubs, usually regarded as enzymatic cells emitting enzymes into the intestinal lumen, are in the author's opinion protein reserve cells (see below).

The details of intracellular digesting are best known for fresh-water triclads where the most complete studies are those of Westblad (1922) and Willier, Hyman, and Rifenburgh (1925). The latter fed planarians, starved 2 weeks, on beef liver and studied sections of worms fixed at varying intervals after the beginning of feeding. The liver torn into minute bits by the suctorial action of the pharynx fills first the anterior, then the posterior rami of the intestine. Some 45 to 60 minutes were required for complete filling of the intestine in the species used (*Dugesia dorotocephala*), but the phagocytic cells of the gastrodermis begin to engulf the food as soon as this comes in contact with them. These cells swell through intake of water, bulge into the lumen, often fuse into syncytia, and take in food particles in amoeboid manner by putting out pseudopodial processes. The food particles together with fluid are formed into typical food vacuoles like those of Protozoa. About 8 hours were required for all the food in the lumen to be phagocytized by the phagocytic cells which at this time are found packed with food vacuoles in various stages of digestion (Fig. 12.3). The food in the vacuoles, at first loose and faintly stainable, gradually condenses to a strongly staining (eosinophilous) homogeneous ball (Fig. 12.4) shown by chemical tests to be of protein in nature. About 8 hours are required for food vacuoles to reach the homogeneous ball stage. These balls then gradually disappear, presumably as a result of the breakdown of their protein into soluble substances; about 5 days are required for the complete disappearance of the food balls from the phagocytic cells. In a few phagocytic cells, the food balls do not disappear but subdivide into small spherules, and such cells then become granular clubs. Westblad, using indicator dyes with the food, was unable to detect any change of hydrogen-ion concentration in the food vacuoles so that digestion appears to take place in a neutral medium. He found evidence of proteolytic and lipolytic enzymes but not of diastatic ones; in fact it appears that planarians are unable to digest carbohydrates.

In *D. dorotocephala,* the basal halves of the phagocytic cells are normally packed with fat, which largely disappears during the early stages of intracellular ingestion and digestion. It is also known that, during this period, the oxygen intake is doubled or tripled. It thus appears that the phagocytic activities of the gastrodermis consume extra energy and that this energy is supplied by the oxidation of the fat reserves. During the later stages of digestion fat reappears in the phagocytic cells and increases to its former amount. The indications are that the protein of the food is transformed into fat, without the intervention of carbohydrate stages, and that this fat is stored in the gastrodermis as a reserve.

Fat droplets also occur sparingly in the mesenchyme of planarians. As already noted, planarians cannot digest carbohydrates and no carbohydrate reserve (glycogen) has been detected in most species studied, although von Brand (1936) found 3 per cent by weight in *Planaria torva* in autumn, less in winter.

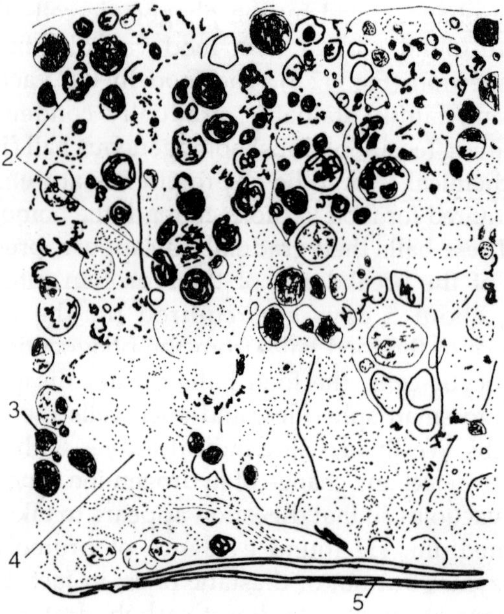

Fig. 12.3 : Intracellular digestion. Gastrodermis of *Dugesia dorotocephala* 6 hours after feeding beef liver showing various stages of food vacuoles. 1, early stage of digestion; 2, lager stage with homogeneous areas forming; 3, final stage, vacuole becomes a homogeneous eosinophilous ball; 4, fat vacuoles; 5, muscle fibers.

The available researches indicate, the, that in fresh-water planarians practically the entire processes of digestion, assimilation, and storage of food occur within the phagocytic cells of the intestine. Nothing is known of the manner in which the food reserves are transported to and utilized by other body tissues.

In those Turbellaria that swallow their food whole one infers that considerable digestion must take place in the lumen before phagocytosis can occur. Metschnikoff (1878) and Westblad (1922) claim that digestion is wholly extracellular in those rhabdocoels having a ciliated gastrodermis (Catenulida, Macrostomida). Westblad saw protein and fat particles undergo digestion in the intestinal lumen of *Stenostomum* while starch grains remained unaltered. However, Jacek (1916), also using *Stenostomum,* found that digestion is wholly intracellular and that protein particles (egg white) and oil droplets are digested in vacuoles in the gastrodermis; during protein digestion the vacuolar fluid changed from an acid to an alkaline reaction and protein was converted into and stored as fat. Starch grains were ejected unaltered. Some observations of the author confirm the occurrence of intracellular digestion in *Macrostomum.* It may be assumed that in rhabdocoels small enough particles are ingested at once by the gastrodermis whereas larger ones must first undergo some extracellular digestion. Extensive storage of glycogen in temnocephalids was reported by Fernando (1945). Nothing is known of digestive processes in polyclads, which, it may be recalled, also have a ciliated gastrodermis.

Defecation has been witnessed by several observers. The worm first takes in water through the pharynx, then by vigorous muscular contractions expels indigestible remnants. Those polyclads that possess them use the anal pores as well as the mouth for defecation.

i. Starvation. - The fresh-water planarians can endure prolonged starvation, from 6 to 14 months, depending on species. During this time they may reduce to 1/10 to 1/13 the original length and 1/300 of the original volume. This reduction results from tissue degeneration and does not involve any alteration of cell size. The nervous system undergoes no alteration so that greatly starved animals have disproportionately large heads. Epidermis and musculature are also practically unaltered, but the mesenchyme degenerates and disappears for the most part. The digestive

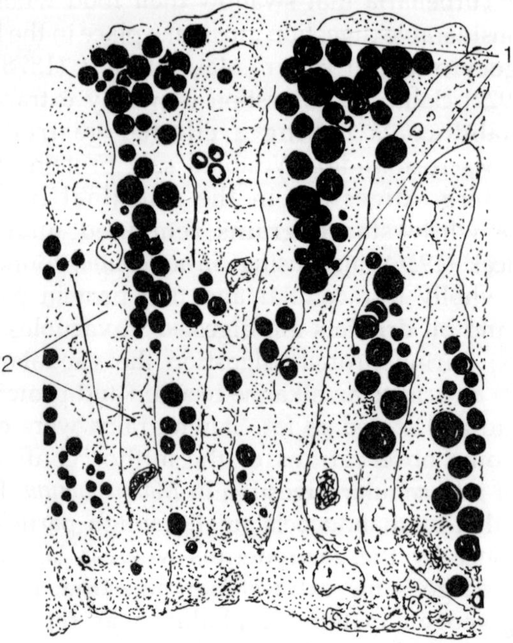

Fig. 12.4 : Intracellular digestion (continued). Gastrodermis of *Dugesia dorotocephala* 3 days after feeding beef liver; all food vacuoles have reached the stage of homogeneous eosinophilous balls 1, food ball; 2, fat vacuoles.

system retains its form, and there appears to be on reduction in the number of diverticula, but the volume of the intestine is greatly diminished through degeneration of large numbers of the phagocytic cells. The gastrodermis loses food inclusion and reduces to a flattened syncytial state. In the later stages of starvation the fat stored in the phagocytic cells begins to be utilized but Willier, Human and Rifenburgh (1925) found that the surviving gastrodermis still contained considerable fat after 3 months' starvation in *D. dorotocephala*. However, the complete breakdown of most of the gastrodermis must furnish a large amount of fat and protein to the planarian. The greatest degeneration in starving worms is seen in the reproductive system. The copulatory apparatus, sex ducts, and yolk glands entirely disappear, and the gonads reduce to small clusters of cells. The oxygen consumption of planarians per unit weight declines during early starvation but rises continuously later, and indication that starvation in these worms acts as a rejuvenating factor.

The ability of other Turbellaria to endure prolonged starvation is not well known, but the smaller forms are probably able to survive for only short periods in the absence of food.

j. Respiration.- The Turbellaria are devoid of any special respiratory structures; the exchange of respiratory gases takes places by simple diffusion through the body surface. So far as known, respiration is of the ordinary aerobic type, comprising the intake of free oxygen and the emission of carbon dioxide. Exact measurements of respiratory rate have been made chiefly on fresh-water planarians. The rate of oxygen consumption of *Dugesia dorotocephala* is about 0.2 to 0.3 *cc*. per gram per hour in adult worms; the rate per unit weight higher the smaller (younger) the worms, and is also increased by regeneration, during the early stages of intracellular digestion (see above), and during later stages of starvation. The oxygen consumed is also affected by the osmotic pressure of the medium as well as by the kind of salts in the medium. The rate of oxygen consumption of *Dugesia* species is constant at different oxygen concentrations in the water until this concentration falls below about one-third saturation. The cutting of planarians into pieces accelerates the rate of oxygen consumption, and this increase is greater the smaller the pieces, *i.e.*, if planarians are cut into six or eight pieces the rate of oxygen take of the combined pieces so greater than that of intact worms. This stimulatory effect of cutting declines with time and disappears within 24 hours. After allowance has been made for this effect, an axial series of pieces of *D. dorotocephala* shows an anterio-posterior decline in rate of oxygen intake, another expression of the axial gradation already mentioned several times.[1] Some experiments on the carbon dioxide production of the polyclad *Stylochus ijimai* are also available. Here, too, the rate of carbon dioxide production was found the decline with increasing size and to show an axial gradation in the form of a U, *i.e.*, a decrease from the anterior end toward the middle and an increase again toward the posterior end. Unfortunately sexual worms were used, and as the reproductive system is not evently distributed throughout the worm, a factor incapable of evaluation was present.

The oxygen consumption of planarians is depressed in the presence of potassium cyanide. It can be depressed as much as 90 per cent without injury to the worms. This indicates that the oxidative mechanism of planarians is chiefly of the cyanide-sensitive type, that it belongs to the "Warburg-Keilin" system, involving an iron-containing oxidase, a cytochrome complex, and a dehydrogenase. The dehydrogenase enzyme attacks the substance furnishing the energy, oxidizing it, and reducing the cytochrome complex which in turn becomes reoxidized by reducing the oxidase enzyme. The latter is restored by means of the free oxygen in the air or water, and so the system continues indefinitely as long as free oxygen is available. Cyanide acts by preventing the oxidase from taking up free oxygen. The cyanide-stable respiration, which constitutes only a small part of the respiration of planarians, operates by way of enzymes other than oxidases.

k. Excretion.- Nothing whatever is known of the form in which turbellarians eliminate their nitrogenous wastes. From the fact cited above that protein food is stored chiefly as fat in planarians, it seems evident that a process of deaminization must occur and hence that ammonia or some related substance must be given off. Several attempts have been made by use of vital dyes to determine what tissue is involved in elimination. Such experiments rest on analogy with conditions in vertebrates where it is known that vital dyes introduced into the body are treated like waste products and ejected by way of the kidneys.

When the Acoela (which have no protenephridial system) are exposed to vital dyes, the dyes collect in the central or digesting mesenchyme and are ejected through the mouth. In triclads, also, the gastrodermis serves as the chief means of elimination of vital dyes, the dyes granules collect in mesenchymal vacuoles which are then passed to the intestine. In a number of kalyptorhynchid and typholoplanid rahabdocoels, also in *Stenostomum,* the dye collects in certain portions of the protonephridia, mostly the more proximal portions, and in the paranephrocytes which also pass it into the protonephridia so that in these forms some evidence is available of an excretory function of the protonephridia, aided by paranephrocytes. No evidence has been seen of any passage of substances through the flame bulbs in any turbellarian. It seems probable that in Turbellaria the necessity of eliminating nitrogenous wastes is avoided by converting them into insoluble granules ("concrements") that are held permanently in the mesenchyme and in pigment cells.

Regulation of water content is in all probability one of the chief functions of the protonephridial system and the only function of the flame bulbs. This view is supported by the fact that the protonephridia are much better developed in fresh-water than in marine Turbellaria. Atrophy of the protonephridium of *Stenostomum* is followed by edema and general degenerating. New flame bulbs were formed by the evagination of capillaries in response to the accumulation of water in the worm. In a comparison of specimens of *Gyratrix hermaphroditus* from fresh, brackish, and salt water, Kromhout (1943) noted a gradual reduction in the protonephridial system with increasing salinity of the medium. Probably influx of water into the turbellarian body evokes activity of the flame bulbs, resulting in a current of water along the protenephridial tubule.

l. Chemotaxis.- This was already discussed in connection with finding of food, which is in effect a chemotactic reaction. Experiments have also been performed on the behavior of planarians to chemical substances in capillary tubes placed in the same dish with them. Pearl (1903) tested in this manner various concentrations of several acids, alkalies, and salts, and found that in general a positive response is given to weak concentrations of all the substances tried, and a negative one to stronger concentrations. A positive reaction is the same as that toward food, *i.e.*, the planarian on perceiving the diffusing chemical turns towards and then proceeds directly to the capillary. In the negative response, the animal turns away from the capillary. Weak acids proved especially attractive, and planarians would often grip the capillary and extend the pharynx into its orifice. Koehlar (1932) also employed the capillary method, using snail blood or crushed planarians, and found highly positive responses in some species, which can be caused to follow such as capillary around. Reactive species will also take a median course between double capillaries. Planarians with one auricle removed may show "circus movement" in a dish of diluted snail blood, circling toward the intact side. By adding India ink to snail blood diffusing from a capillary, it can be seen that response occurs when the carbon particles have passed along the auricular sense groove.

m. Reactions to Contact; Thigmotaxis.- The reactions of fresh-water planarians to mechanical stimuli have been investigated in detail by Pearl (1903). To a light touch on the side of the head, the worm may give a positive response, pausing, turning the raised head toward the stimulus, and moving to it. A similar stimulus along the body sides causes usually a small local contraction and may evoke the positive turning if not too far posterior. A positive reaction is never exhibited by postpharyngeal regions. Stronger tactile stimulation of the anterior end evokes the typical negative response, a turning of the head away from the stimulus and a gliding off in the new direction. Very strong or repeated prodding of the head causes jerking back by means of longitudinal contraction and turning through a larger angle. Moderate to strong mechanical stimulation of the sides of the body or the posterior region evokes local contraction and the hurried type of locomotion for a short time, followed by the usually gliding.

If a planarian is cut in two, the anterior piece reacts as to strong posterior stimulation, *i.e.*, moves by the hurried method for a short time, then returns to the usual gliding. The posterior piece generally crawls backward immediately after the cut and continues for some time to give this response to mechanical stimulation of the cut surface. To such stimulus elsewhere it gives either the negative turning away or the hurried crawling. A positive reaction to contact is not obtained in the absence of a brain.

The Turbellaria are markedly thigmotactic; the dorsal surface is negatively, the ventral surface positively thigmotactic. This is another way of saying that flatworms strive to keep their ventral surface in contact with a substratum, regardless of gravity, and their dorsal surface freely exposed to the medium. Pearl (1903) notes that if a light object is gently places on the dorsal surface of a resting or moving planarian, the worm will promptly move away from under the object. Planarians will not come to rest in a situation where both surfaces are in contact. When placed on their backs, planarians, polyclads, and probably all flattened Turbellaria give a characteristic *righting reaction*. The worm twists into a spiral in such a way that the ventral surface of the head finally comes in contact with a substratum. The head then glides away and the spiral unwinds as the rest of the body follows. Planarians dropped through water (planarians cannot swim) give the righting reaction, usually several times, as they fall. Pieces of planarians also give the righting reaction although more slowly than when the head is present. Brainless pieces of polyclads also right themselves, but the reaction differs from the normal in that any part of the piece, not always the anterior end as in the case of intact worms, may attach first. However, this may also occur in pieces of planarians.

The entire body of planarians is sensitive to contact, but in general tangoreceptivity is highest on the auricles or sides of the head and decreases in an anteroposterior direction. In the absence of the brain, the auricles are no longer more sensitive than other parts of the body margin.

n. Phototaxis.-Although some Turbellaria are positive to light, especially pelagic types, larval and juvenile stages of polyclads, acoels, and rhabdocoels that contain holophytic symbionts, and the rhabdocoels of Lake Baikal, the members of the class as a whole avoid light and unless other factors are operative seek the darkest areas of their environment. Many studies have been made on the response of turbellarians, chiefly fresh-water planarians, to light. Exposure to light increases the activity of planarians and causes resting individuals to start locomotion; the rate of locomotion may also be more rapid in the light. This stimulating actions of nondirective light has been termed *photokinesis*. After a time the effect decreases, and the worms slow down and finally come to rest in the least illuminated region of the container, if such exists. The avoidance of light by planarians is an example of a phobotactic or phobic response, that is, the darkened area is found by random movements, a negative reaction when passing into more lighted areas and a lack of reaction when passing into more darkened regions. The most careful test of the phobic nature of light avoidance in planarians is that of Ullyott (1936). In a light gradient in which all possibility of directed rays from reflections, etc., was avoided, the planarians eventually came to rest in the darker part of the gradient. This phobic response to light is given equally well by specimens from which the eyes have been removed; these also show no change in rate of locomotion. Decapitated specimens also avoid light, although they move more slowly. Inmost species light is perceived by the entire surface, although in some the posterior region shows no reaction to light. Both dorsal and ventral surfaces are sensitive of light, although planarians tend to keep their dorsal surfaces toward the light.

Planarians perceive changes in light intensity and may give evidence of such perception by hesitation, head, swingings, and change of path when encountering a new intensity. Various experiments that have been performed on the possibility of color perception by planarians are really tests of intensity discrimination according to the more critical papers. On contrasting backgrounds of black and white or dark and light colors, planarians come to rest on the darker ground and in general on colors of longer wave lengths. Such reactions are probably merely avoidance of light of higher intensity. A very exact phobic reaction was reported by Lemke (1935) for *Dugesia lugubris* on a background of alternate black and white circles; the planarian follows the black circles, turning back at the boundary with the white circles (Fig. 12.5). Eyeless specimens lose the reaction. Other species may not give such exact response to intensity boundaries and may in fact exhibit no change of behaviour to considerable changes of intensity.

To directed light, planarians give a fairly precise negative response, turning away from the light and proceeding in a direction approximately opposite to the incidence of the light rays. The path of planarian can therefore be directed by means of light rays. This phototopic reaction is mediated primarily by the eyes, for in worms from which the eyes have been removed (without injury to the brain) the response

Fig. 12.5 : Reaction to light. Continuous line gives patn normal *Dugesia lugubris* on background of alternate white and black circles; worm gives avoiding reaction to white circles. Broken line gives path of individual of which the eyes have been extripated.

is much less definite and precise, although still elicitable. If a spot of light is shone upon the eyes or head the negative turning away results, but if it is shone on middle or posterior regions there is not such response. Hence the topic reaction is mediated by the anterior end only, whereas the phobic response to light is mediate by the entire body in most species, with sensitivity decreasing in an anter-oposterior direction. The sense organs responsible for the phobic reaction have not been identified, but the eyes are evidently the chief receptors for the topic response.

According to the generally accepted analysis of Taliaferro (1920), directed light affects only those retinal cells whose long axes are parallel to the light rays. Because of the curvature of the planarian eye, only a few such cells would be affected in any given position of the planarian. The pigment cup also acts to shade retinal cells except those parallel to the light rays. This mechanism accounts for the highly directive effect of light on planarians, *i.e.*, their ability to perceive rather exactly the direction of light, and it also makes it rather probable that planarians can see movement, although not objects. The decrease in intensity of light as a moving object interrupts light rays would affect a succession of retinal cells. Probably perception of movement is the function of the numerous eyes of land planarians and polyclads since a moving decrease of intensity would affect successive eyes rather than successive retinal cells of the same eye. According to Kawaguti (1932) the rear part of *Bipalium* is insensitive to light because it lacks eyespots.

Planarians can perceive ultraviolet radiations; they react negatively to them but move slowly than in visible light. The behavior is the same in normal and eyeless specimens. Short exposures to ultraviolet and to direct sunlight are fatal to planarians. White species appear to be more sensitive to sunlight than colored ones.

Some data are available on the behaviour to light of cave planarians that live in perpetual darkness. *Sphalloplana percoeca* from Kentucky caves showed no avoidance of light or seeking of dark areas but would writhe on exposure to direct sunlight and might be killed by less than 2 minutes of such exposure. *Phagocata subterrannea* from an Indian cave, however, was negative to light but less exact and more given to wandering than epigean species. Whether this different behaviour is correlated with the fact that the first species is eyeless and the second had eyes cannot be stated.

o. Rheotaxis. In general fresh-water turbellarians inhabiting still waters fail to react to water currents. Strong currents usually elicit only a "clamping" reaction, contraction and increased adhesion to the substratum. Weak currents may, however, call forth a positive rheotactic response, a turning of the anterior end towards the current and placing of the body in line with it. This has been reported for planarians and the rhabdocoel *Bothromestoms*. In the latter animal, extirpation experiments indicated that the eight groups of sensory bristles along the lateral body grooves are rheoceptors. The reaction of still-water species to *weak* currents is probably part of the food-catching mechanism, since turbellarians are able to detect the presence of intact prey by water disturbances created by them. Pond planarians may develop a rheotactic response after being kept for some time in a rather strong water current. Planarians that habitually live in flowing water usually exhibit a positive rheotaxis, for obviously without such a reaction they could not maintain themselves in their habitat. Thus, *Dugesia dorotocephala*, a spring-dwelling species, was found to show positive rheotaxis (travelling upstream) in nature and in the laboratory; but the positive reaction could be altered to a negative one (travelling downstream) by change of water or lowering the temperature. The outstanding example of a positively rheotactic planarian is *Crenobia alpina*, inhabitant of rapid alpine streams. As already noted, this animal is more or less incapable of finding food in the absence of a water current. It is highly positive to currents and will follow a tube from which a water current is issuing. In case of two currents of equal velocity opposite or at right angles to each other, it will take a resultant path between them, and of two currents of unequal velocity it will select the faster one. The rheotactic reaction is given by all parts of the body, although most readily and to weaker currents by the head. Loss of the entire body margin does not eliminate the reaction so that rheoreceoptors must be distributed over the entire body surface being, however, most numerous on the head. After being kept for long periods (6 months to a year) in still water this species retains its rheotactic response undiminished. Beauchamp (1933) worked with a British stream planarian[2] in which the rheotactic response showed a definite seasonal cycle. In winter when temperatures were low and the reproductive system developing, the animal was positively rheotactic and accumulated at the head of hill streams; after the breeding season, negavite rheotaxis ensued and the worms travelled downstream and even into the lake into which the streams flow.

p. Thermotaxis and Temperature Relations.- The majority of the Turbellaria and undoubtedly eurythermous, enduring wide ranges of temperature, but some species appear to be somewhat stenothermous. The behavior to temperature changes is considerably dependent on the temperature at which the worms have been living prior to the tests. Available data are limited to fresh-water planarians. For local application of temperature change, a narrow hairpin-shaped tube is used through which water of the desired temperature is circulated. Planarians react to temperature change of 2 to 3°C. from that at which they have been living. To moderate increase applied to the anterior end a positive reaction may be given at first; the worm turns towards the tube and may test it with the anterior end, before turning away. Usually a negative reaction is eventually given. If the tube is very hot, a negative reaction may occur at once. Reactions to lowered temperatures are similar but less definite, and often no reaction is given. Localized heat applied to posterior regions results in local contraction, more rapid gliding away, or the hurried type of crawling. The lower limit of reaction to increased temperature was found by Koehler to be 24°C. for the anterior end of *Crenobia alpine,* 32 for its posterior part, 28-30°C and 40, respectively, for *Dugesia lugubris.* The ventral surface of planarians is also sensitive to temperature, so that thermal perception is evidently distributed over the whole body. When placed in a temperature gradient, *D. dorotocephala* which had been living at 22°C. aggregated in a zone ranging from 17 to 26; *Phagocata gracilis* aggregated at 0 to 10°C., whether taken directly from its natural habitat at 9.5°C. or kept for a month the laboratory at 20 to 22°C. The thermal death point of the former species is about 42°C, of the latter 30°C.

The most interesting behaviour to temperature is that of *Crenobia alpina*. This animal when presented with two streams of different temperature selects the cooler one and follows this about, being able to distinguish a difference of 3°C.

Planarians possess a high capacity for acclimation to temperature, being able when kept at high temperately (27 to 30°C.) to lower their rate of metabolic activity below and when kept at low temperature (8 to 10°C.) to raise to above that expected for such temperatures.

q. Geotaxis.- The Turbellaria in general show perception of gravity for when the oxygen supply of the medium decreases below requirements they come to the surface, *i.e.*, exhibit negative geotaxis. This is not a response to an oxygen or other gradient for its is given by worms in a closed vessel completely filled with water of uniform constitution and turned upside down. Unfed planarians in water of good oxygen content in the dark tend to seek the bottom of the vessel, but after feeding they tend to come to the surface probably because of the increased oxygen demand during intracellular digestion. The outstanding example of geotactic response among Turbellaria is seen in the acoel *Convoluta roscoffensis*. This animal is negatively geotactic when undisturbed, positively so when disturbed. On the sandy tidal zone on the channel coast of France, where it lives in countless numbers, it comes to the surface of the sand at low tide, and disappears into the sand when the tide returns or an any other disturbance. This rhythmic geotactic behaviour with reference to the tides is retained for about a week by *Convoluta* when kept in the laboratory in vessels of still water. It is independent of day and night. That the tidal response is really a geotactic one was proved by Fraenkel (1929) who showed that when *Convoluta* is simultaneously exposed to centrifugal force and gravitation it takes a resultant path between the two forces.

Although Turbellaria without statocysts react to gravity, it appears that the statocyst is of some importance in the geotactic reactions of the Acoela. The anterior tip of *Convoluta,* which contains the statocyst, reacts normally to gravity, but the rest of the body has lost the geotactic response. In regeneration experiments on Acoela it has been noticed that pieces without the statocyst have difficulty in righting themselves and in general react poorly to stimuli.

r. Galvanotaxis.- The reaction of turbellarians to the direct constant electric current has been ascertained by a number of workers. All forms tested are cathodic, *i.e.,* they turn their anterior ends towards the cathode and may swim or crawl towards this pole. *Stenostomum* in an electric current behaves exactly like ciliate Protozoa and its cilia also beat with reference to anode and cathode as in ciliates. *Convoluta* is highly cathodic. Fresh -water planarians turn the anterior toward the cathode and may place the body in line with the current and proceed toward the cathode or the body may curve into a C-, U, or W-shape with ends and often ventral surface facing the cathode. This cathodic orientation of posterior as well as anterior end may be attributed to the presence in such species of a fission region or some region of especial activity, such as a adhesive area, in the posterior part of the animal. Decapitated planarians show the same response as entire ones, and middle pieces move anterior end first to the cathode or backward to the anode. The land planarian *Bipalium* is likewise cathodic, moving by loops to the cathode, and pieces obtained by cutting the body in thirds also loop to the cathode. The one polyclad tested behaved like planarians, turning toward the cathode and usually also curving its body into a U-shape toward this pole.

The various explanations of galvanotaxis that have been advanced are based chiefly on the idea of compulsory effect of the current directly on the ciliary or neuromuscular mechanism. This type of explanation appears unsatisfactory to the author, for it fails to explain reversals that occur under different physiological conditions. It can hardly be supposed that the neuromuscular mechanism also undergoes reversal during temporary changes of conditions. The fact that individuals facing the anode at the time the current is made also turn and proceed to the cathode is further difficult to explain on any such theory. According to the author's theory of galvanotaxis also expressed independently by Robertson (1927), the flat-worm takes that position with reference to the electric current which best corresponds with its own electrical currents. For at least the lower animals, as has been shown by numerous experiments have inherent electrical currents related to their anteroposterior axiation. These currents presumably originate from different rates of chemical activity along this axis. In general, the anterior or head end is electropositive to lower levels, and hence it orients so as to face the cathode. The rear end of flatworms through the presence of a fission plane or other active region may also be electropositive with reference to the middle; hence the C or U attitude taken by such species the current.

s. Desiccation.- The Turbellaria have no surface protection against evaporation and hence require an aquatic or humid environment. Some species secrete a mucous cyst or covering when exposed to desiccation and some land planarians can endure relatively dry environments as the South American pampas by habitually dwelling under cattle dung. In general, however, exposure to desiccation results in death. Fresh-water planarians when subjected to drying curl up so as to expose as little surface as possible and from time to time attempt to escape by extending the head in seeking movements or giving backward crawling waves but are unable to progress on a dry surface. When placed on a wet spot in a dry environment they give a phobic reaction whenever the head extends outside the wet spot. The land

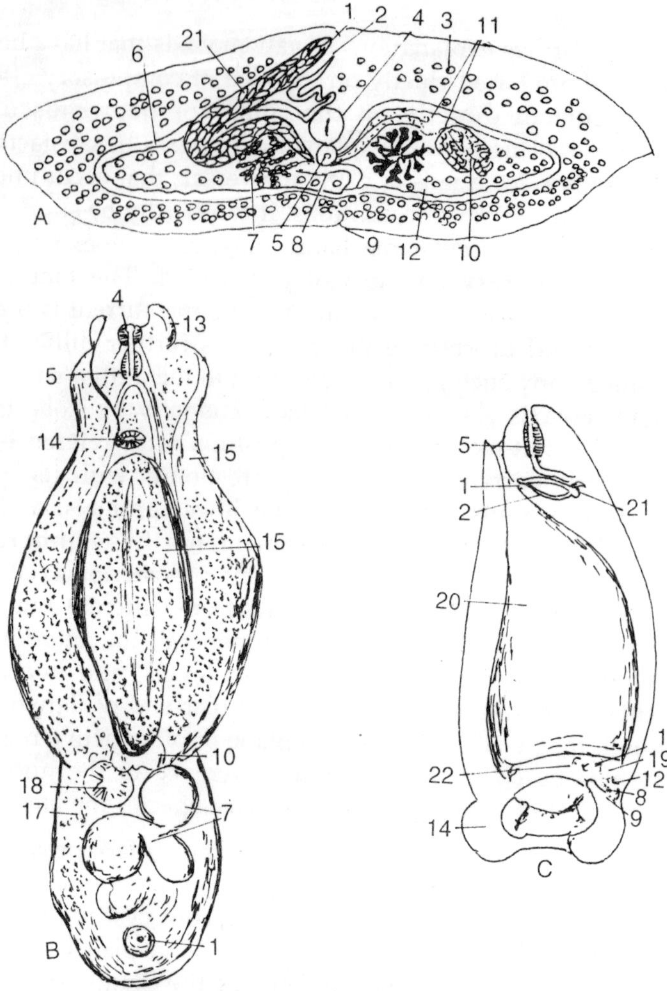

Fig. 12.6 : Variants of trematode shape. *A, Transversotrema* fish trematode from the Red Sea. *B.* Strigeid fluke *Alaria alata* C. Longitudinal section of the amphistome fluke *Gastrothylax* showing deep ventral pit. 1, gonopore; 2, sperm duct; 3, yolk glands; 4, oral sucker; 5, pharynx; 6, intestine; 7, testis; 8, excretory bladder; 9, nephridiopore; 10, ovary, 11; yolk ducts; 12; Laurer's canal; 13, lateral suckers; 14, acetabulum; 15, forebody; 16, hold-fast; 17, hindbody; 18 ootype; 19, seminal receptacle; 20 ventral pit; 21, uterus; 22, beginning of uterus.

planarian *Bipalium* was found by Kawaguti (1932) to lose water according to the simple laws of evaporation in relation to the humidity and temperature of the environment. Water was lost even in an atmosphere of 100 per cent humidity. Recovery occurred if the water loss did not exceed 45 per cent of the fresh weight.

t. Mass Protection.- Following the discovery by Bohn and Drzewina (1920) that a number of *Convoluta* are more resistant to fresh water or much diluted sea water than single specimens in the same (limited) volume of fluid, much work was done by these authors and also by Allee and his students on such mass protection in various turbellarians (and other animals) against adverse conditions and agents, such as toxic chemicals, ultraviolet radiation, salinity changes. In general, resistance to such agents and conditions is much greater when there are a number individuals as contrasted with single or a few individuals in the same quantity of medium. The protection is not caused primarily by the fact that, in the case of a toxic chemical, for instance, there is less of the chemical per worm in the mass experiment, but is does result from the greater quantity of animal in a limited amount of medium. The French authors postulate the secretion under these conditions of a special protective substance of which of course more would be emitted by the larger number of animals, but the experiments of Allee indicate rather that the normal secretions and exudates of the animals furnish the protective materials. Protective action is furnished by disintegrated animals or the water in which the same or other animals had lived. In the case of turbellarians the mucous and other glandular secretions emitted from the surface are probably the chief means of protection. Such secretions, exudates, or disintegrated remains neutralize toxic chemicals and, in some unknown way, probably not by adding salts to the medium, also protect marine turbellarians against fresh water.

u. Learning.- The Turbellaria are capable of associative memory. *Stenostomum* can be trained to turn back into the dark at a light-dark boundary if punished by an electric shock when entering the light area. As *Stenostomum* is negative to light only a very weak light to which it does not react can be employed. The reverse experiment, training it to avoid the dark and remain in the light, requires longer training and gives a less definite result. Retention is poor. Similar experiments have been performed with the fresh-water planarian *Dugesia gonocephala* using shaking and electric shock as punishments, and rough versus smooth substratum, weak light versus darkness, and vertical versus horizontal surfaces as contrasting conditions. In general, the planarian could be taught by repeated punishment to avoid that condition to which it normally gives a negative response, as light and rough substratum *i.e.*, to turn back at the boundary in the absence of punishment, but the reverse attempt, to teach it to turn back into conditions that it normally avoids, gave poor or no results. The experiments of Soest and Dilk came from the same laboratory as the experiments with the formation of associations in Protozoa reported in P–C. Much more convincing is the work of Hovey (1929) on the formation of an association in a leptoplanid polyclad. These worms were subjected for hours to alternating periods of light lasting 5 minutes and darkness lasting 30 minutes. Light induces movement, darkness a quiescent state as is general in Turbellaria. Whenever a worm started to move on exposure to light it was stopped by being touched on the anterior margin. The number of touches required to cause the worms to cease movement in the light decreased notably with repeated trials although no worm learned to remain entirely motionless when exposed to light. Possibility of motor fatigue, injury to the anterior margin, or adaptation to light was ruled out. Following extirpation of the cerebral ganglia, worms did not learn to move less in the light, and repeated hindrance to such movement had not effect. Evidently the brain is involved in associative learning in polyclads.

Reference:-

1. The existence of such an axial gradient in respiratory rate in planarians has been disputed by others (e.g., Parker, 1929) for other species, but the author feels that in their experiments insufficient account was taken of the stimulating effect of section and of the sexual condition of the animals. Only nonsexual animals should be employed.
2. Said to be *Crenobia alpina* but this seems improbable.

Chapter 13

Physiology of Trematodes

The physiology of trematodes has been much less studied than that of cestodes or nematodes, probably on account of the difficulty of obtaining suitable experimental material.

Before considering the physiology of trematodes, it is as well to review certain features typical of the group which directly or indirectly impose certain limitations or requirements on their physiology.

(*a*) All trematodes possess at least one sucker, which means that they are brought into intimate contact with host tissues. The definitive habitat must thus present a suitable surface to which attachment can be made and on which feeding can take place.

(*b*) Without exception, they possess a well-developed alimentary canal usually with a muscular pharynx. This kind of system is especially suitable for ingesting semiliquid or viscous food such as intestinal contents, mucus, blood and bile, and with the exception of a few progenetic forms distribution of adults is limited to habitats, chiefly in vertebrates, where such food materials are available.

(*c*) Their body covering (tegument) bears a striking resemblance to that of cestodes and, as in that group, is absorptive in nature; nutrition may thus be taken in both through the gut and through the tegument.

(*d*) Their egg capsules are generally delicate structures, incapable of withstanding desiccation, and normally require water for development.

(*e*) Their life cycles are complex (excepting the Monogenea) and always involve a molluscan intermediate host, sometimes a second intermediate host and, more rarely, a third intermediate host. The physico-chemical conditions and the nutritional levels of the different environments encountered may thus vary profoundly, so that striking changes in the metabolism of various stages may be expected.

Chemical composition

As is stressed elsewhere the chemical analysis of parasitic organisms seldom provides significant information unless it is related to the nutritional state of the host, and the degree of maturity of the parasite. Thus the carbohydrate content of an intestinal parasite may vary within considerable limits, depending on whether the host has been actively feeding or starving before autopsy. This point is illustrated particularly by experiments with *Fasciola*. Chemical analysis data are usually expressed as a percentage of the dry weight. The dry/fresh weight ratio may vary appreciably with the maturity of the fluke. Thus in the case of *Echinostoma revolutum*, experimentally developed in chicks, the dry weight after 10 days was approximately 10 per cent and after 30 days 20 per cent.

Carbohydrate. As in other parasitic helminths, the main polysaccharide found in trematodes is glycogen, and that obtained from *Fasciola* is a highly dextrorotatory compound which does not appear to differ significantly from glycogens from vertebrate tissues. The glycogen predominantly occurs in the parenchyma with small quantities in the muscles and reproductive system. Species living under conditions of low oxygen tension generally appear to contain higher levels of glycogen than those inhabiting more aerobic environments. As might be expected, starvation produces a marked loss of glycogen and in *Fasciola* parenchymal glycogen virtually disappears after 24 hours' starvation. Comparable results have been obtained with *Echinostoma revolutum* after 72 hours. The amount of stored carbohydrate present at any one time will, of course, be further complicated by the demand of the parasite in relation to the phase of growth and reproduction

Table 13.1 Chemical composition of Trematodes*

Species	Host	Location	Chemical composition % dry wt.					References
			dry matter as % fresh wt.	inorganic ash	carbo-hydrate (glycogen)	protein	lipid	
Haplometra cylindracea	frog	lungs	16.7	5.3	9.5	63	—	Halton (1964)
Haematoloechus medioplexus	frog	lung	17.5	6.1	9.0	66	—	Halton (1964)
Diplodiscus subclavatus	frog	rectum	16.5	4.0	15.5	62	—	Halton (1964)
Gorgoderina vitelliloba	frog	bladder	19.3	4.5	8.7	—	—	Halton (1964)
Fasciola hepatica	sheep	bile ducts	19.5	5.5	16.0	59	12-13	Halton (1964), von Brand (1964)
Fasciola gigantica	buffalo	bile ducts	22.4	—	25.4	66	12.8	Goil(1958a, 1958b, 1961)
Paramphistomum explanatum	buffalo	bile ducts	25.7	—	25.0	53	4.5	Goil (1957, 1958a, 1958b)
Gastrothylax crumenifer	buffalo	reticulum	27.9	—	30.3	49	1.3	Goil (1957, 1958a, 1958b)

Table Continued

Species	Host	Location	Chemical composition % dry wt. dry matter as % fresh wt.	inorganic ash	carbo-hydrate (glycogen)	protein	lipid	References
Schistosoma mansoni (male)	mouse	venous system	—	—	14-29	50	—	von Brand (1966)
Schistosoma mansoni (male)	mouse	venous system	—	—	3-5	65	—	von Brand (1966)
Schistosoma mansoni (female)	mouse	venous system	—	—	3-5	65	—	von Brand (1966)
Gynaecotyla adunca	bird	intestine	—	—	1.6-7 (wet wt.)	—	—	Vernberg & Hunter (1956)
Dicrocoelium dendriticum	sheep	bile duct	20.2	6.0	28.5	43	6.5	Eckert & Lehner (1971)

* Factors such as age, degree of maturity and nutritional condition of host must be known for these figures to be of any real significance.

Lipid. Only *Fasciola* and *Gastrothylax* appear to have been studied in any detail. In *Fasciola* much of the lipid appears to be concentrated in the excretory vessels and expulsion of fat droplets through the excretory pore has been described. That much of the lipid in trematodes represents excretory products is supported by the common occurrence of fatty material in the excretory bladders of encysted metacercariae. The lipids in *Fasciola* and *Gastrothylax* consist of phospholipids, unsaponifiable matter (possibly cholesterol), unsaturated and saturated fatty acids.

Protein. The composition of the tissue proteins of trematodes has been very little studied and a major gap in our knowledge occurs here. These figures are based on a total nitrogen conversion factor of 6.25 adopted from vertebrate tissues, and this figure may need some modification in the light of further experimental work. The proteins of the eggs and metacercarial cysts, however, have been studied in some detail; the main structural proteins involved appear to be chiefly sclerotin and keratin.

Water relationships

The osmotic relationships of trematodes with their environments have been little studied. This is surprising, for their complex life cycles may involve in turn passage through fresh or sea water (by the miracidium and the cercaria), an intermediate molluscan host, and often a second intermediate host (usually an arthropod or a vertebrate) before reaching the definitive host. A study of the osmotic changes encountered and how they are overcome might produce some interesting results. Both miracidia and cercariae of *Fasciola hepatica* face considerable osmotic changes on passing to or from tissue (Δ= 0.58°C for cattle) to fresh (Δ = 0°C) or sea water (Δ = 2.2°C), or to snail tissues (Δ = 0.22°C for *Lymnaea stagnalis*). In a miracidium, except for the simple flame-cell system, there is no special device for osmotic control. Rather surprisingly, in *Fasciola* large changes in osmotic pressure (0-1.0 per cent NaCI) produce no change in the flame cell activity of the miracidia. The few studies carried out on adult trematodes suggest that they are poor regulators and relatively permeable to water; data is available for *Fasciola gigantica Gastrothylax crumenifer Haematoloechus medioplexus*.

Table 13.2 Nature of food and enzymes present in some species of *Digenea*. + = present; = not tested for.

Species	*Host*	*Location*	*Food*	Enzymes present					
				Protease	*Lipase*	*Alkaline phosph-atase*	*Acid phosph-atase*	*Esterase*	*Amino peptidase*
Haplometra cylindracea	frog	lungs	blood	+	.	+	+	+	.
Haematoloechus medioplexus	frog	lungs	blood	+	.	+	+	.	.
Diplodiscus subclavatus	frog	rectum	blood	.	.	+	+	.	.
Opisthioglyphe ranae	frog	intestine	tissue mucus blood	.	.	+	+	+	.
Gorgoderina vitelliloba	frog	bladder	tissue blood	.	.	+	+	+	.
Fasciola hepatica	sheep	bile duct	tissue blood	+	+	+	+	+	.
Schistosoma mansoni	mouse	venous system	blood	+	.	+	+	.	.
Cyathocotyle bushiensis	duck	rectum	tissue	.	.	+	.	+	+

Nutrition

The nutrition of a number of species has been examined by Halton (1967a) who, together with Jennings (1968), have given excellent reviews of this problem.

Nature of food materials

The habitats of trematodes are so diverse that it is not surprising to find that the food materials utilized cover a wide range and include blood, bile, mucus, intestinal contents. The question is further complicated by the fact that some species can migrate through the body and are exposed to different kinds of food materials during this process.

Probably the most controversial species studied has been *Fasciola*. According to Dawes (1963*a*, *b*, *c*) this species is essentially a tissue feeder utilizing epithelium and connective tissue and only incidentally takes in some blood. Other workers, notably Halton (1967*a*) have produced evidence that *Fasciola* feeds predominantly on blood but that tissue and mucus is also ingested. The relative amounts of blood and tissue taken in may be a matter of degree, depending on the site and activity of the organism at any time.

The nature of the food materials is further complicated by the fact that absorption can take place through the tegument. Thus schistosomes, which live in the hepatic portal system, probably utilize soluble food materials such as amino acids and monosaccharides via the tegument, in addition to blood. In schistosomes and in some other blood-feeding trematodes a blackishbrown pigment, derived from ingested blood is found; although the exact nature of this pigment has not been studied, there is some evidence that it is identical in composition with the malarial pigment.

The nutrition of larval trematodes has been largely neglected. There is evidence that the rediae of *Parorchis acanthus* and *Cryptocotyle lingua* can also absorb nutrients, such as glucose, through the body surface as well as ingesting particulate matter in the gut.

Digestion

Understanding of the process of digestion in trematodes is complicated by presence of unusual morphological features, the nature of which are not properly understood. Thus, E.M. studies have shown the presence of protoplasmic extensions in the form of lamellate processes, ridges or 'microvilli' or combinations or modifications of these in the gut.

Species studied include: *Schistosoma mansoni; Schistosomatium douthitti; Gorgodera amplicava* and *Haematoloechus medioplexus.*

Various hypotheses have been put forward regarding the function of these processes. Thus it has been suggested that (*a*) at least some digestion occurs in vacuoles formed by these processes, (*b*) absorption of macromolecules occurs across the membranes of the lamellae, (*c*) the processes provide a porous surface on which membrane digestion could operate.

An extensive examination of nine species by Halton (1967a) has revealed that digestion in Digenea appears to be predominantly extracellular. Enzymes make the gut contents soluble, the products are then absorbed and the process of digestion is completed in the gastrodermis. Attempts to demonstrate digestive enzymes have not been very conclusive. In homogenates of *F. hepatica* and *H. cylindracea* a protease, a dipeptidase, a lipase, and acid and alkaline phosphatases have been detected. Few of these have been detected histochemically but problems of technique may be involved. There is, however, evidence of an esterase in the gut. From *S. mansoni,* a globin-digesting enzyme with a pH optimum of 3.9 has been extracted. Direct evidence that haemoglobin is ingested and digested by schistosomes has been obtained by labelling reticulocytes with tritiated L-leucine and showing its later incorporation in the worm tissues. In *Schistosoma rodhaini,* histochemical methods for the exopeptidase, leucine aminopeptidase, failed to demonstrate this enzyme in the gut although it was present in the tegument. This further confirms the important metabolic role played by the tegument.

One additional factor operates in trematode digestion, namely the fact that certain adhesive organs serve not only as organs of attachment but also as extracorporeal organs of digestion and absorption. Thus the tegumental lining of the suckers and associated regions of *Haplometra cylindracea* has been shown to give strong reactions for carboxylic esterases suggesting that extracorporeal digestion takes place in this region. It is in the strigeoid trematodes, however, that extracorporeal digestion appears best developed. This process has been extensively studied by Erasmus and his co-workers. The three bird species, *Cyathocotyle bushiensis, Apatemon gracilis minor* and *Diplostomum phoxini* have been particularly studied. In these species three main host-parasite interfaces are found: (*a*) the general tegument, (*b*) the lappets and (*c*) the adhesive organ lobes. Well-developed glands are associated with the lappets in some species. There is now abundant evidence that both the lappets and the adhesive organs release histolytic enzymes which digest host tissues. This results in the production of finely fragmented material which may then be ingested. In *A. gracilis* and *C. bushiensis* the oral sucker does not appear to be permanently attached and the organism is free to move and digest material. In *A. gracilis* the host-parasite interface is so intimate that it is 'impossible to distinguish precisely the extent of the parasite tegument.'

The adhesive organ is not only proteolytic in function but also appears to be 'placental' and serves an important role in the absorption of materials. Apart from the strigeids, little work appears to have been carried out on the host-parasite interface and this is an area in which much work remains to be done.

Respiration

Respiration metabolism of trematodes has been reviewed by Vernberg (1968).

Eggs. The eggs of most trematodes embryonate in an aerobic environment but the extent to which oxygen is necessary for embryonation is not known. In *Fasciola,* the oxygen consumption is 0.3 mm^3 /mg (d.w.) but rises to five times this figure after about 12 days' development at 25°C.

Table 13.3. O_2 consumption of Trematodes in atmospheric air

Species	*Location*	*Temp. °C*	*Qo_2 ($\mu 1 0_2$/mg dry wt. hr)*	*References*
Cotylophoron indicum	Reticulum	37.5	0.023	Goil (1961)‡
Dicrocoelium dendriticum	Liver	37	1.2	Eckert & Lehner (1971)
Echinostoma revolutum	Intestine	40	2.8-4.9	Taft & Fried (1968)
Fasciola gigantica	Bile ducts	37.5	0.037	Goil (1961)‡
Fasciola hepatica	Bile ducts	37.5	0.17†	Prichard & Schofield (1971)
Gynaecotyla adunca	Intestine	41	0.05-0.13*	Vernberg & Hunter (1961)
Gastrothylax crumenifer	Reticulum	37.5	0.30-0.33	Goil (1958)‡
Paragonimus westermani	Lungs	37.5	0.74-0.86	Read & Yogore (1955)‡
Paramphistomum cervi	Reticulum	38	0.03	Lazarus (1950)‡
Paramphistomum explanatum	Bile ducts	37.5	0.84-1'34	Goil (1958)‡
Pleurogonimus malaclemys	Intestine	36	0.03-0.07*	Vernberg & Hunter (1961)‡
Proctoeces subtenuis	Organ of Bojanus	15-17	0.168	Freeman (1962)‡
Saccocoelium beauforiti	Intestine	36	0.02-0.04*	Vernberg & Hunter (1961)‡
Schistosoma mansoni, males	Mesenteric veins	37.5	9.1	Bueding (1950)‡
Schistosoma mansoni females	Mesenteric veins	37.5	10.7	Bueding (1950)‡
Clonorchis sinensis	Liver	37	2.280	Bueding *et. al.* (1971*b*)
Paragonimus ohirai	Lungs	37	2.238	Bruce *et al.* (1971*b*)

*μl C_2/ hr/μg N. †Wet weight.

In several species - *S. mansoni, S. japonicum* and *Paragonimus ohirai* - miracidia show an 'ageing effect' and consume more oxygen after five hours than after one hour; differences between strains do, however, occur. In *Schistosoma* but not *Paragonimus* respiration was stimulated by glucose, a result suggesting the presence of larger glycogen reserves in the latter species.

Larvae; rediae and sporocysts. Work on larval respiration has been reviewed by Vernberg (1963) and Smyth (1966). Little work has been carried out on rediae and sporocysts; these in general utilize oxygen if it is offered to them. In the starving sporocysts of *Cercaria dichotoma,* oxygen consumption appears to be related to the number of contained cercariae.

Cercariae. As might be expected of predominantly free-living organisms, cercariae have a marked aerobic metabolism and cannot survive long under anaerobic conditions. The cercariae of *S. mansoni,* for example, become inactive without oxygen after one hour and some 75 per cent die within four hours.

Some cercaria (e.g. *Himasthla quissetensis*) are 'conformers' i.e. their respiration decreases in direct proportion to the decrease in oxygen tension. Other species, however (e.g. *Zoogonus rubellus*), are conformers at some range of oxygen tensions only.

Adults. As in the case of most other helminth parasites it is extremely difficult to determine the significance of these figures. The fact that trematodes utilize oxygen when it is offered to them *in vitro* has little bearing on the *in vivo* situation unless the level of oxygen available *in vivo* is also known.

The influence of different substrates, such as those occurring in the Krebs or Embden-Meyerhof pathways, has been studied in a number of trematodes and there is some evidence of the operation of part or all of these pathways. The influence of a number of recognized inhibitors of mammalian metabolism on trematode respiration has been examined. The effects of various inhibitors on the oxygen consumption of *Fasciola* in the presence of succinate or NADH are shown in Table 13.4 The succinoxidase present in *Fasciola* appears to be relatively insensitive to amytal, requiring 10-100 times the concentration normally required to inhibit the corresponding mammalian enzyme. An even more striking result was obtained with Antimycin A for which a concentration of 10^{-5} was required - about 1000 times that required by mammalian systems.

Table 13.4. The effect of various respiratory inhibitors on the oxygen uptake of *Fasciola hepatica*.

Substrate	*Inhibitor*	*Concen-tration*	*Oxygen uptake nmole/ min per mg protein*	*Inhibition %*
Succinate	–	–	5.4	–
	Amytal	2×10^{-3}M	5.5	0
	Amytal	2×10^{-2}M	6.8	+
	Antimycin A	10^{-5}M	3.6	34
	KCN	10^{-3}M	7.3	+
	KCN	10^{-2}M	11.4	+
	Azide	10^{-4}M	5.0	8
	Azide	10^{-3}M	3.5	35
	Malonate	10^{-4}M	2.6	63
NADH	–	–	4.1	–
	Amytal	2×10^{-2}M	3.4	16
	Rotemone	10^{-4}M	2.1	48

* The oxygen uptakes were measured polarigraphically. Succinate or NADH concentration was 4×10^{-4}M. The sign + indicates an apparent stimulation of oxygen uptake.

In contrast to mammalian systems (in which the electron transport system can be inhibited by 10^{-3} cyanide), appeared to *stimulate* oxygen uptake in *Fasciola*. A similar stimulation of respiration by cyanide has been reported for *Paramphistomum cervi* and some cestodes and nematodes. Thus, although *Fasciola* appears to possess enzymes similar to those of the mammalian chain - NADH-cytochrome *c* oxidoreductase, succinate-cytochrome *c* oxidoreductase, NADH oxidase and cytochrome *c*-oxygen oxidoreductase - it is apparent that major differences in the electron transport chain of these two systems occur. Cytochrome *c* does not appear to be present in *Fasciola* although it has been reported in *Echinostoma revolutum*. In *Schistosoma mansoni,* however, cyanide does have an inhibitory effect on oxygen uptake which may indicate the presence of a cytochrome system.

Respiratory pigments with absorption spectra approaching those of mammalian haemoglobin have been described in several species of trematodes but there is no evidence that this substance acts as a respiratory pigment.

Metabolism

Carbohydrate metabolism

Like most other parasitic helminths, trematodes have a pronounced carbohydrate metabolism. This is evidenced by the substantial amount of stored carbohydrate, the extensive use of endogenous carbohydrate, the high rate of uptake of exogenous glucose and production of fatty acids as metabolic by-products.

Under starvation conditions, endogenous glycogen is rapidly used but its reserves are built up again when exogenous sources become available. Adult trematodes can make use of hexoses, such as glucose and mannose. Experiments with gut-ligatured *Fasciola* have demonstrated that absorption can take place not only via the gut but also through the tegument; this result is in keeping with the metabolic activity implied from ultrastructure studies.

Although much further remains to be done in this area, the few studies carried out - largely on *Schistosoma mansoni,*

Dicrocoelium dentdriticum and *Fasciola hepatica*-indicate that trematodes generally follow a Embden-Meyerhof pathway of phosphorylating glycolysis.

The end-product of carbohydrate metabolism in *Schistosoma* is lactic acid; up to 91 per cent of the glucose utilized being converted to this acid. In contrast, in *Fasciola*, up to 95 per cent of the carbohydrate is converted to the volatile fatty acids, acetic and propionic acids. The proportion of these acids produced is a function of the oxygen tension, ranging from 2.5:1 in nitrogen/carbon dioxide to approximately 1:1 in oxygen.

There is also evidence of a complete Krebs cycle in some trematodes and for a partial cycle in others, but the terminal oxidation appears to be deficient. In *Fasciola*, however, the Krebs cycle appears to be of relatively minor importance in metabolism, but the presence of the enzymes isocitrate lyase and malate synthetase suggests that the glyoxylate cycle is functioning in this species. There is evidence, too, that the pentose-phosphate pathway operates in this species.

In the majority of helminths, the period immediately following anaerobiosis is characterized by an increase in the rate of respiration. This is often referred to as the repayment of an 'oxygen debt.' In trematodes, however, there appears to be no oxygen debt, and the fatty acids are almost entirely excreted without further metabolizing.

Protein metabolism

It has already been pointed out that trematodes in general can ingest and digest complex proteins. An efficient protein metabolism must occur, for in some species astonishing numbers of eggs can be produced; *Fasciolophsis buski*, for example, produces up to 25 000 daily. The potential protein synthesis of larval trematodes is even greater - a singel miracidium of *S. mansoni* may produce more than 200 000 cercariae.

It is interesting to note that an aminopeptidase has been detected in the rediae and cercariae of *Philophthalmus gralli*.

This could explain the lytic action of these larvae in the snail tissue.

There is very little direct evidence on the amino acid requirements of adults as investigation of this depends largely on *in vitro* studies and these, in turn, depend on knowledge of metabolism so that somewhat of a dilemma exists.

As in the case of carbohydrates, amino acids may be absorbed through the tegument and, as in cestodes may even 'leak' back through this structure. Experiments on the uptake of *S. mansoni* appear to substantiate this view.

Transamination. A number of transaminases have been detected especially in *Fasciola* and *Schistosoma* (Kurelec & Ehrlich, 1963; Huang *et al.*, 1962). In *S. japonicum*, glutamicpyruvic and glutamic-oxaloacetic transaminases have been demonstrated, and in *Fasciola* additionally, a pyruvic acid/alanine transaminase has been found. Much more, however, remains to be done in this field.

End-products of metabolism. Ammonia and uric acid appear to form the nitrogenous compounds most commonly excreted, although urea has also been reported. The materials excreted may, however, be affected by the host diet. Thus, *Fasciola* appears to be largely a protein feeder and excretes some ten times the amount of ammonia excreted by the cestode *Hymenolepis diminuta,* which utilizes largely carbohydrate. *In vitro,* the addition of glucose to media in which *Fasciola* is cultured greatly decreases the amount of ammonia excreted but increases the lactic acid released.

Lipid metabolism

The lipid metabolism has been very little studied, although the lipid analysis of several forms is known.

In *Schistosoma mansoni,* the major lipids are cholesterol, triglyceride, phosphatidylcholine, phosphatidylethanolamine and phosphatidylinositol. Like cestodes this species has lost its ability to synthesize *de novo* sterols and fatty acids but when supplied with dietary sources it has retained the capacity to synthesize all its complex lipids.

During starvation, the fat content rises, presumably as the result of accumulated fatty acids as by-products of carbohydrate metabolism. Lipases have been detected in a number of species, often in the lappets and adhesive organs of strigeids. As could be expected, acetylcholinesterase and pseudo-cholin-esterase have been detected in a number of species, because histochemical methods for these are well developed. It was first shown in Monogenea that such methods provide elegant techniques for the demonstraticn of nervous tissue in whole mount preparations. These methods are readily applied to most species, such as *Schistosoma* spp. and will undoubtedly be of considerable value in further studies on the trematode nervous system.

Chapter—14

Physiology of Cestodes

General account

Cestodes may be considered to be biological models of exceptional physiological interest on account of (*a*) the unusual habitat in which they live, (*b*) their basic morphology, which is characterized by the lack of an alimentary canal, and (*c*) their pattern of strobilar growth.

Expanding these points further, it can be seen that:

(*a*) cestodes live in an environment whose physicochemical properties and nutritional level may vary in relation to the feeding pattern and nutritional condition of the host; some species (e.g. *Hymenolepis diminuta*) undergo diurnal migration patterns in relation to the host feeding pattern.

(*b*) Cestodes lack a dlgestive tract and a defined circulatory system; their outer body covering (the tegument, is essentially a 'naked' protoplasmic surface through which all substances must both enter and leave the body; *active* rather than passive transport appears to be generally involved in these processes. The occurrence of this naked surface implies that cestodes would be strongly antigenic.

(*c*) It has generally been assumed that, for nutritional purposes, cestodes utilize small molecules derived from the digested food sources of the host. The situation appears to be more complex, however, for it is now known that cestodes are capable of fixing carbon dioxide and that this process serves as a valuable source of carbon atoms. Furthermore, there is some evidence that cestodes may be capable of utilizing larger molecules, such as protein, either by direct uptake or by a process known as membrane (= contact) digestion this hypothesis is, however, by no means proved. In the latter process membrane-bound enzymes are believed to be capable of digesting large molecules with which they come in contact. Pinocyytosis has not, however, been demonstrated in cestodes, although several attempts have been made to do so.

(*d*) Possibly in relation to the process of membrane digestion, mentioned above, it is likely that in some cestodes (*e.g.*, *Echinococcus granulosus*) the scolex acts as an organ of nutrition (*i.e.*, has a placental-like function) as well as of attachment.

(*e*) The growth of cestodes is unusual in that it results in the continuous production of *embryonic* tissue (*i.e.* the strobila) which differentiates into a strring of sexual individuals which eventually producef fertile eggs. Certain species (*E. granulosus, Taenia serialis,* and probably *T. crassiceps*) are capable of dedifferentiation and, under certain conditions, cystic forms can give rise to either strobilar or cystic forms depending on which stimulus has been applied.

The physiology of this group of parasitic helminths has been more extensively studied than any other and a large literature exists. Various aspects of cestode physiology have been reviewed by Read & Simmons (1963), von Brand (1966) and Smyth (1969). The majority of experimental work has been carried out on those species which are comparatively easily maintained in the laboratory, *e.g.*, *Hymenolepis* spp;

Taenia spp. and *Echinococcus granulosus* amongst the Cyclophyllidea and *Spirometra mansonoides, Diphyllobothrium* spp., *Schistocephalus solidus* and *Ligula intestinalis* amongst the Pseudophyllidea. With rare exceptions, the physiology of the remaining cestode orders has been little investigated and this area offers much scope for future work.

Chemical composition

General comment

As emphasized elsewhere, data on the chemical analysis of parasitic worms are of little value unless qualifying information is provided on the nutritional state of the host at the time of autopsy. The carbohydrate content especially is liable to fluctuate, and as this affects the dry weight, changes in other constituents, such as protein when expressed as a percentage, may be more apparent than real. The lipid content is also directly related to that of the host. Moreover, the chemical composition may vary wity the 'strain' of both parasite and host. It can also vary in specimens from different regions of the intestine.

The data must also be accepted with caution on technical grounds, because some of the older analytical methods have been shown, by more modern workers, to be unreliable. Some of the problems inherent in analysis have been reviewed by von Brand (1960) and Hopkins (1960).

Major chemical constituents. The proportions of the main tissue constituents - protein, lipid and carbohydrate - show a somewhat different pattern from most other invertebrates in that the carbohydrate content tends to be high and the protein relatively low. Larval cestodes are particularly rich in glycogen which may reach astounding levels (>50 per cent in the plerocercoids of *Ligula* or *Diphyllobothrium*) Table 58.

Table 14.1. Analyses of Cestodes, expressed as a percentage of dry wight. For figures to be of value, the nutritional condition of the host and the analytical method used, should be known.

Species	Dry wt. as % of fresh wt.	Glycogen	Lipid	Protein	Inorganic substances	Stage	References*
Anoplocephala magna (*Taenia plicata*)	27.5	6	33.1	—	1.22	adult	von Brand (1952)
Calliobothrium verticillatum	27.6	—	—	—	—	adult	Read *et al*. (1959)
Cittotaenia perplexa	27.6	—	—	20.60	—	adult	Campbell (1960b)
Diphyllobothrium latum	—	17.9	—	—	—	plerocercoid	Markov (1939)
Diphyllobothrium latum	9.0	20	16.6	60	4.8	adult	von Brand (1952)
Diphyllobothrium sp.	29.9	31.5	—	41	—	plerocercoid	Archer & Hopkins (1958b)
Diphyllobothrium sp.	30.8	36.2	—	48	—	adult	Archer & Hopkins (1958b)
Diphyllobothrium dendriticum	27.0	36.2	—	—	—	plerocercoid	Reuter (1967)
Dipylidium caninum	20.4	—	—	—	—	adult	von Brand (1952)
Echinococcus granulosus	14.8	19.8	13.6	62.5	13.5	protoscoleces	Agosin *et al*. (1957)

Table Continued

Eubothrium rugosum	—	22.8	—	—	—	adult	von Brand (1952)
Hydatigera (Taenia) taeniaeformis	22.3	19.7	4.2	27.1	29.0	strobilocercus (mice)	von Brand & Bowman (1961)
Hydatigera (Taenia) taeniaeformis	28	43.3	5.3	26.3	18.1	strobilocercus (mice)	Hopkins (1960), Hopkins & Hutchison (1960)
Hydatigera (Taenia) taeniaeformis	29.5	24.9	3.1	28.9	28.4	strobilocercus (rats)	von Brand & Bowman (1961)
Hydatigera (Taenia) taeniaeformis	20	—	6.9	40.6	—	adult	Hopkins (1960), Hopkins & Hutchison(1958)
Hydatigera (Taenia) taeniaeformis	26.7	23.2	6.3	45.0	22.0	adult	von Brand & Bowman (1961)
Hymenolepis diminuta	22.4	45.7	20.1	31.0	—	adult	Fairbairn *et al.* (1961)
Hymenolepis citelli	—	—	16.1	—	—	adult	Harrington (1965)
Ligula intestinalis	29.0	38-52	35-45	—	—	plerocercoid	Markov (1939)
Moniezia expansa	9.2-11.0	24-32	30.1	21.8	10.5	adult	von Brand (1952, 1966); Campbell (1960b)
Multiceps multiceps (Coenurus cerebralis)	25.3	—	—	—	27.4	scolex	von Brand (1952)
Multiceps multiceps	12.4	—	—	—	4.1	membranes	von Brand (1952)

Table Continued

Raillietina cesticillus	20.5	31.8	15.5	36.4	11.5	adult	Reid (1942)
Schistocephalus solidus	31.8	50.9	—	35.8	5.8	plerocercoid	Hopkins (1950)
Schistocephalus solidus	38	28.0	—	—	—	adult	Hopkins (1960)
Spirometra mansonoides	—	—	16	—	—	plerocercoid	Meyer *et al.* (1960)
Spirometra mansonoides	—	—	24	—	—	adult	Meyer *et al.* (1966)
Taenia crassiceps	20.0	27.5	—	—	—	larva	Taylor *et al.* (1966)
Taenia hydatigena (*marginata*)	16-26	28	4.9	—	—	adult	Featherston (1969) von Brand (1952)
Taenia saginata	12.2	48.8	11.2	32.0	5.3	adult	von Brand (1952) Machnicka-Roguska (1961)
Taenia solium	8.7	25.4	16.2	46	6.4	adult	von Brand (1952)
Thysanosoma actinioides	16.3	—	—	29.0	—	adult	Campbell (1960b)
Triaenophorus nodulosus	—	13.8	—	—	—	adult	von Brand (1952)

Calcareous corpuscles. Cestodes (and trematodes) often contain enormous numbers of curious bodies termed *calcareous corpuscles,* made up of an organic base together with inorganic material. They vary much in size; in some species being very large - 16 - 32 µm (*Spirometra mansonoides, Echinococcus granulosus*), in most species, they may be as small as 12 µm. The organic material contains RNA, DNA, proteins, polysaccharides and alkaline phosphatase. The inorganic materials are mainly calcium, magnesium, phosphorus and carbon dioxide with traces of other metals, but these constituents, especially phosphate, can vary considerably in relation to metabolic conditions and to origin. Ultrastructural studies have shown that in *T. taeniaeformis,* the corpuscles are formed intracellularly and a single corpuscle is formed in one cell, which is apparently destroyed in the process.

The role of calcareous corpuscles is not clear but it has been speculated that they act as major reserves for (*a*) phosphates and for (*b*) other organic ions together with carbon dioxide. These materials may be called on suddenly - phosphates for phosphorylation, the ions to act as enzyme catalysts and the carbon dioxide for CO_2-fixation - such as when a larva enters a host intestine and immediately requires a large amount of energy for its establishment process; such processes frequently involve muscular attachment.

Carbohydrates

As with trematodes and nematodes the main carbohydrate reserve in cestodes is 'glycogen', which is a typical energy reserve of helminths inhabiting biotopes with a low oxygen tension. The properties of cestode 'glycogens' have not been well characterize and their physical characteristics vary somewhat with their methods of extraction. A few workers have distinguished between *lyo-glycogen* and *desmo*-glycogen (precipitated by TCA) but this distinction has not been widely made. Glycogen is distributed largely in the arenchyma and muscles.

The data presented in Table 58 shows that the total glycogen content of cestodes may vary in the range 6-48 per cent of the dry weight. Larval cestodes generally show a higher and more constant glycogen content than the corresponding adults; this may reflect the more stable intermediate host environment - usually the coelomic cavity or tissues. The glycogen content can show marked variation within any species in relation to the nutrition of the host and the degree of maturity of the strobila (orpart of it). Thus, in *Raillietina cesticillus,* after 20 hours of host starvation, the glycogen level falls from 4.6 to 0.25 per cent (wet weight). Again, the glycogen, protein and lipid content of *Hymenolepis diminuta* varies very greatly with the age of the worm and the degree of maturity.

It is thus clear that the glycogen content of a strobila, or a particular part of it, may depend on a number of factors, only some of which are understood. Any quoted figures for the carbohydrate content should thus be accepted with caution, unless the previous metabolic history is known, and the region of the strobila defined.

Protein

Like the glycogen content, the figures are meaningful only against the age, degree of maturation and previous metabolic history of the worm. Cestodes are unusual in the animal kingdom in that in many species the protein content is less than the sum of the carbohydrate and fat. This appears to be related to the high level of glycogen reserves in some species.

Figures for protein content are based on multiplying the total nitrogen content by 6.25. It must be emphasized, however, that this figure is based on average analysis of *mammalian* tissues and may not hold for invertebrate tissues generally. There is evidence that some species of cestodes contain significant quantities of non-protein nitrogen so that early analyses for 'protein' in cestodes should be accepted with some reservation.

Fractionation of tissue proteins has been carried out in a few species, and in some, protein has been found to be conjugated with substances such as bile acids, glycogen or cerbrosides.

The chief structual proteins in cestodes are the sulphur-containing keratin (hooks and embryophores) and sclerotin (egg-capsules); the latter has already been considered.

Analysis (both in the scolex and in the oncosphere) of the keratin in hooks and embryophore of several cyclophyllid species suggests that its structure and composition do not differ substantially from that of vertebrate keratin. An exceptionally detailed examination of the physical and chemical characteristics of the keratin of the hooks in the oncosphere of *H. microstoma* and strobilocercus of *H. taeniaeformis* has been made by Dvorak (1969*a*, *b*).

The amino acids in the tissues of a number of species have been examined but, in general, no unusual information has emerged.

Table 14.2. The chemical components of the immature, mature and gravid regions of the strobila of *Hymenolepis diminuta* from 5 to 16 days postinfection, expressed as a percentage of dry weight per worm.

Age of worms (days)	Total dry wt./ worm(mg)	*Non-Protein nitrogen*			*Protein nitrogen*			*Lyo-glycogen*			*Desmo-glycogen*			*Total Lipid*		
		Immature	Mature	Gravid	Immature	Mature	Gravid	Immature	Mature	Gravid	Immature	Mature	Gravid	Immature	Mature	Gravid
5	0.151	1.76	—	—	4.22	—	—	12.62	—	—	2.38	—	—	23.63	—	—
7	2.88	0.70	0.73	—	1.66	2.88	—	9.58	24.98	—	1.22	2.83	—	4.90	15.53	—
9	11.02	0.22	0.75	—	0.5	3.17	—	3.07	39.86	—	0.39	5.47	—	1.81	19.07	—
11	30.17	0.07	1.24	—	0.18	3.04	—	1.19	39.59	—	0.15	3.19	—	0.61	21.57	—
13	60.39	0.05	1.16	0.09	0.12	2.49	0.26	0.97	32.33	3.12	0.13	3.07	0.44	0.52	16.82	2.78
16	82.20	0.20	1.06	0.46	0.11	1.74	1.84	0.59	21.41	21.51	0.11	2.73	2.74	0.47	11.33	11.72

Table 14.3. Chemical composition of hooks of *Hydatigera taeniaeformis* (Strobilocerci) and *Echinococcus granulosus* (Protoscoleces).

	% composition	
Substance	*H. taeniaeformis*	*E. granulosus*
Carbon	41.81	47.48
Hydrogen	6.13	6.24
Nitrogen	15.43	15.63
Sulphur	11.57	5.58
Phosphorus	0.34	0.49
Ash	1.04	1.38

Lipid

The lipid content of cestodes clearly varies considerably from species to species. Moreover, the lipid content may vary considerably even in the same species grown in the same host species fed on different diets. This is related to the fact that host intestinal fatty acids and sterols are *directly* (*i.e.*, without further digestion) absorbed by cestodes and the qualitative composition of cestode lipids generally follows that of the host. The lipid content can also vary with the age of the proglottid.

Not many detailed fractionations of cestode lipids have been made. Some older methods of analysis for some lipid components have been replaced by methods, such as gas chromatography for fatty acids, which are now acepted as being more reliable; earlier data may thus need re-evaluation.

All the usual lipid fractions have been identified in cestodes - phospholipids, glycolipids, fatty acids, glycerol and unsaponifiable material. The unsaponifiable material, which is largely cholesterol, may account for more than 20 per cent of the total lipids. Perhaps the most detailed lipid analyses are those for *H. diminuta*.

Table 14.4 Changes in glycogen content of adult and larval *Taenia taeniaeformis* after incubation under various conditions.*

State	Condition	Substrate	No.of expts.	Consumption of Glucose	Consumption of Glycerol	Percentage glycogen content of fresh tissues Initial	Percentage glycogen content of fresh tissues Final	Change in glycogen content µg/ 100 mg/ 1 hour
Adult†	Anaerobic	Glucose	15	1.90±0.17	—	2.51±0.27	2.32±0.24	— 63
Adult †	Aerobic	Glucose	13	1.37± 0.24	—	3.52± 0.45	3.43± 0.43	— 30
Adult	Anaerobic	Mixture	36	1.72± 0.15	3.83± 0.22	4.11± 0.40	3.91± 0.36	— 100
Adult	Aerobic	Mixture	35	1.49±0.19	4.91±0.31	4.69±0.42	471±0.40	+ 10
Adult	Anaerobic	Glycerol	19	—	3.30±0.33	4.60± 0.48	4.47±0.47	— 65
Adult	Aerobic	Glycerol	19	—	4.62± 0.40	5.57± 0.58	5.42± 0.53	—75
Larva‡	Anaerobic	Glucose	12	0"51± 0.05	—	5.58± 0.18	5.55± 0.19	— 15
Larva	Aerobic	Glucose	12	0.36± 0.05	—	5.58± 0.18	5.57± 0.17	— 5
Larva	Anaerobic	Mixture	17	1.30±0.14	2.44±0.13	5.90±0.22	5.85±0.18	— 25
Larva	Aerobic	Mixture	17	1.39±0.10	4.07± 0.12	5.90± 0.22	6.48± 0.16	± 290
Larve	Anaerobic	Glycerol	12	—	2.00±0.37	6.87±0.31	6.84± 0.31	-15
Larva	Aerobic	Glycerol	12	—	4.33+0.51	6.87± 0.31	6.93± 0.45	± 30

* The substrate concentration was as follows: glucose 11 mM, glycerol 22mM, or a mxture of glucose 11mM and glycerol 22mM. All consumption figures are µmoles/ 100 mg fresh weight/1 hour.

† In these two series, taken from von Brand *et al.* (1964) the experiments lasted 3 hours, in all others 2 hours.

‡ In experiments with larvae, the same initial glycogen values are given, because in each individual experiment the larvae were divided equally for aerobic or anaerobic incubation, and for determination of the initial glycogen values.

Carbohydrate Metabolism

From a series of studies carried out on both cyclophyllidean and pseudophyllidean cestodes it is possible to build up a general picture of the carbohydrate metabolism. These experiments fall into two groups: (*a*) *in vivo* experiments: carried out by feeding the host on experimental diets and observing the effect on the chemical composition of a worm; (*b*) *in vitro* experiments: concerned largely with the kind and quantity of carbohydrate utilized and the end-products of carbohydrate metabolism. The carbohydrate metabolism dominates other aspects of metabolism and this aspect of cestode physiology has been extensively studied. Only a broad review will be presented here.

In vivo carbohydrate utilization

It has long been recognized that carbohydrate must be present in the host diet for the normal establishment and growth of cestodes. If rats already infected with *H. diminuta* are placed on carbohydrate-deficient diets or one deficient in sucrose, there is a decrease in the number and size of strobila and the numbers of gravid and mature proglottids. A number of other experiments have been carried out on the effect of restricting the carbohydrate in the host diet to other substances such as glucose, galactose, maltose and dextrin. Care must be taken with such experiments, however, or diets with different calorific intakes may result. Evidence from experiments, in which the carbohydrate content of the diet was varied calorifically, appear to indicate clearly that the factor limiting growth of cestodes *in vivo* is the actual availability of glucose in the intestinal lumen. With this species, taking glucose as a standard, dry weight of worms from rats fed on a maltose diet showed a decrease of 3.9 per cent, dextrin a decrease of 11.8 per cent and sucrose a decrease of 22.8 per cent. In these experiments galactose failed to support worm growth although some limited growth has been obtained by other workers. These results are in keeping with the sequence of events after the carbohydrates, mentioned above, are ingested by a mammal, *i.e.*, both sucrose and dextrin require further enzymatic degradation before being available (as glucose) for worm growth. Starch, which also produces glucose on hydrolysis in the intestine, supports cestode growth *in vivo*.

In vitro carbohydrate utilization

Extensive experimental work on the *in vitro* utilization of carbohydrates by cestodes has been carried out. There is increasing evidence that the conditions under which some of these early experiments were carried out were not sufficiently close to the *in vivo* situation to be reliable. Hence, some early work may need re-evaluation. For example, the PCO_2 level can have a profound effect on the utilization of glucose and synthesis of glycogen. This is probably related to CO_2-fixation and its role in the production of fumarate used for the reoxidation of NADH under anaerobic conditions. The presence or absence of ions (such as Na^+) can also influence uptake of glucose

Most species studied have been found to utilize glucose and many use galatose (contrast *in vivo*-see above); a few species can utilize maltose and one species (*Cittotaenia*), sucrose. The mechanism of uptake for all kinds of carbohdrates is not known in detail but experiments using ^{14}C have already shown that the uptake of glucose is by active transport and not by simple diffusion. Such a mechanism would permit the uptake of glucose from low levels in the gut. Utilization of glucose in many species results in a build up of glycogen in the parenchyma. In larval *Taenia taeniaeformis*, however, significant glycogen synthesis only took place under aerobic conditions when *both* glucose *and* glycerol were present in the media. When either of these were present alone some sparing of tissue glycogen utilization took place, but no glycogen synthesis occurred. The significance of the interaction between the metabolisms of glucose and glycerol is not yet clear but there is evidence that the α-glycerophosphate (Bücher) cycle is involved. Glyceol is known to stimulate respiration in other species, e.g. *Mesocestoides corti*.

It is interesting to note also that glucose uptake and glycogenesis are stimulated *in vitro* by the presence of insulin in the medium; hence the endocrine balance of the host may have an influence on the carbohydrate metabolism of a cestode.

Intermediary carbohydrate metabolism

Most of the work on this has been carried out on the genera *Hymenolepis, Echinococcus, Taenia* and *Monezia*. It generally can be said that the three main metabolic pathways best known in animal tissues, also occur in cestodes although often in a modified form. These are: (*a*) Embden-Meyerhof pathway (glycolysis) (*b*) Krebs cycle or citric acid cycle (oxidative decarboxylation). (*c*) oxidative phosphorylation (the electron transport system: the cytochrome system).

Lesser well-known pathways, the pentose-phosphate pathway (the hexose-mono-phosphate shunt) and the α-glycerophosphate cycle also occur in a few species.

Table 14.5. Embden-Meyerhof pathway of anaerobic glycolysis*

1. Glycogen + nH_3PO_4 $\xrightleftharpoons{\alpha\text{-glucan phosphorylase}}$ *n* glucose 1-phosphate
2. Glucose 1-phosphate + glucose 1, 6-diphosphate $\xrightleftharpoons{\text{Phosphoglucomutase}}$ glucose 1, 6-diphosphate + glucose 6-phosphate

Alternatively, starting with glucose, 1 and 2 are replaced by 3.

3. Glucose + ATP $\xrightleftharpoons{\text{Hexokinase}}$ glucose 6-phosphate + ADP
4. Glucose 6-phosphate $\xrightleftharpoons{\text{Glucosephosphate isomerase}}$ fructose 6-phosphate
5. Fructose 6-phosphate + ATP $\xrightleftharpoons{\text{Phosphofructokinase}}$ fructose 1, 6-diphosphate + ADP
6. Fructose, 1, 6-diphosphate $\xrightleftharpoons{\text{Aldolase}}$ dihydroxyacetone phosphate + D-glyceraldehyde 3-phosphate
7. Dihydroxyacetone phosphate $\xrightleftharpoons{\text{Triosephosphate isomerase}}$ D-glyceraldehyde 3-phosphate

8. D-glyceraldehyde 3-phosphate + dehydrogenase + NAD $\underset{\text{dehydrogenase}}{\overset{\text{Glyceraldehydephosphate}}{\rightleftharpoons}}$ 3-phospho-D-glyceric acid-enzyme complex + $NADH_2$

9. 3-phospho-D-glyceric acid-enzyme complex + $H_3 PO_4$ $\underset{\text{dehydrogenase}}{\overset{\text{Glyceraldehydephosphate}}{\rightleftharpoons}}$ 1, 3-diphospho-D-glycerate + dehydrogenase

10. 1, 3-diphospho-D-glycerate + ADP $\overset{\text{Phosphoglycerate kinase}}{\rightleftharpoons}$ 3-phospho-D-glycerate + ATP

11. 3-phospho-D-glycerate + 2, 3-diphospho-D-glycerate $\overset{\text{Phosphoglyceromutase}}{\rightleftharpoons}$ 2, 3-diphospho-D-glycerate + 2-phospho-D-glycerate

12. 2-phospho-D-glycerate $\overset{\text{Phosphopyruvate hydratase}}{\rightleftharpoons}$ phosphoenol-pyruvate + H_2O

13. Phosphoenol-pyruvvate + ADP $\overset{\text{Pyruvate kinase}}{\rightleftharpoons}$ pyruvate + ATP

If O_2 is absent, the $NADH_2$ from Reaction 8 is oxidized in muscle by Reaction 14 and the end-product is lactate:

14. Pyruvate $\overset{\text{Lactate dehydrogenase}}{\rightleftharpoons}$ L-lactate

Sum of reactions 3-14; glucose + 2 ADP + 2 $H_3 PO_4$ = 2 lactic acid + 2 ATP + $2H_2 O$

* Enzyme terminology as recommended by the Enzyme Commission.

Table 14.6. The citric acid or Krebs cycle*

1. Acetyl-CoA + oxaloacetate $\overset{\text{Citrate synthase}}{\rightleftharpoons}$ citrate + CoA
2. Citrate $\overset{\text{Aconitate hydratase}}{\rightleftharpoons}$ *cis*-aconitate + H_2O
3. *cis*-aconitate + H_2O $\overset{\text{Aconitate hydratase}}{\rightleftharpoons}$ Ls-isocitrate
4. Ls-isocitrate + NADP $\overset{\text{Isocitrate dehydrogenase}}{\rightleftharpoons}$ oxalosuccinate + $NADPH_2$
5. $NADPH_2$ + ½ O_2 $\underset{\text{and cytochrome oxidase}}{\overset{NADPH_2\text{ cytochrome } c \text{ reductase}}{\rightleftharpoons}}$ NADP + H_2O
6. Oxalosuccinate $\overset{\text{Isocitrate dehydrogenase}}{\rightleftharpoons}$ 2-oxoglutarate + CO_2

7. 2-oxoglutarate + oxidized lipoate $\xrightleftharpoons[\text{pyrophosphate}]{\text{Oxoglutarate dehydrogenase with thiamine}}$ 6-S-succinyl-hydrolipoate

8. 6-S-succinyl-hydrolipoate $\xrightleftharpoons{\text{Lipoate acetyltransferase}}$ succinly-CoA + dihydrolipoate

9. Succinyl-CoA + H_2O $\xrightleftharpoons{\text{Succinyl-CoA hydrolase}}$ succinate + CoA

10. Succinate + ½ O_2 $\xrightleftharpoons{\text{Succinate dehydrogenase and cytochrome oxidase}}$ fumarate + H_2O

11. Fumarate + H_2O $\xrightleftharpoons{\text{Fumarate hydratase}}$ L-malate

12. L-malate + NAD $\xrightleftharpoons{\text{Malate dehydrogenase}}$ oxaloacetate + $NADH_2$

Sum of reactions: Acetyl-CoA + $2O_2$ = $2CO_2$ + H_2O + CoA

* Enzyme terminology as recommended by the Enzyme Commission.

It is beyond the scope of this book to discuss these pathways in detail, but the extent to which they are utilized by cestodes is summarized, briefly, below.

The Embden-Meyerhof pathway. A number of studies-especially on *H. diminuta* and *E. granulosus*- have demonstrated most, but not all, of the enzymes of this pathway. Thus, myokinase, phosphorylase, phosphohexomutase, hexokinase, aldolase, phosphoglyceraldehyde dehydrogenase and lactic dehydrogenase have been detected.

Although all the enzymes of the Embden-Meyerhof sequence have not been identified, the results strongly suggest that this pathway is followed in these species and is probably the general pattern in cestodes.

The Krebs (citric acid) cycle. This cycle is essentially the metabolic centre at which the carbohydrate, fat and protein metabolisms make contact and are enabled to exchange intermediate compounds.

In most cestodes, evidence for the occurrence of the enzymes involved in the Krebs cycle is incomplete. Although the existence of many enzymes of the cycle, such as succinic dehydrogenase, is well established, it yet remains to be demonstrated (with a few exceptions) that the cycle is actually operative in the intact organism (*i.e.,* not just in homogenates).

A species in which the Krebs cycle has been studied in great detail is *Echinococus granulosus* in which it has been shown that a number of Krebs cycle intermediates were oxidized not only in homogenates but also in intact protoscoleces. There seems no doubt that in *Echinococcus* a Krebs cycle, in whole or in part, operates. Only very incomplete evidfence is available for the existence of the cycle in other genera such as *Taenia, Moniezia* and *Hymenolepis.*

The pentose phosphate pathway. Although the Embden-Meyerhof pathway is the main conversion route for carbohydrates in some species of cestodes a small proportion of carbohydrate may be utilized via the pentose phosphate pathway. In *Echinococcus,* although 60 per cent of the glucose utilized is metabolized via the Embden-Meyerhof pathway, some 20 per cent may be utilized by the pentose phosphate pathways. Evidence for the pentose phosphate pathway in other species is very incomplete.

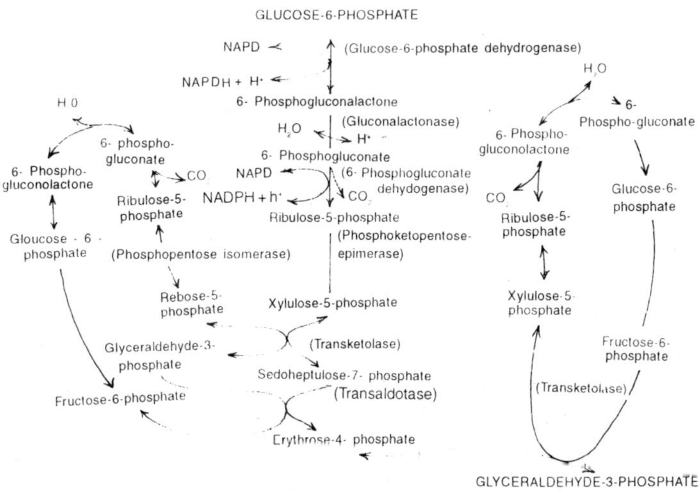

Fig. 14.2 : Pentose phosphate pathway.

CO_2-fixation in cestodes. Although CO_2-fixation was not thought to be common in animals, experimental work provides increaseing evidence that the process is more common in parasitic organisms than formerly thought, and occurs extensively in cestodes.

In terms of the host-parasite relationship it is clearly economical for an intestinal parasite to utilize carbon dioxide because its use has two advantages, (*a*) carbon dioxide is a waste-product of the host metabolism and thus freely available and (*b*) it penetrates cells very readily and thus can be rapidly utilized by an intestinal parasite.

The most active CO_2-fixing enzyme appears to be phosphoenolpyruvic acid carboxylase according to the equation:

Phosphenol pyruvic acid (PEP) + CO_2 + guanosine diphosphate (GDP) = oxaloacetic acid (OAA) + guanosine triphosphate (GTP)

The main metabolic end-product in *Echinococcus* is succinic acid (see below). This is often indicative that CO_2-fixation is taking place and following a pathway from pyruvic to succinic acid according to the equation:

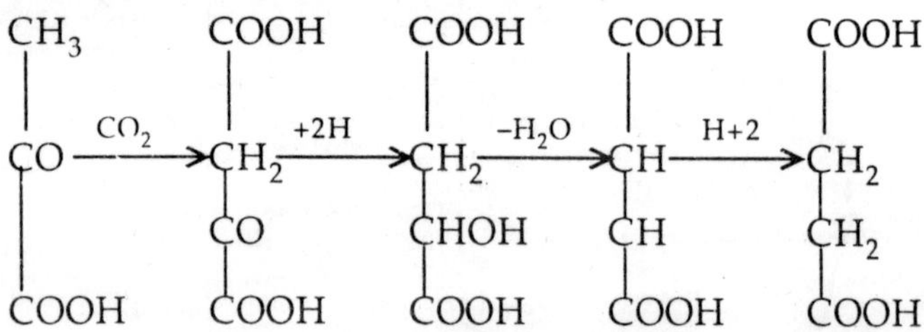

pyruvic acid — oxalo-acetic acid — malic acid — fumaric acid — succinic acid

Note that in some nematodes (*e.g., Ascaris,* p. 358) the succinic acid is further decarboxylated and reduced to other organic acids.

It is known that fumarate can act as a hydrogen acceptor for the reoxidation of NADH, a substance involved in glycogenesis and glycolysis. Thus the fumarate is essentially substituting for oxygen in terminal respiration in an anaerobic environment. This would explain the fact that in certain cestodes (e.g. *H. diminuta*) the presence of carbon dioxide has a marked effect on the synthesis of glycogen and utilization of glucose (Fairbairn *et al.*, 1961). This stimulating effect of carbon dioxide suggests that early work on carbohydrate metabolism carried out in the absence of carbon dioxide is open to re-evaluation.

End-products of carbohydrate metabolism

In considering this question, it is important to note that the nature of the end-products is greatly influenced by the PCO_2 and PO_2 of the gas phase. Thus under aerobic conditions, the majority of cestodes studied (e.g. *Echinococcus granulosus; Hydatigera taeniaeformis; Moniezia expansa*) secrete largely lactic acid, but succinic acid under anaerobic conditions. Other organic acids and ethanol are also produced but in much smaller quantities. The situation in *H. diminuta* is not clear, for some workers have reported lactic acid as the main end-product, whereas others have reported mainly succinic and acetic acids. These differences are probably due to differences in the gas phase (often unstated) by different workers.

Respiration

Oxygen consumption

Although a number of studies on oxygen consumption of cestodes has been made, it is difficult to assess the significance of such data. This is largely because - with a few exceptions the oxygen tension at the establishment site is frequently not known and in some species, at least, the consumption is very dependent on the oxygen tension. Thus, in *Echinococcus,* at tensions >5 per cent oxygen, consumption is directly proportional to the tension, yet below the 5 per cent level the consumption ceases completely. Again, in larval

Mesocestoides corti, the respiration rate declined at a tension of 2 per cent oxygen.

Two general conclusions can be drawn from the data available.

(*a*) all species of cestodes examined have been found to utilize oxyten *in vitro* when available.

(*b*) consumption of glucose in the absence of oxygen (*i.e.*, fermentation) proceeds at almost the same rate under anaerobic conditions as under atmospheric oxygen.

As pointed out above, the fact that oxygen is consumed *in vitro* does not imply that oxygen is utilized *in vivo*, unless we hae evidence that a similar level of oxygen is available *in vivo*. Several species, such as *H. nana*, are capable of growing to maturity in the complete absence of oxygen. Again, a careful study of the development *in vitro* of *H. diminuta* by Roberts & Mong (1969) found that this species appeared to be 'completely facultative with respect to oxygen presence or absence, under the conditions of the experiments.' This ability to develop under aerobic or anaerobic conditions would clearly be of considerable adaptive value.

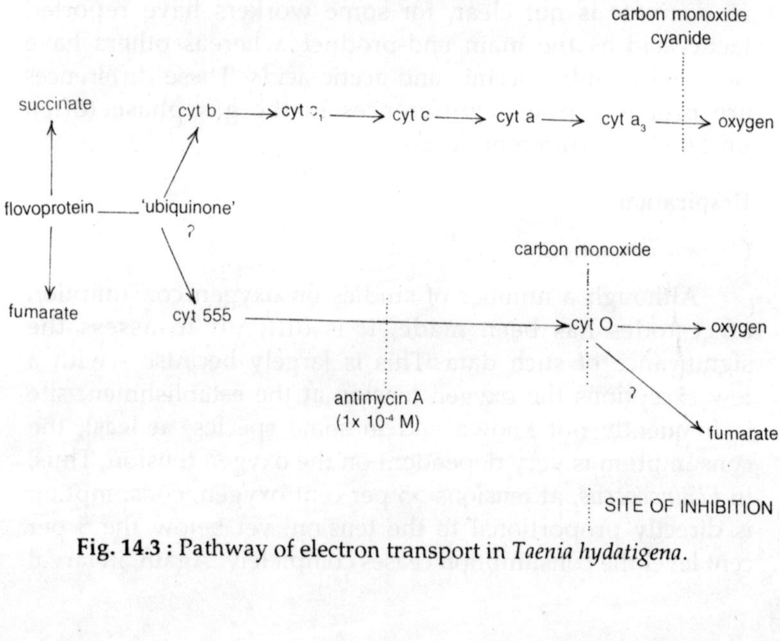

Fig. 14.3 : Pathway of electron transport in *Taenia hydatigena*.

Terminal respiration: electron transport

Although a substantial amount of work has been carried out on electron transport of cestodes, very little unequivocal information is at present available. This is related largely to the difficulties of obtaining cestode tissues in sufficient quantities and of making mitochrondrial preparations of sufficient purity. Progress in this field has been reviewed by Bryant (1970) and Smyth (1969). It is beyond the scope of this book to deal with this matter in great detail and only a brief account will be given here.

Probably two pathways operate: (*a*) the conventional one utilizing oxygen and the cytochrome system and (*b*) the pathway utilizing fumarate derived from carbon dioxide as indicated above.

A 'classical' cytochrome system of the mammalian type involving cytochrome oxidase can only operate efficiently at minimum oxygen tensions of the order of 5 mm. The very few observations made on the PO_2 of the gut suggest that at least on the surface of the mucosa tensions as high as 30 mm may be encountered. In the crypts of Lieberkühn, in which some species (e.g. *Echinococcus*) become embedded, tensions may be considerably higher, it is of interest, therefore, to note that this species appears to require oxygen for evagination.

There is evidence from cyanide inhibition of aerobic respiration and from spectra studies, that components of a 'classical' cytochrome system may be present in at least some species of cestodes. Thus, direct or indirect evidence suggests that a cytochrome system occurs in *D. latum, Echinococcus granulosus, Triaenophorus lucii, H. diminuta, Hydatigera taeniaeformis, Taenia hydatigena* and *Moniezia expansa*. The most complete study appears to have been carried out on *Moniezia expansa* and *T. hydatigena* in which the respiratory chain components contained two terminal oxidases one of the cytochrome *a* type and one containing cytochrome.

In *Moniezia*, cytochromes *b*, *c* and aa_3 have been positively identified and the experimental evidence indicates that they do participate in electron transport. This, together with the fact that cristate mitochondria have been isolated from this species indicates that *Moniezia* can undergo oxidative phosphorylation. The fact that an aerobic system can operate does not exclude anaerobic metabolic processes, because, as discussed earlier, in the absence of oxygen fumarate can act as an electron acceptor; in this situation succinic dehydrogenase acts as a fumarate reductase. The extent to which aerobic or anaerobic pathways are followed will clearly depend on the availability of oxygen in the particular intestinal site in question. A possible mechanism controlling these alternative pathways is described below.

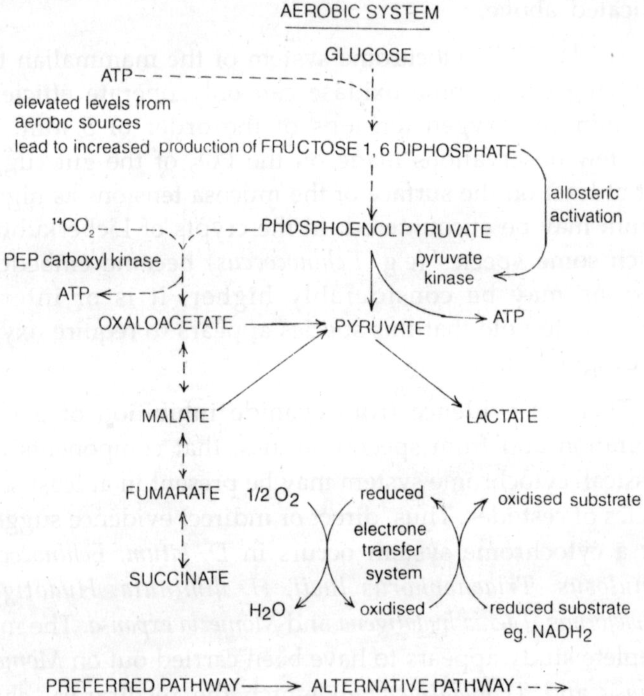

Fig 14.4 : Hypothesis to account for succinate and lactate production in *Moniezia expansa:* the possible *aerobic* pathway resulting in *lactate* production.

In *Moniezia*, hydrogen peroxide appears *in vitro*, apparently as a result of the action of cytochrome o, but it is questionable if this substance ever appears *in vivo*. The occurrence of a peroxidase in the tegument of *Moniezia* and *Hymenolepis* may suggest that this enzyme functions to remove peroxide - a substance which is highly toxic.

Metabolic regulation under aerobic and anaerobic conditions

Based largely on work with *Moniezia expansa*, Bryant (1972*a*, *b*) has produced an ingenious hypothesis to account for the switch from lactate production to succinate production in the absence of oxygen. This is based on the fact that phospho-enolpyruvate (PEP), a product of the Embden-Meyerhof glycolysis pathway, is normally converted to lactate in a reaction, the first step of which is mediated by pyruvate kinase. However, as already shown, in the presence of carbon dioxide, PEP is also capable of being acted on by the enzyme phosphoenolpyruvate carboxykinase (PEPC).

According to Bryant (1972*a*, *b*), under *aerobic* conditions, increased availabiltiy of ATP would lead to increased levels of fructose-1, 6-diphosphate (FDP). The latter has an activating effect on pyruvate kinase, thus leading to increased lactate production. Under *anaerobic* conditions i.e. with oxygen not available, an alternative electron transport system, with fumarate as the terminal acceptor becomes functional. This alternative system produces less ATP, which in turn leads to the lowering of concentration of FDP, thus removing the activation effect on the pyruvate kinase. The result is that the enzyme PEPC can now compete more favourably for PEP and increased succinate production takes place. Moreover, malate inhibits the action of FDP, so that activity along one pathway would inhibit the activity along another.

This ability to switch pathways, has two advantages. Firstly, the oxidation of glucose can continue in the absence of oxygen. Secondly, the NADH generated during the oxidation of glucose to PEP provides electrons for the reduction of fumarate to succinate instead of pyruvate to lactate. The occurrence of these alternative pathways thus makes cestodes exceptionally adaptable to environmental conditions. The efficiency of this adaptation in *Echinococcus* is evident from the fact that consumption of glycogen is almost the same under anaerobic as aerobic conditions.

Protein metabolism

General account

It is evident fromt he remarkable growth rate of some species that rapid protein synthesis takes place. It must be

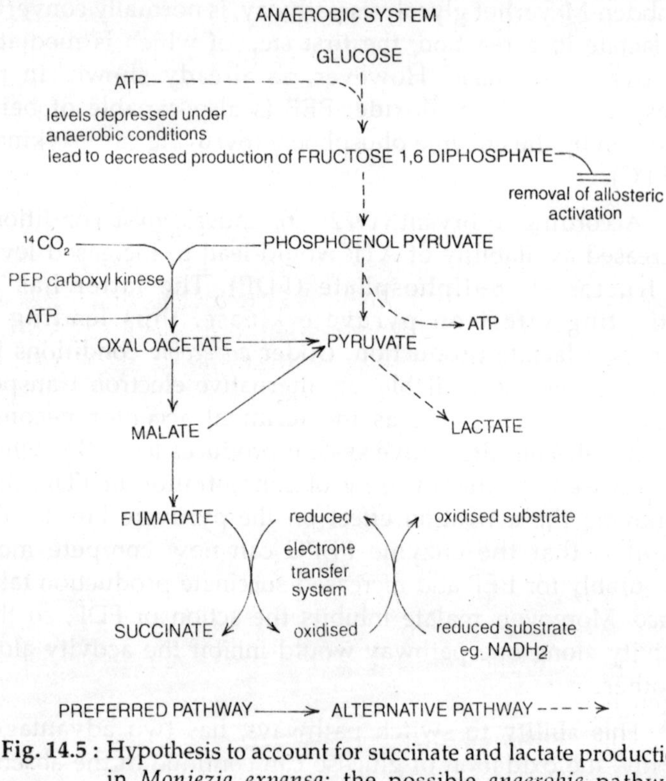

Fig. 14.5 : Hypothesis to account for succinate and lactate production in *Moniezia expansa*: the possible *anaerobic* pathway resulting in *succinate* production.

borne in mind, however, that cestodes are essentially made up of a string of embryos and hence different parts of a strobila may not only be synthesizing at a different rate, they may have different nutritional demands depending on the degree of maturation of the proglottids in question.

Our concept of the way in which protein or its breakdown products are taken up by an absorptive surface, such as the intestinal mucosa, has undergone drastic revision within recent years. It is now known that not only are small molecules, such as amiino acids, absorbed, but also larger ones such as dipeptides or even whole protein molecules may be ingested. The same processes may take place in the cestode tegument, although there is, as yet, no unequivocal evidence that pinocytosis - which would be necessary for the ingestion of large molecules - actually occurs in cestodes.

As discussed earlier the processes of absorption and/or digestion in cestodes are further complicated by the possibility that a process known as 'membrane' (= contact) digestion can occur. This concept developed by Ugolev (1965; 1968), envisages digestive enzymes *bound* in or on the plasma membrane covering the surface in question. There is some evidence from *in vitro* studies on *Echinococcus* that the scolex in this species can digest proteins at the host/parasite interface by this proces. There is also some indirect evidence, from experiments with *Hymenolepis* spp. and *Moniezia expansa* that starch hydrolysis in these species takes place by membrane digestion.

The occurrence of proteolytic enzymes in cestodes has not been much investigated, but they have been reported in *Taenia, Diphyllobothrium* and *Echinococcus.*

Amino acid uptake

The uptake of amino acids by cestodes has been extensively studied and it is now known to take place largely by active transport, although a diffusion component may also be involved. This may generally be recognized experimentally by the fact that the reciprocal of the velocity

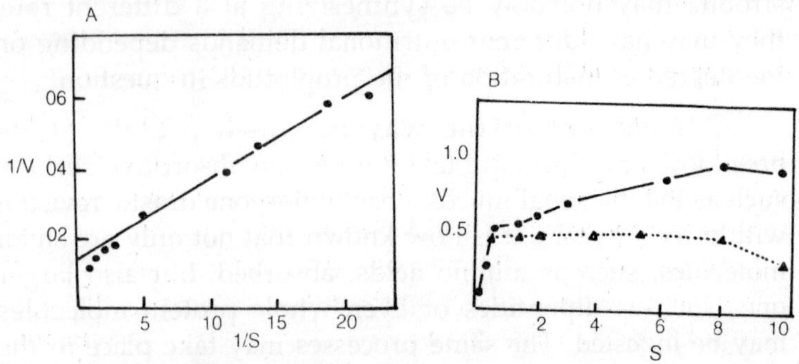

Fig. 14.6 : Uptake of proline by *Hymenolepis diminuta*. *A.* after 2 minutes' incubation. V, velocity in μM/hr/g ethanol extracted dry wt. S. proline concentration in mM. *B.* after 1 minute incubation. V. velocity in μM/min/g worm water. S. proline concentration in mM. Solid circles = observed values; open circles = values corrected for 'diffusion'.

of uptake $\left(\frac{I}{V}\right)$ against the reciprocal of the concentration $\left(\frac{I}{S}\right)$ (a Lineweaver-Burk plot) is a straight line (Fig. 14.6A). The evidence for this is based on:

(*a*) cestodes can accumulate amino acids aginst a gradient

(*b*) the uptake of one amino acid can inhibit the uptake of another (i.e. there is 'competition' for the carrier locus in the amino acid transport system).

Early work has been reviewed by Smyth (1969) and Read & Simmons (1963); later representative studies have been those carried out on *Taenia crassiceps; Hymenolepis citelli* and *H. diminuta*. In *H. diminuta,* a species which exhibits a diurnal migration pattern the uptake of methionine was found to differ in the anterior and the posterior parts of the worm. This can be accounted for by the fact that the level of dietary methionine in the anterior region reaches a peak 1.3 hours after feeding, whereas the level in the posterior region of the gut remains relatively stable. The latter fact is related to a homeostatic mechanism which maintains the molar ratios (but not necessarily the concentrations) relatively constant. The mechanism whereby this is achieved has already been discussed. The amino acid 'pools' (i.e. the free amino acids in the tissues) in cestodes of elasmobranch fish have been examined in detail by Simmons (1969). The pools showed remarkable consistency in some cases. For example, *Lacistorhynchus tenuis* and *Calliobothrium verticillatum,* which are parasites of the elasmobranch *Mustelus canis,* have almost identical amino acid pools. This suggests that the amino acid pool of the host intestine and the ability of the parasite to adapt to it, plays an important role in host specificity.

Intermediary metabolism

Little is known of the precise amino acid requirements of cestodes, i.e. which amino acids are 'essential' and cannot be synthesized by the host. *In vitro* studies should ultimately provide the answer to this question.

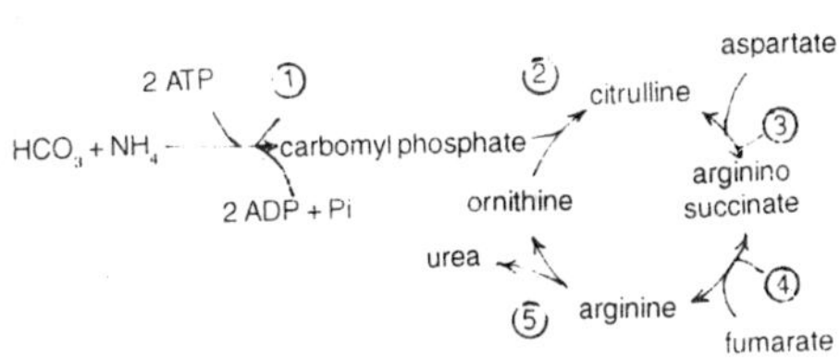

Fig. 14.7 : The (Krebs-Henseleit) ornithine cycle. Numbers refer to enzymes as follows: (1) Carbamyl phosphate synthetase (E.C.2.7.a). (2) Ornithine transcarbamylase (E.C.2.1.3.3). (3) Arginino-succinate synthetase (E.C.6.3.4.5). (4) Arginino-succinate lyase (E.C.4.3.2.1). (5) Arginase (E.C.3.5.3.1).

The chief mechanism of amino acid formation is by transmination and the α-ketoglutaric acid/glutamic acid and/or the pyruvic acid/alanine systems have been demonstrated in *Hymenolepis* spp., *Anoplocephala magna,* and *Raillietina cesticillus.* In *Hymenolepis,* some 17 amino acids failed to act as amino donors; this suggests that this species is largely metabolically dependent on the host for these materials, which would, of course, nomally be freely available in the intestine. It has already been pointed out above that the amino acid pool in some cestode species in the same host have almost identical amino acid pools.

Cestodes have also been shwon to take up purines and pyrimidines either by active transport or by facilitated diffustion there appear to be at least two binding sites on the purine-pyrimidine permease.

End-products of nitrogen metabolism

The intermediary metabolism of cestodes has been little studied. Although the nitrogenous compounds excroeted are known for some species, the metabolic pathways whereby these products are produced are poorly known. The end-products most commonly reported are urea, uric acid and ammonia but numerous other compounds have been found.

The production of urea points to the occurrence of an ornithine cycle and two important enzymes of this cycel ornithine transcarbamylase and arginase have been detected. Despite the presence of these enzymes, a *functional* ornithine cycle could not be detected in *M. expansa, D. caninum, T. pisiformis* and *E. granulosus.*

A number of cestodes have also been shown to excrete amino acids (at least *in vitro*) but it is not known whether or not this represent 'abnormal' leakage from the tissues. The same phenomenon occurs in trematodes and nematodes.

Lipid Metabolism

Althought he lipid metabolism of only a very few species has been investigated, a generalized idea of the pattern is beginning to emerge. Studies, based largely on experiments with labelled materials, carried out on *H. diminuta Spirometra mansonoides* and *R. cesticillus* have led to the following general conclusions about the lipid metabolism of cestodes:

(*a*) the ability to synthesize both fatty acids and sterols *de novo* has been completely lost. Cestodes are thus entirely metabolically dependent on the hosts for these materials; more complex lipids may be synthesized from these exogenous sources.

(*b*) the mechanism of uptake (at least for acetate) is by meditated transport at low levels, but diffusion also operates at higher concentrations.

(*c*) the sterol and fatty acids of the parasite and host are thus (not surprisingly!) qualitatively almost identical; they are not, however, quantitatively identical and some control mechanisms in lipid synthesis must exist in cestodes.

(*d*) The synthetic ability (utilizing exogenously provided fatty acids and sterols) retained, includes the ability to

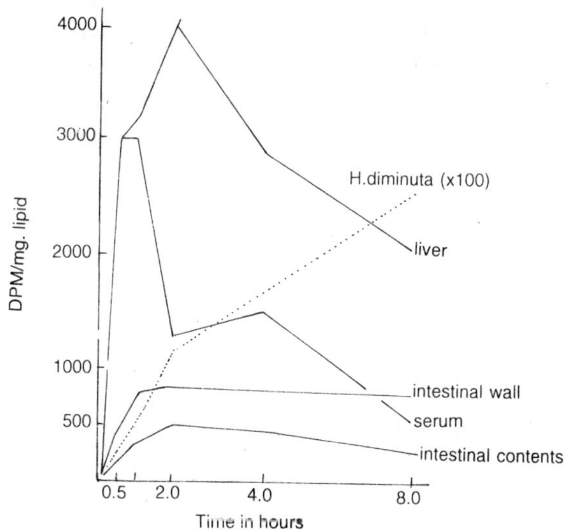

Fig. 14.8 : Incorporation of 8.2 x 10^6 dpm of intravenously injected U-^{14}C-palmitate into rat tissues and *H. diminuta*.

synthesize its own triglycerides, sterol esters, phospholipids and glycerolipids. Cholesterol, however, cannot be synthesized and must be absorbed from the host. Since labelled atoms from ^{14}-C-glucose and ^{32}p can be incorporated with these molecules it is evident that the synthetic ability retained includes that to synthesize the non-fatty portion of lipids.

(*e*) the synthesis of triglycerides, mentioned above probably takes place via the α-glycerophosphate pathway; it is envisaged that monoglycerides are hydrolysed after absorption probably at the surface of the worm and the heterogenous free fatty acids are converted to triglycerides by this pathway.

As emphasized above, most of the conclusions have been based on work on *H. diminuta* and *S. mansonoides*, and it would be interesting to extend similar investigations to other species.

From the above results, it is evident that the lipid content of any cestode will depend largely on the lipid content of the host diet. This has been strikingly demonstrated in *R. cesticillus*. In worms from chickens fed on a natural balanced

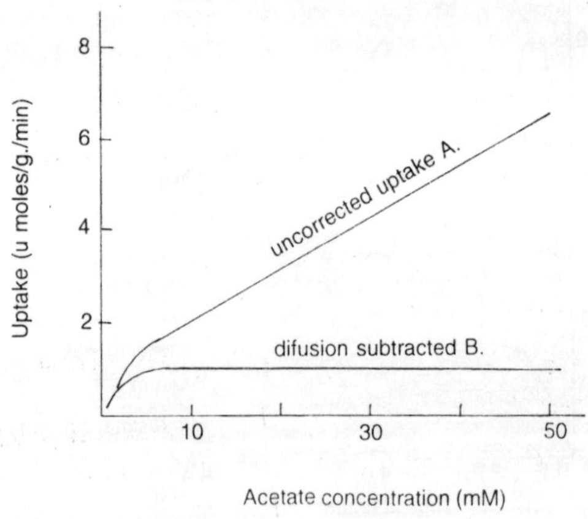

Fig. 14.9 : Uptake of ^{14}C-acetate by *H. diminuta*, as a function of acetate concentration.

diet, the lipid content was found to be 20.1-21.6 per cent. In worms from chickens fed on a low-fat diet the lipid content fell to 10.1 per cent compared to 29.6 per cent in worms rrom chickens fed on a similar per cent diet with added corn oil.

It is well known that the uptake of lipids is assisted by the presence of surface-active agents and it is interesting to speculate if the well-known ability of bile or bile salts (such as sodium taurocholate) to induce evagination of the scolex in cestodes could be due to its presence assisting the uptake of a lipid or sterol by the worm. This absorbed material could then act as a hormone or a trigger substance which could initiate evagination. It should also be noted, however, that bile itself is an important source of lipids and in bileless hosts (experimentally prepared) the uptake of fatty acids such as palmitate is greatly reduced.

Chapter—15

Physiological Chemistry

It is not yet possible to present a satisfactory general account of nematode biochemistry and physiolgy, for the simple reason that nearly all the information available relates to the animal-parasitic species only. As many of these species are highly specialized, it would be presumptuous to supose that details of their physiology would apply without serious modification to the less specialized plant-parasitic and freeliving forms. Nevertheless, the historical development of parasitology has been such that a reviewer attempting to deal with nematodes generally has no alternative but to accept the existing knowledge of physiological processes in the animal parasites as his basic material and from this to compare and predict, however modestly, whatever he feels may be applicable to the free-living and plant-parasitic forms.

Those who work with nematodes are in a particularly fortunate position to study the development and nature of parasitism. In no class of animals is the gradation of adaptations from the free-living to the facultagtive and oblighate modes of parasitism, whether plant or animal, more readily apparent or more widespread. It is therfore unfortunate that the freeliving species, which are the obvious starting point for physiological study, have received so little attention. However, there are indications, among which discussions in the present context are not the least, that biologists are now aware of the problem facing them. It is hardly necessary, therefore, to defend the attempt which is made in the following pages to present the subject of nematode physiology as a whole, nor to apologize for the repeated emphasis on the need for the broadly conceived studies of all nematodes, regardless of their ecological classification as free-living, plant-parasitic, or animal-parasitic species.

There are many approaches to any discussion of nematode physiology, not all of which are entirely suitable to a given occasion. In the present account of the plan which has been adopted deals first with the processes of digestion, absorption, and composition of the perienteric fluid. These lead naturally to the discussion of assimilation of the various foodstuffs, to the composition of the organs and tissues, to the biosynthesis of tissue components, when known, and to the consideration of organ and tissue function. A general account of respiration, both aerobic and anaerobic, is followed by a more detailed account of the metabolism of carbohydrates, fats, and proteins. This is succeeded by a discussion of excretion, water balance, permeability, and egg-hatching.

No attempt will be made to deal comprehensively with all aspects of nematode physiology, nor to make mention all the literature. Large and complex subjects such as immunity in animal hosts, resistance in plant hosts, and the effect of alterations in the host's physiology upon the establishment of infections have been regretfully excluded. References to most of the literature dealine with nematode physiology are to be found in reviews which are readily available.

DIGESTION AND ABSORPTION

Nematodes feed on a wide variety of substrates, including bacteria, fungi, algae, protozoa, other nematodes and soil microfauna, and living and decaying higher plants and animals. Whether their buccal armament features stylet, teeth, or lips, the ingestion of food appears to depend ultimately upon the powerful sucking action of the esophageal muscles. The food of many species, notably those in which the diameter of the stylet is usually too small to admit the passage even of bacterria, is primarily fluid, whereas in others such as the predators, the food is solid. Considering the complexity and diversity of the natural food, it is probable that in one or another species most of the known digestive enzymes are to be found, although there is little concrete evidence supporting this supposition. Similarly, it is not unreasonable to suppose that the nutritional requirements may very widely from one species to another, and here it is gratifying to know that some success has been achieved by studying nutrition in axenic cultures. This phase of nematode physiology will not be discussed as it is given special attention by Dougherty.

Digestion of foodstuffs by nematodes may be extracorporeal, intracorporeal, or a combination of the two, but is always, so far as is known, extracellular. There appears to be no authentic description of intracellular digestion following phagocytic engulfment of food particles, such as occurs in more primitive metazoans.

Extracorporeal digestion has long been thought to occur in nematodes whose food must enter the body by way of a hollow stylet whose dimensions are such that even bacteria would traverse it with difficulty. This stylet is of course employed in piercing the hard parts of the food material, and as a rule it receives the duct of the dorsal esophageal gland somewhere near its base. Students of plant-parasitic nematodes have grown accustomed to the obvious inference that the dorsal gland produces a secretion which is used to predigest the plant food. It is curious to find, therefore, that Linford (1937) appears to be the only investigator who has actually described the outward flow of the secretion, the organisms observed being *Aphelenchoides* sp. and *Meloidogyne* spp. In these species secretion is followed by extracorporeal liquifaction of the affected tissues, and then by ingestion of the liquid. Linford suggested also that the predaceous *Aphelenchoides* produced a paralysant which immobilized its prey while digestive secretions were pumped into it.

This early study by Linford did not include any observations on the actual nature of the secretion, and twenty years later not much has been added to our knowledge. Recent work by Myuge (1957a) and Zinovev (1957) has revealed that the bulb nematode (*Ditylenchus allii*), the potato tuber nematode (*D. destructor*) and the root-knot nematode (*Meloidogyne* spp.) all secrete protease and amylase into the distilled water in which they are suspended. *D. allii,* which is specific for the onion, does not penetrate the onion cell, the fluids of which are toxic to it, but feeds on the inter cellular substances, and may therefore be supposed to secrete protopectinases or pectinases which digest the middle lamella. This species secretes only 14% as much amylase as *D. destructor,* which feeds on potato starch. Moreover, the thermophilic *Meloidogyne* spp. secrete increasing amounts of enzymes as the temperature is raised to 24°, whereas the secretion of the non-thermophilic *D. allii* does not change. These observations of the Russian workers seem to provide the only clear evidence supporting the specific enzymic nature of the salivary secretions of plant-parasitic nematodes. In passing, however, it is worth noting that Tracey (1958) was able to detect both chitinase and cellulase in homogenates of three species of *Ditylenchus* (*D. destructor, D. dipsaci, D. myceliophogus*). Whether these enzymes play a part in the extracorporeal digestion of the cell walls of higher plants and fungi has not been shown. Nor is it possible to distinguish clearly between the secretions which may be responsible for tissue penetration and those which are directly concerned with the predigestion of foodstuffs. There is no reason to believe, in fact, that the dissolution of a cellulose cell wall in the process of nematode penetration into tissues does not result in the creation of soluble sugars which are of dietary value, and in the case of *D. allii* already mentioned, which digests the middle lamella and thus produces maceration of the onion cells without actually penetrating them, it is tempting to suppose that the products of protopectin digestion are used in part at least as nutrients.

This uncertainty regarding the function of extracorporeal digestion is even more striking when one turns to a consideration of the animal-parasitic nematodes. Many non-migratory parasites produce extensive cytolysis of the host tissues which has been thought to be due to secretions of the esophageal glands. A striking example is to be found in the dog kidneyworm, *Dioctophyme renale,* which has a large and highly dendritic dorsal gland opening for forward in the esophagus. Similarly, Thorson (1956) found that esophageal extrascts of the dog hookworm (*Ancylostoma caninum*) contain a protease, and that the sera of dogs refractory to further infection with hookworm stongly inhibit this protease. Thus, it appeared that the protease was part of the glandular secretion, and that it possessed antigenic activity. It is in the migratory larvae of various species, however, that the difficulty inherent in distinguishing between enzymes responsible for tissue penetration and predigestion of foodstuffs are clearest. Lewert and Lee (1954, 1956) have shown that several species of skinpenetrating filariform larvae cause definite changes in the acellular ground substance of the host dermis, and that these effects are associated chiefly with the production of a collagenase by the larvae. It is not difficult to conceive that the extracorporeal digestion of collagen which seems essential for effective penetration of the parasite, produces amino acids or soluble peptides which may also be eminently suitable as larval food. The whole subject of extracorporeal digestion, therefore, is one which will need much more study before its general features are clearly outlined.

Intracorporeal digestion is probably more common in nematodes than the extracorporeal type, and might be expected to occur even in those species in which a preliminary external solubilization of foods occurs. Many nematodes feed by ingestion of microorganisms, decaying organic matter, intestinal contents, or tissue particles. Saprophytic species, including the free-swimming intestinal parasites, may of course feed in part on predigested material, but there is good evidence that solid particles are also ingested and digested. Hobson (1948) has thorougly reviewed the evidence in the case of intestinal parasites. As for free-living saprophytes, McCoy (1929) showed that the larval stages of *Ancylostoma caninum,* which normally develops in the host feces, can be grown successfully in pure cultures of various coliform bacteria, on which they feed; and Overgaard Nielson (1949), in his extensive study of free-living nematodes observed that saprophytic species feed on the microorganisms associated with the decaying plant tissues, rather than on the tissues themselves. Thus, the existence of intracorporeal digestion in many, and perhaps most, nematodes seems assured.

The convincing observations which have just been cited have not yet been well-supported by detailed biochemical studies. Hirsch and Bretshneider (1937), and Janicki (1939) described granules in the foregut of *Ascaris lumbricoides* and *Dioctophyme renale* which were regarded as being secretory, although the discharge of the granules into the intestine was not seen. The first thorough study of digestion in nematcdes was made by Rogers (1941a), using the combined intestine and intestinal contents of two species, *Strongylus edentatus* and *A. lumbricoides.* Rogers found that amylase protease and lipase were present in both species, and that they were well adapted in their properties to the natural habitats of the two parasites, that is the horse caecum and the pig jejunum and ileum, respectively. The investigation of *Ascaris* digestive enzymes was later continued in more detail by Carpenter (1952), but with reference only two enzymes present in the esophagus and intestinal cells. Carpenter's experiments, which were exceptionally complete, revealed that the maximum enzyme activity occurred in the foregut, which also contained most of the socalled secretory granules. Most of the enzymes, which included an amylase; lipase, protease and four different peptidases, were present also in the esophagus and to a lesser extent in the mid-and hindgut. Savel (1955b) reported that more protease and peptidase activity were to be found in *Ascaris* intestine than in other tissues, and several workers, *e.g.* Rogers (1940a), have found that blood feeders digest hemoglobin, as shown by the formation of hematin in the intestinal lumen.

In summary, there can be no doubt that nematodes are likely to contain an array of digestive enzymes. Some of these may be secreted to the exterior by the dorsal esophageal gland. Less certainly, the esophageal glands may contribute secretions which pass to the intestine, where they are augmented by secretions from the foregut. In some nematodes, extracorporeal digestion is essential; in some it may be useful but unessential, and in many others it may play no part in the normal digestive processes, which are then entirely intacorporeal.

Considering the doubts which still exist concerning the mechanism of absorption of digested foodstuffs in higher animals, it is not surprising to find that absorption by nematodes is scarcely understodd at all. It sems certian, however, that absorption occurs by way of the intestine, and not through the cuticel. Intestinal cells of enmatodes are apt to contain a number of inclusions, some of which have been reported to consist of fats, carbohydrates, proteins, or minerals, which tend to disappear during starvation. It is not known whether such inclusions are synthesized in the intestine from absorbed foods. A number of workers, including Rogers (1947), Yamao (1951) and Chowdhury (1955) have pointed out that the nematode intestine contains an abundance of acid or alkaline phosphatases, which have often been thought to be essential for the absorption of simple sugars against a concentration gradient. This hypothesis has never been strongly supported by experimental evidence, and it may be significant that in several nematodes the tissues contain very little glucose, but contain instead considerable quantities of the disaccharide trehalose. As trehalose is not a common dietary carbohydrate there exists an as yet unexplored possibility that in nematodes, as in the desert locust *Schistocercagregaria,* the concentration gradient in the intestine normally favours the absorption of glucose, the absorbed glucose being efficiently converted to trehalose in intestinal or other tissues.

HEMOLYMPH

All nematode tissue are bathed in the perienteric fluid, or hemolymph, which fills the body cavity completely and is always under pressure owing to the tonicity of the longitudinal muscles of the body wall. Obviously, the hemolymph must be used to transport solutes from one tissue to another, whether these be the products of intestinal digestion and absorption, mobilized storage products, or metabolic wastes. On the other hand, hemolymph is essentially cell-free, and it would be surprising, in the light of modern knowledge, to find that an appreciable synthesis of substances such as proteins or fats occurred in it. While there is as yet no evidence on this point, it is nevertheless true that *Ascaris* hemolymph, which has been studied is some detail, is a very complex solution. Nearly all our present information concerning hemolymph stems from studies on *Ascaris* and *Parascaris,* and as it is improbable that knowledge concerning more normally sized nematodes can be secured in detail, its composition in *Ascaris* warrants close consideration.

In *Ascaris,* hemolymph is a clear, pink, slightly sticky solution, somewhat acid (pH 6.2 to 6.4), with a rather high osmotic pressure (equivalent to 0.192 M sodium chloride) which is, however, consistently lower than the osmotic pressure of the host gut fluids. This high osmotic pressure is the reflection of a large ash content (0.9%) in which the ions of sodium, potassium, calcium and magnesium predominate. There is much less chloride than is required to balance these cations, the anion deficiency being made up in large part by volatile and non volatile organic acids and possibly by bicarbonate. Although the resemblance between the ion content of *Ascaris* hemolymph and the host fluids is remarkable this should not, for reasons which will appear later, be regarded as a general feature of nematode hemolymph.

Proteins, carbohydrates, and fats are all present. Proteins, comprising 4.9% of the fresh weight and 66% of the total solids, consist primarily of albumins and gloulins which differ little in their amino acid composition from those of mammalian plasma. Hemoglobin is present in small amounts and several enzymes have been identified, among which are invertase, maltase, amylase, esterase, and protease. Non-protein nitrogen consists chiefly of free amino acids, peptides, urea and ammonia. As already mentioned, it is improbable that synthesis of proteins occurs in the hemolymph itself, although no evidence to show their origin in other organs or tissues has been reported. Possibly the intestine, whose cytology is complex may be the source of the hemolymph proteins.

Table 15.1 Ionic Composition of *Ascarfs* Hemolymph, and Pig Intestinal Fluids.

	No	K	*Ca*	*Mg*	*Cl*	P	*Dry Wt.*	*Ahs*
			Millimole/litre				Per-cent	
Hemolymph. . .	129	25	5.9	4.9	53	17	7.4	0.91
Intestinal Fluid. . .	124	27	4.5	5.5	61	24	6.5	1.1

Carbohydrates are abundant in hemolymph, consisting of glycogen (0.40%) and trehalose (0.77%). Glucose and other free carbohydrates occur only in traces. Fats are less abundant; but significant amounts of phospholipids (0.18%) and triglycerides (0.13%) are present.

Although it is probable that the composition of the hemolymph of other species in general is similar to that of *Ascaris,* there may also be important differences. Thus, the ionic composition might be expected to reflect the conditions under which a particular species lives, and might change from one developmental stage to another, depending upon the extent of the ecological variation encountered. Similarly glycogen has been thoroughly identified in *Ascaris,* where it occurs in most tissues and less thoroughly in many other species; and trehalose occurs in the tissues of a number of nematoides. The species examined, however, have all been animal parasites. The food of free-living and plant-parasitic forms is so varied that there is no reason to expect that soluble sugars such as trehalose, or polysaccharides like glycogen, will be found to be the main carbohydrates in all nematode. The dietary carbohydrate of many species, in fact, probably contains very little of the glucose which is required for the synthesis of glycogen and trehalose. Investigation of the carbohadrates of free-living and plantparasitic nematodes is, therefore, a matter of importance.

Table 15.2 Glucose and Trehalose in Animal-Parasitic Nematodes.

	Trehalose	*Glucose*
	% of tissue solids	
Porrocaecum decipiens (larvae)	2.18	0.16
Trichinella spiralis (larvae)	1.76	0.04
Uncinaria stenocephala	0.91	0.77
Trichuris ovis	0.48	0.09
Ascaridia galli	0.38	0.78
Heterakis gallinae	0.10	0.43
Litomosoides carinii	0.06	0.01

ASSIMILATION AND CHEMICAL COMPOSITION

Assimilation is the means by which simple food-stuffs or products of digestion of foodstuffs are converted into the complex constituents of the organism, and in the widest sense includes biosynthetic processes in general. It presupposes, therefore, some knowledge of the complex molecules which are its end result. In the present discussion it is proposed to combine, where possible, a review of the chemical composition of nematodes with the means by which complex molecules are synthesized in these animals. This arrangement is not burdensome, as for the most part there is no exhaustive information concerning either chemistry or biosynthetic mechanisms.

It is worth remembering that nematodes possess a few biological characteristics which much strongly affect the quantitative aspects of assimilation. Their growth rate, for example, is not exceptionally fast, and indeed in many parasitic species is rather slow. At various stages in the life cycle, moreover, metabolism may be greatly depressed, as in larvae of plantparasites living temporarily in the soil, filariform larvae of animal parasities, and surviving infective eggs of plant or animal parasites. In all nematodes the amounts of energy required for motion and locomotion are probably small. On the other hand the needs of the female reproductive organs are impressively high. Every day, for example, a mature female *Ascaris* lays eggs and accompanying secretions which amount to some 10% of her own weight; and in 22 days a minute free-living *cephalobus elongatus* weighing approximately on microgram was observed to lay eggs weighing about 6 μg. This seemingly disproportionate amount of energy expended on egg production is, of course, related to the large numbers of eggs laid, but is also due to the fact that the nematode egg is cleidoic. So far as is known the embryonating eggs absorb no nourishment from the environment, and as a result all their caloric requirements must be provided by the adult female in the form of yolk materials which are deposlted in the mature ooyte. Moreover, in a numebr of plant and animal species the fully embryonated and infective egg survives for long periods in the absence of a suitable hatching stimulus, and must therefore contain additional caloric reserves for this purpose. As the yolk material is comprised primarily of proteins, carbohydrates, and fats, the adult female must possess adequate systems for the extensive biosynthesis of all three chemical groups.

Carbohydrates

As already mentioned, information concerning carbohydrates in nematodes is largely confined to the animal parasites, whose metabolism might be expected to be centered around glucose and glycogen. It is not safe to assume that these compounds are of equal importance in free-living or plant-parasitic species.

Until recently, glucose in animal parasites was estimated by means of relatively unspecific reducing agents. In one investigation, however, it has been determined by oxidation with glucose oxidase, which is extremely specific for glucose. The total glucose in seven species ranged widely from 0.01% of the dry weight in the filarial worm, *Litomosoides carinii,* to 0.7% in the fowl ascarid, *Ascaridia galli.* In a parallel series of analyses, trehalose, which is a dimer of glucose, ranged from a high value of 2.18% of the dry weight in codworm (*Porrocaecum decipiens*) larvae to 0.0% in *L. carinii.* In all two species (*A. galli, Heterakis gallinae*) there was more trehalose than glucose

Table 15.3 Chemical composition of nematodes

	Glycogen		*Lipid*		*Protein*	
	% *f.w.*	% *d.w.*	% *f.w.*	% *d.w.*	% *f.w.*	% *d.w.*
Ancyclostoma caninum	1.6	—	—	—	—	—
*Ascaridia galli**	4.2*	17.9*	1.2	—	—	—
*Ascaridia galli**	4.8'	19.8	1.3	—	—	—
*Ascaris lumbricoides**	3.3-8.7	14-24	1.3	—	—	48-57
*Ascaris umbricoides**	—	—	1.6	8.9	—	—
Dipetalonema gracilis	0.2	—	—	—	—	—
Dirofilaria immitis	1.9	10	—	—	—	—
Dirofilaria uniformis	1.7	—	—	—	—	—
Eustrongylides ignotus (larvae)	6.9	28	1.1	4.4	—	—
Heterakis gallinae	2.7	11.8	—	—	—	—
Litomosoides carinii	0.8	5	—	—	—	—
Parascaris equorum	2.1-3.8	10-23	—	—	—	—
Porrocaecum decipiens (larvae)	—	55	—	3.5	—	—
Strongylus vulgaris	3.5	—	—	—	—	—
Trichuris vulpis	2.3	—	—	—	—	—
Nippostrongylus brasiliensis larva	—	5-6†	—	15-20†	—	73-76†
Nippostrongylus brasiliensis adult	—	—	—	11.9	—	—
Trichinella spiralis	—	—	—	5.5	—	—

and no other low molecular weight carbohydrate seemed to be present in appreciable amount. The distribution of glucose and trehalose in the different organs and tissues is known only in *Ascaris* in which, comparatively, the intestine and cuticle contain more glucose, and the muscle, hemolymph and reproductive tissues an overwhelming amount of trehalose.

Reference has already been made to the well-characterized *Ascaris* glycogen which has its counterpart in many other animal-parasitic species and which differs little in detailed structure from mammalian glycogen. Intestine, muscle and reproductive organs are rich in glycogen, whereas cuticle and hemolymph are not. Sometimes glycogen is stored in immense amounts, as in *P. decipiens* larvae, where it comprises 55% of the dry weight, and *Ascaris* muscle, which contains 70%. These amounts, however, are exceptional, and many species contani much less. Apart from *Ditylenchus dipsaci*, whose solids contain 10% of unidentified carbohydrate no plant parasites have been examined.

Glycogen synthesis in mammals occurs by phosphorylation of glocose to glucose-1-phosphate (hexokinase phosphoglucomutase), followed by polymerization of glucose-1-phosphate (phosphorylase) with liberation of phosphate. Adenosine triphosphate is also required. Although clear evidence is lacking, this may also be the pathway followed in nematodes, for several tissues of *Ascaris* may contain phosphorylyse, hexokinase and **ATP.** In some species, most notably in *L. carinii* addition of glucose to a saline medium containing serviving worms reduces the rate of utilization of tissue glycogen or leads to an actual synthesis of glycogen.

The mechanism by which trehalose is synthesized in plants and animals which normally contain it has not yet been learned. Trehalose is of course formed by linkage between the first carbon atom in each of two glucose molecules, whereas glycogen, including *Ascaris* glycogen consists primarily of 1→4 linkages between glucose units. Clearly, trehalose cannot be directly converted to glycogen by simple polymerization, but must, if it contributes to glycogen synthesis, first submit to hydrolysis or rearrangement from 1→1 to a 1→4 type of linkage. It is not necessary of course, for either carbohydrate to be formed directly from dietary glucose. Many amino acids are freely convertible to 3-carbon compounds from which glucose may be synthesized and there is also evidence that in embryonating *Ascaris* eggs higher fatty acids give rise to similar 3-carbon compounds from which both trehalose and glycogen are synthesized. The latter represents a conversion of fat to carbohydrate which has not been conclusively demonstrated in other animals, although it is well-known in germinating seeds, which bring about the conversion by adaptation in the tricarboxylic acid cycle.

Apart from the carbohydrates discussed above, which are primarily concerned in energy metabolism, others of a more or less complex nature might be expected to occur in nematodes. These include the ribose and deoxyribose of nucleic acids and the amino sugars and uronic acids of glycoproteins and mucoproteins, which have not been much studied, although a number of reports have appeared describing the histochemical distribution of nuclei acids and glycoproteins in nematode tissues and eggs. Two structural carbohydrates, however, have been investigated in some detail, and therefore deserve mention.

The first of these is ascarylose, a distinctive hexose occur ring in the glycosides which are components of the yolk in *Parascaris* and *Ascaris* oocytes, and later make an important contribution to the vitelline membrane of the fertilised eggs. These glycosides will be discussed later. Ascarylose itself is 3, 6—di-deoxyaldodexose a unique carbohydrate not known to occur naturally except in nematodes. No details of its biosynthesis are available.

The second structural carbohydrate requiring mention is chitin, which occurs in many nematodes only in the hard part of the shell. No actual isolation has been reported, but the results of chemical and histochemical tests, and enzymic hydrolysis on eggs of many plant- and animal-parasitic species are convincing. Free-living species have not been examined, and the presence of chitin in some animal-parasitic species seems doubtful. It cannot be claimed, therefore, that chitin is universally present in nematode egg-shells. As the hard shell, like the vitelline membrane lying close to its inner surface, is a primary egg envelope, the chitin in it must be synthesized from yolk material. This synthesis and extrusion of chitin or its precursors from the fertilized oocytes of *Parascaris* and *Ascaris* has been described cytologically, and it is probable that glycogen and trehalose in the yolk provide the glucose from which the N-acetylglucosamine units of chitin are formed. The conversion of glucose-6-phosphate to glucosamine in rat liver proceeds in the presence of glutamine as a source of amino groups. *Ascaris* oocytes contain considerable glutamine or related substance, as well as a source of acetate needed for the acylation of glucosamine to N-acetylglucosamine. There is no question, therefore, that the oocyte possesses appropriate substrates for chitin synthesis, although the way in which the synthesis takes place has not yet been studied.

Lipids

In some respects the nematode lipids, first studied in detail by Flury (1912), are better known than the carbohydrates, although as usual, detailed studies have been made only on *Ascaris* and *parascaris*. Generally speaking, they can be said to resemble the lipids of higher animals, but with important differences. Before discussing these in detail, however, the results of analyses for total lipids in several species are worth noting. Viewed generally, the amounts of lipids in five species of animal parasites, ranging from 1% to 8% of the dry weight, are not exceptional. In contrast, the values for three species of *Ditylenchus* and for the vinegar eelworm, *Turbatrix aceti*, are extremely high.

Table 15.4 Total Lipids in Nematodes (% of dry weight)

Animal Parasites		
Ascaris lumbricoides	8.0	(von Brand, 1941)
Dioctophyme renale	1.0	(von Brand, 1957)
Trichinella spiralis (larvae)	5.5	(von Brand *et. al.*, 1952)
Eustrongylides ignotus (larvae)	4.4	(von Brand, 1938)
Porrocaecum decipiens (larvae)	3.5	(Fairbairn, 1958b)
Plant Parasites		
Ditylenchus dipsaci		
Ditylenchus destructor	33	(Tracey, 1958)
Ditylenchus myceliophagus		
Free-living		
Turbatrix aceti	41	(Comenga, 1955)

Unfortunately, details of the lipids in *Ditylenchus* and *Turbatrix* have not been secured. Investigations of *Ascaris* lipids, however, have established clearly that, if the over-all concentration is not unduly high, the distribution in various organs and tissues is interesting. Thus, lipids comprise 17.5% of the fresh weight of eggs, 6.0% of the ovaries, and considerably less than 1% in other tissues. Clearly, the reproductive organs are the most important sites for lipid deposition. In all tissues there are appreciable quantities of triglycerides, phospholipids and unsaponifiables, although the relative concentration of phospholipids in the reproductive tissues and egg is low.

The composition of the triglyceride and phospholipid fractions is not dissimilar to that of higher animals and need not be discussed further, except to say that the menas by which *Ascaris* synthesizes these substances are unknown. The unsaponifiable fraction, on the other hand, is particularly interesting, for it contains unique substances of great importance in the parasite's economy.

In higher animals, most of the unsaponifiable fraction is composed of sterols, and chiefly of cholesterol. *Ascaris* unsaponifiables contain cholesterol, and also a second, saturated sterol, but the bulk of this fraction consists of a substance isolated many years ago by Flury (1912) and Faure-Fremiet (1913) and named by them ascaryl alcohol. Quite recently Fouquey *et. al.* (1957) established the structure of this compound which is not a true lipid, but a series of closely related glycosides with the solubility characteristics of true lipids. The generic name, ascaroside, which was suggested for these glycosides should now be accepted as correct. One component of ascarosides, ascarylose, has already been mentioned in the discussion of carbohydrate chemistry. This hexose is combined in glycosidic linkage with a high molecular weight, aliphatic, straight-chain diol to form ascaroside B, the main ascaroside in *Parascaris* ovaries.

Table 15.5 Distribution of Lipids in Tissues of Female *Ascaris* (Fairbairn, 1957)

	Total lipids	*Trigly-cerides*	*Phospho-lipids*	*Unsaponi-fiables*
	% wet weight	% of total lipids		
Hemolymph	0.32		62	
Cuticle	0.61	43	41	16
Muscle	0.75	40	49	11
Reproductive system	6.0	70	8.3	22
Fertilized Eggs	17.5	69	9.2	22

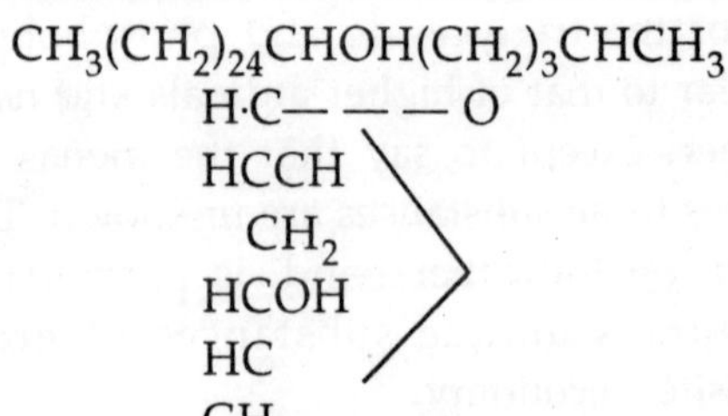

Ascarosides A and C contain the same hexose, but differ slightly with respect to the higher alcohol. Ascarosides occur in *Ascaris* muscle and cuticle and in *Eustrongylides* males, but are concentrated mainly in the oocytes and in the fertilised eggs, and it is in these cells that their function is clearly understood.

Ascarosides first appear in the developing oocytes as refringent granules which are finally discharged from the fertilized eggs to become the main component of the innermost, or vitelline, memberane. The process of shell formation in *Ascaris* and *Parascaris,* which lies beyond the boundaries of the present discussion, has been reviewed recently; and the function of the vitelline membrane will be mentioned later. At this point it is appropriate to suggest, however, that although ascarosides have been identified only in *Ascaris* and *Parascaris,* it is likely that they, or closely related substances, are constituents of other nematode eggs. Thus, a vitelline membrane is present, and when more closely examined is found to be soluble in lipid solvents, highly impermeable, chemically resistant and to melt at about 70°C. These are the outstanding properties of *Ascaris* and *Parascaris* membranes, and are sufficiently unusual to suggest similarities in chemical structure. However, much more work will be necessary in ordre to establish the validity of this supposition.

It is clear from this brief discussion that in many, perhaps in most, nematodes there is an extensive synthesis of lipids, some of which are incorporated in the somatic tissues, but most of which, in the female, are required for the production of eggs. It is probably safe to assume that as in other animals, the phospholipids and sterols are structural components of cells, concerned in the formation of cellular and subcellular membranes. In the embryonating fertilized egg, for example, these lipids, unlike the triglycerides, are apparently not metabolized. The part played by ascarosides in the somatic tissues is not clear. Triglycerides, on the other hand, serve as energy stores, as will become apparent later.

Proteins

The proteins of nematodes are of exceptional interest but have been studied by modern analytical methods only in a few instances. The total protein in *Ascaris* makes up eight to nine per cent of the wet weight and the amino-acid composition generally resembles that of its host. The haemolymph of *Ascaris* contains 4.9 per cent proteins of which 2.8 per cent are albumins and 2.1 per cent globulins. The fibrous proteins of the cuticle have been most thoroughly studied but complete agreement has not been reached on their composition.

X-ray and chemical analysis have inducated the presence of keratin, collagen and tanned protein although there are differences between species. There is little evidence in nematodes that protein is stored for subsequent use in energy metabolism. It is true that nitrogen catabolism during prolonged starvation *in vitro* is fairly extensive, as will appear later, but the large amount of protein deposited in *Ascaris* oocytes is apparently not used at any stage of embryonation for energy. Although protein granules in the larval intestines of *Agamermis decaudota, Ditylenchus dipsaci* and *Meloidogyne javanica* disappear during development to the adult stage, it is not clear whether these are mobilized for incorporation into other tissues, or are catabolized. In general then, it seems likely that the bulk of nematode proteins are structural, or potentially structural, tissue components.

Mention was made previously of the large amounts of albumins and globulins in *Ascaris* hemolymph, whose origin and function are unknown, but which must surely originate in some tissue other than the cell-free hemolymph itself. A third hemolymph protein is hemoglobin, which in *Ascaris* comprises about 2% of the total. Muscle and epidermis, contain a second, and somewhat different, hemoglobin. In animal parasites, the presence of hemoglobin is the rule, for von Brand (1952) listed 13 species in which it occurs, and several others have been added since. More important than these minor statistics is the fact that no species has yet been shown to lack this protein. It is, therefore particularly necessary to examine freeliving and plant-parasitic species, concerning which no report has yet appeared.

Hemoglobin is widespread amongst parasitic nematodes, only three species having been reported to lack haemoglobin it is localized differently in different species. There is evidence that it plays a respiratoryy role in some, but not all species. The haemoglobin in the body wall may also serve to transfer oxygen from the cuticle to the body fluid. Hemoglobins of *Heterakis gallinae* and trichostrongyles have characteristic a-and ß-absorption bands which differ little in position from those of the host, whereas the differences in *Ascaris* hemoglobin are distinct. The fundamental distinction between host and parasite proteins, however, lies in the extremely high affinity of the parasite proteins for oxygen, exen at very low partial pressure of this gas. As a result, no dissociation of oxyhemoglobin occurs except at very low tissue concentrations of oxygen. After quantitative study of this phenomenon, both Rogers and Davenport are agreed that an oxygen transporting function of hemoglobin in *Ascaris* and *Nippostrongylus muris* is improbable.

As no alternative function for hemoglobins has been suggested it is tempting to think that these are simply pigments derived by absorption of modified host hemoglobins. A serious objection here is that some species, such as *Ascaris,* do not feed on the host tissues, and as a result have limited access to host hemoglobins.

There has been no chemical investigation of the proteins in nematode muscle or epidermis. In recent years, however, a good deal of interest has been taken in the cuticle, which is largely protein, and is of well-recognized importance both in the protection of nematodes against their normal environment, and in their destruction by nematocides. The organization and composition of the cuticle is therefore worth considering in some details, particularly since it has the same fundamental structure in all nematodes.

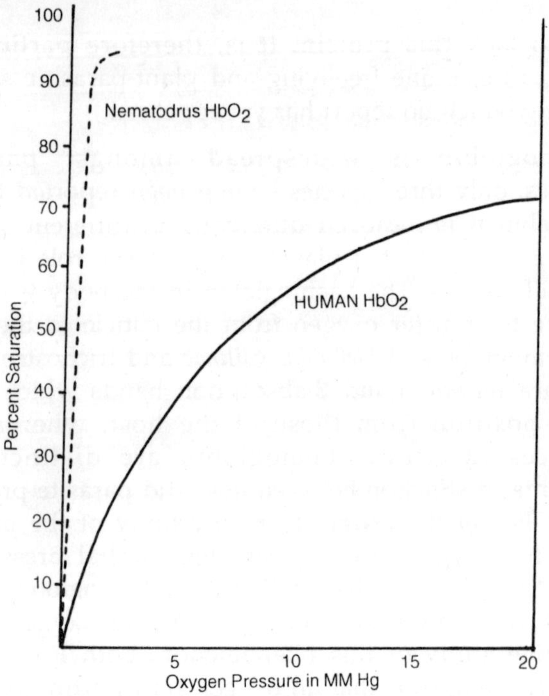

Fig. 15.1 : Comparison of the affinity of *Nematodirus* and human hemoglobins for oxygen.

The cuticle is, so far as is known, a secretion product of the underlying epidermis, but is nevertheless highly organized. It is soft and flexible, and contains about 75% water. When isolated, it can be stretched very considerably in the direction of the longitudinal axis, but very little in the lateral axis. Most of its solids are proteins. However, small amounts of various lipids and carbohydrates are to be found in it whose nature differs little from those occcurring in the muscle. Whether these contribute to its physiological functions or have merely "strayed" into it during the secretion of proteins has not been determined.

Recent work by Bird and Deutsch (1957) and Bird (1958) has considerably increased our understanding of the structural complexity of the cuticle. For present purposes, a simplified diagram provides a suitable basis for the discussion of

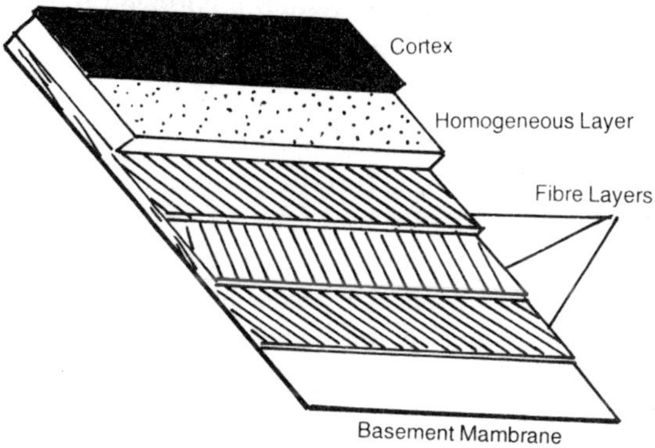

Fig. 15.2 : Schematic representation of *Ascaris* cuticle.

cuticular proteins. Here the basement membrane which lies next to the secreting epidermis is overlaid (in *Ascaris*) with three fibre layers. In many species there are apparently only two fibre layers. These are crossed at an angle of about 70° with the long axis of the worm, and if there is a third layer it lies parallel to the first. An apparently structureless layer lying externally to the fibre layers was called the homogeneous layer by Chitwood (1936) who was the first to make detailed chemical studies of the various parts. This is succeeded by a cortex of varying complexity. Bird (1958) has produced evidence, based on electron microscopy, that in *Ascaris* culticle the cortex contains a radiating network of pore canals which extend into the external cortex, almost to the cuticular surface. The surface itself is covered with an osmiophilic layer less than 1000 Å thick which may be a lipoid.

If whole *Ascaris* cuticle is extracted with water about 20% of its proteins dissolve without seriously modifying the cuticular structure. The extract contains at least five different proteins, all of which seem to be albumins of similar, low molecular weight. Their initial distributionin the cuticle is unknown, although the homogeneous layer would seem to be the most probable locus. This layer also contains sulphur-rich fibrous proteins which resemble fibroin or elastin in their chemical and histochemical properties. It is the fibre layers and the external cortex which have been most studied, partly because of their great intrinsic interest, but equally because they are relatively easy to isolate.

The fibre layers have the staining properties and X-ray diffraction pattern characteristic of collagen which is of course the characteristic protein of tendons and other connective tissues. Amino acid analyses show that hydroxyproline, an amino acid peculiar to collagen, is present in the fibre layerse and in the basal lamella. The collagenous nature of these layers is therefore demonstrable by several techiques. The nature of the protein in the external cortex, which is responsible for the protection of nematodes against potentially damaging environmental influences, is less certain. Early work by Flury (1912), since confirmed established the presence of large amounts of sulfurbearing amino acids, which, considered in relation to its great chemical resistance, suggested a keratin-like structure. Such a structure is not supported by the X-ray diffraction patterns which are those of a collagen, although the presence of hydroxyproline was not reported in amino acid analyses of this protein. To further confuse the picture, evidence has been presented for the presence in the external cortex of polyphenols and polyphenol oxidase, substances which are commonly associated with tanning processes in the arthropod cuticle. Bird (1957, 1958) suggested that these phenolic substances necessary for tanning are transported to the external cuticle through the pore canals which, as already mentioned, were revealed by electron microsocpy. As tanning seems always to be associated with the development of colour, and as the nematode cuticle is not highly coloured, it is probable that if tanning occurs, it is moderate in extent. This view is supported by the fact that well-tanned proteins are resistant to the action of the proteolytic enzyme papain, whereas the adult nematode cuticle is not. Papain and related enzymes have, in fact, been fairly widely used for this reason as antihelminthic agents against intestinal parasites. It seems desirable, therefore, to remain uncommitted or the chemical nature of the external cortex until additional studies have been made. The one possible exception is the aged cyst of

Heterodera rostochiensis, which is deeply coloured, and is of course, a modified cuticle. Evidence for tanning in this cyst was presented several years ago by Ellenby (1946).

The chemical properties of other nematode cuticles, in so far as they are known, are similar to those of *Ascaris.* Chitwood (1936) noted the resemblance between the cuticles of *Ascaris* and *D. dipsaci,* and Bird (1956) detected no important differences in protein amino acids obtained from the entire cuticles of *Ascaris, Toxocara mystax,* and *Strongylus eguinus.* Similarly, histochemical examination of *T. canis, Subulura, Haemonchus, Uncinaria, Chabertia,* and *Echinuria* suggested the presence of collagenous fibre layers, an elastin-containing homogeneous layer, and a polyphenol containing cortex. It seems clear, also that the exteroal critical layer lines the surface of the stomodaeum and esophagus in nematodes generally and that tanning may occur in this lining.

The discussion of cuticle proteins has up to this point been concerned only with adult nematodes. Much less work has been done with the larval cuticle, which is naturally harder to obtain in quantities sufficient for chemical analysis. Monne (1955*b*) reported that the larval cuticles of the lungworms *Metastrongylus* and *Dictyocaulus* were microscopically structureless, but stained in a manner suggesting the presence of tanned protein and were not attacked by papain, which, as noted above, is a fairly efficient agent for the digestion of the adult cuticles. Despite their fragility, therefore, the larval cuticles may be sufficiently well tanned to offer considerable resistance to an adverse environment. Bird and Rogers (1956) in a study of the cast shealths of third stage *Haemonchus, Trichostrongylus* and *Nippostrongylus* larvae found no aromatic amino acids in the proteins, and no evidence for quinonetanning, although considerable hydroxyproline was present. Interpretation of their results is uncertain, as sodium hypochlorite, a strong oxidizing agent, was used to induce exsheathment of the larvae, and the destructive effect of this solution on cuticular proteins must be considered, as the authors realized. A more physiological procedure was employed by Simmonds (1958), who found that fourth-stage

Nippostrongylus larvae cast off the remains of the third stage cuticle when held overnight in saline solution. Proteins of the sheaths obtained in this way contained an abundance of the aromatic amino acid tyrosine and also of hydroxyproline. Physical properties of the sheaths. were those of a somewhat unusual collagen. Simmonds also obtained strong histochemical reactions for carbohydrates and lipids, confirming the chemical analyses previously noted in *Ascaris* cuticle. It is not possible, naturally to compare the cast sheaths used in these experiments with the whole cuticle, for it is well-known that in many animals such as the arthropods that much of the old cuticle is dissolved during the molting process, with only an insoluble remnant being discarded. The physiology of molting in nematodes has not been investigated histologically, but it is probable that here, too, the cast sheath is only a remnant of the original cuticle. Now that axenic culture of several species is possible, there is no technical reason why molting cannot be studied in detail. From both the physiological and practiacal points of view the secretion and organization of the cuticle, and its partial dissolution and rejection four times during the life of each individual, are matters of great interest.

The last group of proteins requiring mention is that of the reproductive organs and eggs. This is a heterogenous group, with interesting examples of extreme specialization. As already emphasized, most of the nitrogen anabolism in the adult female nematode is directed towards the extensive synthesis of ovarian proteins. For the most part, these proteins are deposited in the oocyte as yolk material, to be incorporated subsequently into the structural proteins of the embryo. It is not surprising therefore, to find that extensive amino acid analyses of *Ascaris* ovaries revealed the presence of most of the common acids in well-distributed amounts. During embryonation the yolk protein is converted to the many different proteins necessary for the various cells and tissues. No waste nitrogen accumulates in the egg as a result of this conversion showing that the formation of cell protein from yolk protein is remarkably efficient, and that the yolk protein is not used by the egg as a source of energy.

Many years ago van Beneden described the appearance in *Parascaris* oocytes of refractile granules which are now known to consists primarily of proteins containing more than 20% proline. The clear-cut segregation of these prolamine-like proteins from the yolk material is due to the fact that following fertilisation, the granules are ejected from the cytoplasm, and, according to Yanagisawa (1954, 1955) become a part of the hard shell. Thus, as has long been suspected, the hard shell, although chitinous, also contains a specialized protein. Similar gram-positive inclusions have been described in a number of animal-parasitic nematodes, and in all cases they appear to be transformed after fertilization into constituents of the hard shell.

The spermatogonia of many nematodes contain similar gram positive refringent granules which ultimately fuse to form the prominent acrosome of the sperm. This protein, called ascaridine, has been isolated from *Parascaris* sperm. It contains no phosphorus, sulfur, or purines, but two amino acids, aspartic acid and tryptophane, account for 35% and 15% of the total nitrogen, respectively. After the sperm penetrates the surface of the oocyte, it moves slowly towards the female pronucleus, and as it does so the acrosome, and hence the ascaridine, gradually disappears. Ahead of it, in the egg cytoplasm, appears a dense and local concentration of ribose nucleic acid. Thus there is some connection between the disappearance of ascaridine and the appearance of newly-formed RNA in the fertilized egg, but whether this represents a direct contribution of ascaridine to RNA synthesis has yet to be determined.

If, as seems clear, there is a continual synthesis of protein in the ovaries, then the sources of the amino acids contributing to this synthesis are of interest. Dietary amino acid transferred directly from the intestine to the ovaries are of course an obvious source, and may normally be the most important one. It is also true, however, that *Ascaris* ovaries contain active transaminases which are able to convert α-keto acids derived from carbohydrate metabolism to the corresponding amino acids, with the equilibrium favouring the formation of amino acids. Moreover, Pollak has shown that ovaries can, in the presence of suitable co-factors, condense pyruvate with ammonia to form the amino acid alanine, which he regards as the key amino acid in ovaries. Thus, although most details have still to be worked out, there is evidence that formation of ovarian proteins occurs in the ovaries themselves, and that the required amino acids may be supplied by synthesis, as well as from the diet.

This concludes the discussion of nematode chemistry, and the means by which these animals synthesize some of the complex molecules which are of importance to them. If the discussion has been brief, the references supplied will be found to cover most of the existing literature. Emphasis has been placed on those aspects of nematode chemistry which are most likely to be of general importance, regardless of the species on which the particular investigation was made. In so doing, there is an ever-present danger of fostering an unwarranted complacence, and it cannot be too strongly urged that many more studies are needed, particularly of the free-living and plant-parastitic species. Until these are made, there is no possibility that a discussion based on fact as distinguished from a discussion based on hopeful generalizations, can be presented.

In the succeeding pages, which will deal for the most part with processes of a catabolic nature, it will quickly become apparent that there is an equally great need for strengthening and broadening the foundation on which present knowledge rests.

Table 15.6 Summary of Oxygen Consumption by Nematodes.

($Q_{O_2}^{air}$ = ml O_2 consumed/gm dry weight/hour in saline equilibrated with air).

	$Q_{O_2}^{air}$	
Animal parasites. . . (15 species, 30-38°C)	0.32-18.4	(von Brand, 1952)
Free-living. . . . (24 species, 16°C)	0.68-5.6	(Overgaard Nielsen, 1949)

RESPIRATION

As a rule oxygen is needed for the efficient metabolism of organic compounds. When aerobic respiration is not possible, the anaerobic metabolism which replaces it is much less efficient, owing to failure to oxidize organic compounds completely to carbon dioxide and water. One result of this anaerobic respiration, or fermentation as it is more often called, is that the end-products of incomplete combustion must be eliminated from the body before they attain toxic concentrations. Successful adaptation to an anaerobic or semianaerobic existence therefore involves the development of mechanisms for the excretion of low molecular weight alcohols, organic acids and so forth as waster products. In some animals, and certainly in many nematodes, fermentation continues even when oxygen is being used, so that some molecules are oxidized completely to carbon dioxide and water simultaneously others are oxidized less completely and are excreted chiefly as organic acids.

The general physiology of adaptations to aerobic and anaerobic environments is dealt with by von Brand elsewhere in this book. Here it is only necessary to mention some details which may be of direct concern to the experimenter, or which may influence the mechanism by means of which tissue oxidations occur.

No nematode yet studied has failed to use oxygen when given the opportunity, *i.e.*, no obligatively anaerobic species has been discovered. Some, *e.g. Ascaris*, are poisoned by high concentrations of oxygen, so that it may eventually be desirable to classify these as microaerophiles, a term used by microbiologists to describe the same phenomenon in bacteria and other microorganisms. Generalizations on the subject of oxygen utilization should certainly be deferred, however, until more studies have been made, particularly on plant-parasitic species and the many free-living species which inhabit the essentially anaerobic fresh-water and ocean muds.

Oxygen utilisation by nematodes varies greatly from one species to another, and has not been successfully correlated with weight, volume, or habitat. The upper and lower limits recorded for a number of animal-parasitic and free-living species are set out in Table 15.6, which is based on a summary by von Brand (1952) and on the original investigation of Overgaard Nielsen (1949). It is immediately apparent that the range of Qo_2 values is wide, and that there seems to be no great difference between free-living and parasitiic forms in the ability to use oxygen. The results, of course, were obtained using isolated animals suspended in a saline solution in air, and do not necessarily, or even probably, represent the rate of oxygen consumption in a more normal environment. The represent, in fact, only the *capacity* for metabolizing oxygen. Direct comparison of the animal parasites with the free-livin species is also complicated by the large difference in the temperature at which the Qo_2 values were determined.

In such analyses, a number of factors in addition to temperature may have an important bearing on the results obtained. As a rule, moderate variations in pH or osmotic pressure have not been found to be highly critical. The ionic composition of the suspending medium, on the other hand, may profoundly affect the respiration. Again, the usual Warburg method for determining oxygen consumption, depending as it does upon the continuous removal of carbon dioxide from the gas atmosphere, may result in Qo_2 measurements different from those found under more natural conditions when carbon dioxide is present. Thus, the oxygen consumed by *Nippostrangylus muris* larvae in their early, freeliving stages is unaffected by the presence of 5% CO_2, whereas the Qo_2 of infective larvae is doubled reflecting, perhaps, an adaptation of the infective larvae to the high CO_2 concentrations in the animal tissues in which they will complete their development. Age can also be important. Rogers (1948) observed a considerable decrease in the Qo_2 of infective *N. muris* larvae after twelve days, and the Qo_2 of free-living *Panagrellus redivivus* adults feel about 50% after 24 hours starvation. Although these observations on *Nippostrongylus* and *Panagrellus* were all made in nutrient-free saline, the results must be interpreted differently. Decrease in the Qo_2 of *Panagrellus* can be attributed to starvation, and would not be expected to occur in a nutrient solution. In *Nippostrongylus,* however, the decrease is physiological *i.e.* the infective larvae do not feed, and yet can survive for many days in the soil. The ability to survive must entail a severe restriction in the basal metabolism, which is reflected in decreased oxygen consumption. The importance of recognizing this experimentally uncontrollable condition can be illustrated by considering the changes which take place in the respiration of *Ascaris eggs* during embryonation. Here it has been shown that the Qo_2 falls at first, possibly as an oxygen debt is repaid, but then rises rapidly during the period of cell-division, reaching a peak when the embryo has become vermiform. A sharp decline then sets in, and by the time the egg is infective (about 20 days) the Qo_2 is once again small. Over a period of months the respiration continues to decline, and is in fact scarcely measurable, although the egg remains infective. Clearly, Qo_2 measurement on these eggs would be of little value if recorded without reference to physiological age. It would be surprising, too, if other eggs failed to display a similar variation, and in more general terms, one might predict that variable oxygen requirements are a physiological feature of all species where eggs, larvae, or adults alternate between metabolically active and quiescent states.

A number of investigations have revealed that nematode Qo_2 measurements may or may not be highly dependent upon oxygen tension. This is not a matter requiring discussion here, except to point out that two factors are of primary importance. These are (1) the size of the animal (2) the nature of its oxygen-activating enzymen system. It has been shown that at oxygen tensions less than 16 mm of Hg the rate of diffusion of oxygen into the tissues will limit the Qo_2 of a nematode more than 0.05 mm in diameter. If the

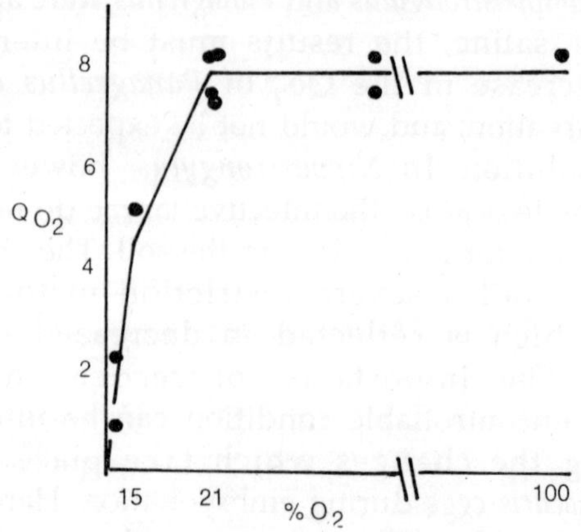

Fig. 15.3 : Oxygen consumption of *Nippostrongylus brasiliensis* at varying concentrations of oxygen in the gas phase.

diameter is 0.14 mm, the limiting tension is 32 mm, and so forth. Assuming that diffusion is not a limiting factor, however, the Qo_2 may still vary according to the oxygen tension if the terminal oxygen activating respiratory enzymes have a relatively low affinity for oxygen. Such is the case in *Ascaris* which seems to depend upon flavin enzymes for oxygen activation. Other species, such as largvae of *Trichinella spiralis,* contain the cytochrome oxidase system, which has an extremely high affinity for oxygen. In these, small size and suitable enzymes create a situation in which respiration is independent of oxygen tension even at 15 mm Hg.

Parasites removed from their normal environment in salime are apt to be very active. *Heterakis gallinae,* whose movements in the fowl caecum must be restricted owing to the thick consistency of the faecal mass, move very actively during the course of Qo_2 measurements in saline, Rogers (1949a), in his pioneering efforts to assess the physiological availability of oxygen to a number of intestinal parasites, suggested that the Qo_2 observed *in vitro* might, on account of the parasites' muscular activity, be considerably higher than the normal *in vivo* requirement. Moreover, repeated observations on several species have led to the conclusion that in the absence of oxygen hitherto active nematodes become quiescent motility being restored upon admission of oxygen. Thus, there appears to be a connection between oxygen consumption and motility which may be of practical importance to the experimenter if motility under unnatural conditions is abnormally high.

Without denying the apparent necesity of oxygen for motility, there is at the same time reason to believe that only a small part of the oxygen consumed is associated with motility. Santmeyer (1956), in observing that the Qo_2 of surviving *P. redivivus* decreased to half its original value after 24 hours, at the same time reported that motility was not affected. Thus, high motility and high Qo_2 are not necessarily directly related. Similarly, in his extensive report on soil nematodes, Overgaard Nielsen (1949) failed to observe a positive correlation between Qo_2 and motility and found that when motility was abolished by anaesthetization of several species with urethane the Qo_2 was reduced only by about 5%. Another pertinent observation was reported by Bueding (1949b), who found that in *Litomosoides carinii* a 97 to 98% suppression of oxygen consumption by cyanide immobilized the parasites, whereas an 85 to 95% suppression by cyanine dyes did not. Thus, there is good evidence that oxygen is necessary for motility, and some evidence that the amount of oxygen used for this purpose is only a small part of the total. This leaves the initial problem, whether abnormal motility of isolated specimens is a serious practical problem in Qo_2 measurement, essentially unresolved.

In addition to measuring oxygen consumption, many workers have determined carbon dioxide production during respiration, and from the results have calculated the molar ratio $CO_2 : O_2$ *i.e.* the respiratory quotient (R.Q.), Carbon dioxide measurements are in themselves legitimate, but the interpretation of R.Q. values even in animals whose metabolism is comparatively well-known is apt to be hazardous, and in nematodes has sometimes led to unwarranted conclusions concerning the kinds of substrates being metabolized. Any interpretation of this sort presupposes that all CO_2 arises from the complete oxidation of the substrate, and that CO_2 fixation into tissue components is not extensive. In nematodes, neither of these suppositions seem justified. As has already been said, oxidative metabolism is often associated with fermentation processes which themselves give rise to CO_2, and in adults of *H. gallinae,* as well as in eggs and muscle of *A. lumbricoides* there is evidennce for a fairly extensive CO_2 fixation. Thus, while measurement of CO_2 output should not be discouraged, interpretation of respiratory quotients should be made with caution.

From the foregoign account it can be said that all nematodes studied so far can use oxygen, that the rate of utilization is dependent upon the physiological stage of development, and that the oxygen consumed is not merely the reflection of an abnormal motility under experimental conditions. To these conclusions one may add another, namely, that the species studied with two dubious exceptions must have oxygen for their continued well-being. The exceptions are *Ascaris,* which while surviving in saline does not seem to favour aerobic over anaerobic conditions, and *Turbatrix aceti,* the free-living vinegar peelworm, some individuals of which have been reported to survive for six months in culture without oxygen. However, experiments of this kind supply no information concerning the possible oxygen requirement for normal growth and maintenance of tissues and production of eggs. Other species including many intestinal parasites quickly succumb to anaerobic conditions, although oxygen tensions in their normal environment are low. It is known that many parasites survive temporary anaerobiosis, and von Brand (1946) has pointed out that flooding of soils must result in oxygen deficiencies to which nematodes must adapt themselves if they are to survive.

There has been general agreement for some years that the metabolism of nematode eggs is primarily, and even obligatively, aerobic. Oxygen requirements of *Ascaris* eggs, which have already been mentioned, may be taken as typical, although no other species has been examined in such detail. These eggs survive anaerobically for some days, but fail to develop unless oxygen is present. Similar observations have been made on a number of other animal-parasitic species. It is interesting to note, however, that no decisive experiment seems to have been performed with the developing eggs of parasites or free-living forms which under natural conditions can have access to very limited supplies of oxygen. An example is to be found in the common pinworm, *Enterobius vermicularis*, which lives in the vermiform appendix and at various levels of the host's caecum, migrating from these loci to the anal and perineal regions where the eggs are laid. Despite the essentially anaerobic character of the natural habitat, the eggs when laid are usually fully embryonated. There is need, therefore, for careful experimental examination of these and similar eggs before generalizations concerning the need of oxygen for development can be accepted.

Oxidative energy needs in animals can be satisfied by proteins, fats or carbohydrates, which in turn may come from the diet, or from body depots. There is no evidence for or against the oxidation of dietary amino acids by nematodes living under more or less natural conditions, nor is there much evidence for any extensive catabolism of tissue proteins. Surviving animal parasites excrete modest amounts of nitrogenous end products but under conditions which can scarcely be called normal, although exactly what is meant by normal is not yet understood. Recent work by Pollak (1957a) has, however, furnished a striking example of how a simple change or addition to the suspending medium can effect metabolism. Pollak found that after two days survival in a balanced saline solution, the ammonia and free amino acid content of *Ascaris* ovaries decreased to less than half its initial value. If, however, the medium contained ammonium ions, the ovarian concentrations of both ammonia and free amino acids was maintained, owing to the ability of this organ to sysnthesize amino acids from keto acids and ammonia. As already mentioned, considerable protein is synthesized in the ovaries and leaves the body in the eggs. During subsequent embryonation this protein is not used as a source of energy, but is concerned solely with the formation of new tissue protein.

The only adult nematode which has been studied appears to have no oxidative fat metabolism, for after several days of starvation *in vitro,* von Brand (1941) could detect no change in the total body fat of *Ascaris.* Under similar conditions fat globules in the intestinal cells disappeared, but it seems likely that these globules represent such a small fraction of the total body fat that their disappearance would not be detected by gravimetric methods. In *Trichinella spiralis* larvae, on the other hand, aerobie *in vitro* survival experimennts resulted in a significant loss of body fat. Histochemical evidence suggests that the surviving infective larvae of several species rely on the metabolism of stored fat for their energy requirements, which are no doubt small, but may extend over a period of several months. In these, and in the infective egg of *Ascaridia galli* loss of infectivity seems to be associated with the reduction of stored fats to a low level. Such observations should naturally be extended to include many more species, using more precise methods.

The most thoroughly studied instance of oxidative fat metabolism is that which occurs in embryonating *Ascaris* eggs. The unembryonated eggs contain a great deal of fat, chiefly as triglycerides which disappears slowly during the first few days of embryonation and then much more rapidly until after 18 to 20 days the egg has become infective. Thereafter the still considerable fat stores decrease much more slowly as the egg enters the semi-dormant respiratory period previously described. Curiously, however, the maximum rate of fat utilization occurs at a time when the Qo_2 is decreasing, the evidence showing that only part of the disappearing fat is completely oxidized, the remainder being partially oxidized and then used for carbohydrate synthesis.

It is tempting to speculate that fat metabolism in adult nematodes is primarily anabolic, with synthesized fats being deposited chiefly in the oocytes, and that fat catabolism occurs chiefly in the eggs and larvae, particularly in eggs which are capable of prolonged survival following embryonation, or in infective larvae which feed little or not at all. Present evidence supports such a view, but is insufficient to permit strong endorsement of it. As for the mechanisms by means of which nematodes oxidize fats, nothing at all is known.

In contrast to proteins, which do not contribute much to energy metabolism, and fats, whose catabolism seems to be restricted to certain developmental stages, carbohydrates are extensively oxidized. As already pointed out, glycogen is stored, sometimes in very large amonunts, in a number of animal-parasitic nematodes. During starvation this glycogen is rapidly depleted in part by fermentative and in part by oxidative processes, the relative proportions depending upon the characteristics of the particular species. This problem will be discussed in more detail later. Here it is sufficient to say that most of the evidence linking oxygen with carbohydrate metabolism is convincing, but indirect. However, in the filarial worm (*Litomosoides carinik*) Bueding (1949b) showed conclusively that at least during the early stages of survival the disappearance of glycogen was linked with the Qo_2. Under natural conditions of course, dietary carbohydrate may be expected to contribute directly to the energy metabolism, any excess being employed for resynthesis of depleted glycogen reserves. Resynthesis of glycogen *in vitro* has been demonstrated in *Litomosoides,* when excess glucose, fructose, or mannose were supplied in the saline medium. In *Ascaris,* sorbose can also be used for this purpose. Glycogen is stored in *Ascaris* chiefly in the bulbs of the muscle cells and in the oocytes of the ovaries, and durinng survival both of these tissues are depleted. Fertilized eggs, with which the uterus is filled, cannot be depleted by starvation as they are not permeable to carbohydrates.

During embryonation of *Ascaris* eggs, there is an extensive disappearance of carbohydrate (both glycogen and trehalose) during the period of active cell division and increasing Qo_2. There can be little doubt that these carbohydrates are a major source of ennergy for the complex biochemical conversions and synthesis taking place at this time. Later, while the embryo is becoming infective, both glycogen and trehalose are resynthesized, so that the infective egg contains just as much of these two carbohydrates as the unembryonated egg. As already indicated, the resynthesis is the result of conversion of egg triglycerides to glycogen and trehalose. It is remarkable that during subsequent survival of the infective egg these large carbohydrate reserves are not used, although, as has been mentioned, there is a gradual disappearnace of fats. The hatched larva therefore begins its new life with a plentiful supply of carbohydrate, whose fate is not yet known.

Nothing is known about the changes occurring in reserve carbohydrates of other nematode eggs, all of which seem to be cleidoic and therefore dependent upon food reserves of some kind. Eggs of most species, of course, hatch spontaneously instead of developing further to an infective, resistance stage, so that it need not be expected that the resynthesis of carbohydrate will be found to be a characteristic feature. However, the eggs of many plant-parasitic nematodes (*Heterodera, Meloidogyne*) are comparable in their development to *Ascaris* eggs and in these it will be of interest to compare the details of carbohydrate biochemistry.

Oxidative Mechanisms

In most animals the first phase of carbohydrate metabolism results in the conversion of glycogen to glucose to pyruvate by way of the well-known Embden-Meyerhof reactions. A number of animal-parasitic nematodes have been found to conform without exception to this metabolic pattern. In many animals alternative pathways for glucose oxidation exist, which have not yet been searched for in nematodes, but which are generally of a secondary nature. The Embden-Meyerhof reactions, of course, do not require oxygen, but merely produce the pyruvate which is then either fermented, or oxidized to carbon dioxide and water. At this point in the metabolic processes, intermediates formed in the catabolism of amino acids and fats also enter the metabolic pool. The general features of intermediary metabolism up to this point are so well-known that they need not be discussed here in detail.

A great many animals proceed with the oxidation of pyruvate by converting it to a two carbon fragment which is then converted to two moles of CO_2 by participation in one complete turn of the Krebs tricarboxylic acid cycle. Hydrogen atoms (protons and electrons) which are liberated during these oxidations are simultaneously oxidized by oxygen to form two moles of water. Oxygen must itself be activated before it can participate in these reactions, the entire conversion of protons, electrons, and activated oxygen to water usually proceeding through the agency of the cytochrome–cytochrome oxidase system.

This concept of biological oxidations is widely accepted, and certainly applies generally to the tissues of higher animals. In lower animals, however, where fewer investigations have been made, it is clear that other oxidative pathways may sometimes be more important, although their precise nature is obscure. The limited knowledge of oxidative metabolism in nematodes certainly leads to the conclusion that some species employ the Krebs cycle-cytochrome oxidase system in tissue oxidations, and that some do not. How general this division is cannot be stated, as only a few animal parasites have been studied.

The Krebs cycle has been identified in whole or in part in *Nematodirus* spp., *Ascaridia galli, Neoaplectana glaseri* and *Trichinella spiralis* (larvae) and unequivocal evidence for the occurrence of cytochrome–cytochrome oxidase in *Trichinella* has been reported. On the other hand neither *Litomosoides* nor *Ascaris* possess a cytochrome system, nor does *Ascoris* make use of the Krebs cycle.

Several methods can be used for the identification of the cytochrome system in tissues. historically, the first of these was based on the spectroscopic detection of cytochrome c, and *Ascaris* muscle was one of the first tissues in which cytochrome *c* was reported. Later work, however, has shown that enzymic assay is a more reliable procedure, and it is on this basis that Bueding and Charms(1952) rejected the idea that cytochromes were concerned in biological oxidatios by *Ascaris*. A third method, which is indirect, relies upon the effect of respiratory inhibitors. The inhibitor most commonly used is cyanide, which is well-known to repress the activity of ferric enzymes almost completely at concentrations approximating 10^{-3}M. Unfortunately, enzymes such as peroxidases are also affected so that cyanide is by no means specific for the cytochrome system. A better inhibitor is carbon monoxide, whose effect in the dark, if reversed by light, appears to be quite specific for ferrous enzymes, of which the components of the reduced cytochrome system are the only known examples. Even this method can lead to puzzling results. Thus, embryonating *Ascaris* eggs are highly sensitive to carbon monoxide and cyanide, but give negative results for the cytochrome system by both enzymic and spectroscopic tests. The only conclusion to be drawn here is that a system is present which at least closely resembles the cytochrome system in its properties.

Nematodes do not seem to differ from other animals in their requirement for high-energy phosphate compounds. Oxidative phosphorylations occur in the species studied by Rogers and Lazarus (1949b) and Chin and Bueding (1954) and several phosphate esters have been tentatively identified. Some details of phosphorus metabolism may be modifications of the pattern normal to higher animals. *Ascaris*, for example, does not appeares to contain appreciable reserves of high energy phosphates such as creatine or arginine phosphates although filariform larvae of *Strongyloides ratti* probably do contain such substances. A reasonably explanation of this difference between species might be that *Ascaris*, which leads a sheltered life in the intestine, has lost

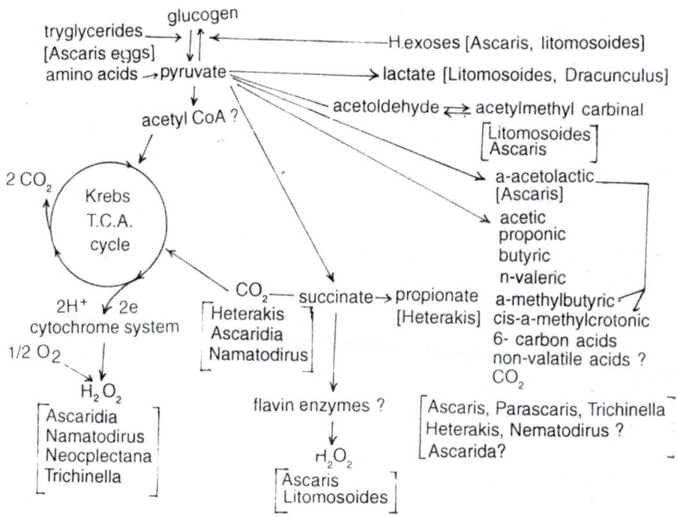

Fig. 15.4 : Oxidation and fermentation of glycogen by animal-parasitic nematodes.

the capacity for storing high energy phosphorus compounds which in other, less-sheltered species, are still necessary to the conduct of a normal life. However, more observations on the phenomena of oxidative metabolism in nematodes must be forthcoming before it will be possible to discern a general pattern, if, indeed, such exists.

As already pointed out, many adult and larval nematodes conduct active fermentations when oxygen is absent, and even when it is present. On the other hand, there is no evidence for the occrrence of fermentations in eggs. This is not surprising, as nematode eggs in general are impermeable to organic molecules, whose accumulation as toxic waste products would be unfavourable to survival.

Investigations of nematode fementations have been confined to a few species, each of which has its own peculiarities. The fact that many, and perhaps most, species are able temporarily to survive anoxic conditions strongly suggests that the ability to ferment is widespread. It is, however, a great deal easier to subject an organism to anaerobic conditions and note its ability to survive than it is to extend the experiment to a study of the excreted waste products. As a result, the widely-held view that fermentations are a more or less characteristic feature of nematode metabolism is still not well supported experimentally.

As already stated, glycolysis of carbohydrate with formation of pyruvate appears to be a normal anaerobic sequance of reactions in nematodes as in other animals, and it is the subsequent metabolism of pyruvate which determines the peculiarly aerobic or anaerobic features of catabolism. As we have just seen, oxidative metabolism of pyruvate by nematodes may or may not follow the same pathways as in higher animals. It is in the fermentation of pyruvate, however, that their divergence is most notable. This divergence is only relative. Most higher animals can survive short periods of anoxia, during which time pyruvate is reduced to lactate, most of which accumulates in the tissues until these are irreversibly poisoned, or until it is metabolized upon readmission of oxygen. Nematodes, on the other hand, are often exposed to more prolonged shortages of oxygen. They do not accumulate fermentation products, but excrete them efficiently. This ability to excrete such waste products provides an explanation for the fact that nematodes under anoxic conditions do not incur a pronounced oxygen debt.

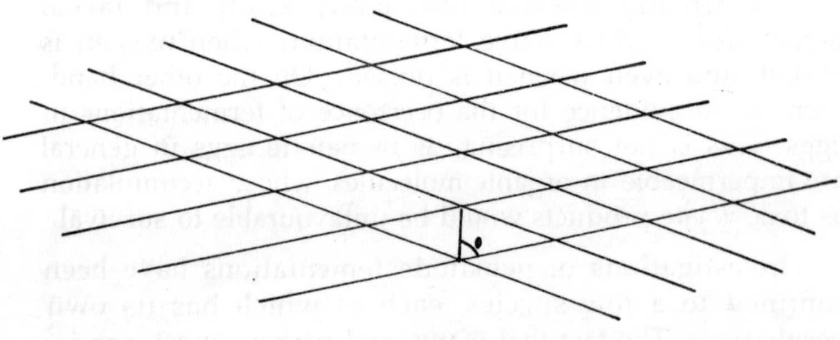

Fig. 15.5 : Geometry of the fibre layers in *Ascaris* cuticle. Θ is normally about 70°. Elongation of the worm results in an increase in volume provided that Θ is greater than 55°. When Θ is less than 55° the volume decrease.

In higher animals, for example, tissue lactate accumulated during anaerobiosis may be metabolized rapidly upon readmission of air, produceing a temporary and quantitative increase in Qo_2 which is spoken of as the repayment of an oxygen debt. If the lactate were excreted instead, there would, presumably, be no oxygen debt.

With some exceptions, the products of nematode fermentation are carbon dioxide and organic acids. Fermentation acids were first studied in *Ascaris* which, as has been shown since, produces a singularly complex mixture. It is probably fortunate that Ascaris was chosen for Weinland's experiments, for the results presented a challenge to biochemists which has proved to be irresistible, and has led to the investigation of other species, most of which, fortunately, seem to excrete somewhat simpler mixtures. The species which have been examined, and the fermentation acids identified are shown in Figure. *Ascaris* adults and *Trichinella* larvae excrete a mixture of acids which is considerably more complex than that of species such as *Heterakis*, Litomosoides, or *Dracunculus*. An immediate resemblance to the end-products of bacterial fermentations is apparent, and indeed there is evidence that the mechanism of propinic acid formation by *Heterakis* and in *Ascaris* muscle is similar to that occurring in *Propionobacterium*. Some of the acids, such as the α-methylbutyric acid produced in large amounts by *Ascaris* are obviusly the result of a long sequence of reactions beginning with pyruvate or a similar preursor. Saz, Vidrine and Hubbard (1958) have proposed a mechanism for this reaction, beginning with the condensation of two moles of pyruvate α-acetolactate and CO_2, which was supported in part by experimental evidence. The *Ascaris* fermentation acids are well tolerated by the parasite itself, as considerable quantities (about 40 mole/litre) are present in the hemolymph. *Heterakis* tissue, however, contain very little acetic and propionic acids, although succinic acid, which is the immediate precursor of propionic acid and carbon dioxide and in this fermentation, occurs in appreciable amounts. Lactic acid, which is the commonest fermentation acid in most animal tissues, is a major excretion product only in the filarial worms *Litomosoides* and *Dracunculus*.

Many microbial fermentation products include non-acid substances, particularly various alcohols. Until recently little attention has been paid to the possibility that nematodes excrete such compounds. However, as Bueding (1951) and Saz, Vidrine and Hubbard (1958) have found that a 4-carbon alcohel, acetylmethycarbinol is a normal excretion product in *Litomosoides* and *Ascaris*, it is possible that non-acid substances occur fairly generally. Carbon dioxide, which is a characteristic product of nematode fermentation, arises from pyruvate or other intermediates in the formation of acid and non-acid fermentation products.

It is noteworthy that none of the fermentations which have just been considered are eliminated during aerobic respiration. In air, the *Litomosoides* fermentation is depressed, and the relative proportions of acetic and lactic acids are altered, but in *Ascaris, Heterakis, Dracunculus* and *Trichinella* there is little change in the total acids excreted. Thus, aerobic respiration in these last four species is simply superimposed upon a constant basic fermentation. *Ascaris* and *Heterokis* are intestinal parasites which probably suffer a chronic oxygen deficiency; *Litomosoides, Dracunculus* and *Trichinella,* on the other hand, live in the pleural cavity, subcutaneous tissue or skeletal muscle of their hosts, in which there is probably no serious lack of oxygen. Moreover, *Trichinella* larvae, owing to their small size and the participation of the cytochrome oxidase system in their aerobic metabolism, make full use of oxygen even at very low partial pressures. Their persistence in maintaining a vigorous anaerobic fermentation is therefore difficult to understand, although it is evidently a fundamental part of their catabolism. Observations such as these illustrate perfectly the need for additional investigation of species occupying widely different habitats.

In concluding this brief account of fermentation, it is legitimate to ask why nematodes indulge in such metabolic acrobatics as the formation of five-and six-carbon acids when the simple reduction of pyruvate to lactate is sufficient for most higher animals. One suggestion might be that more of the energy of the glucose molecule is thus made available. This seems doubtful, since the nematode acids such as α-methylbutric are actually more reduced than lactic acid. Another possibility worth considering is based on the chemical properties of the acids themselves. For the most part, fementation acids are excreted as acids, and not as salts. Tissue cations are conserved in this way, but even more important may be the fact that the permeability of most cell membranes to ions is low. The efficient excretion of acids, therefore, may depend upon their presence in undissociated form. The fermentation acids which have just been discussed fall into two groups with respect to their ionization constants. First lactic acid, and second, all the others, whose respective constants are similar and about one-tenth that of lactic. If *Ascaris* is chosen as an example, we have already seen that its hemolymph is fairly acid (pH 6.4), and can predict that at this pH the proportion of the weaker acids which are present in undissociated form will be very considerably higher than the proportion of any undissociated actic acid which might be present. The formation of very weak acids should therefore favour their efficient excretion.

In the case of parasitic nematodes, the possible effect on the host should alos be considered, as the formation of relatively strong acids would doubtless be highly damaging. *Litomosoides*, as Bueding (1949b) found, excretes both lactic and acetic acids, and under the semi-aerobic conditions which prevail in the pleural cavity, the excretion of acetic acid may be dominant. Nevertheless, the pH of the pleural cavity of infected rats falls to 6.5, and, as lactic acid is about ten times as effective as acetic in its contribution to the final pH, its relative concentration in the mixture might be an important factor in determining the ability of the host to withstand the infection. It must be understood, however, that this kind of reasoning, to be effective, must be supported by a great deal more factual information.

It is perhaps advisable at this point to provide a brief summary of aerobic and anaerobic respiration and biological oxidations. No nematode yet studied has been shown by conclusive evidence to be capable of surviving and reproducing indefinitely under strictly anaerobic conditions, and all species examined have shown some capacity for using oxygen. The evidence suggests that oxygen may be necessary for muscular contraction, but that it is not used exclusively for this purpose. The properties of the hemolymph and tissue hemolymph are such that it has not been possible to assign to them an oxygen-carrying function. Oxidative metabolism may proceed by way of the Krebs tricarboxylic acid cycle, with cytochrome oxidase as the terminal hydrogen acceptor, but this is not always the case, and the general features of cell oxidations are still not clear. Both anaerobic and aerobic fermentations occur in all of the few species which have been investigated, which are by no means representative. Fermentations are typically complex, with excretion of carbon dioxide and at least two organic acids. Lactic acid has been recognized as a major acid in only two species. Nematode eggs seem to require oxygen for development, and no fermentation has been observed in them. However, this view may require modification as more observations are made.

Non-Protein Nitrogen

Although the extent to which proteins are used by nematodes as a source of energy is debatbale, ther is, as in all animals, a continuing catabolism of protein and other nitrogenous compounds which results in the formation and excretion of products classified collectively as the non-protein nitrogen fraction (NPN). Most of the NPN originates from structural components of tissues which are continually being broken down and replaced, although when the diet contains much protein, this may be responsible for a more direct formation of NPN. Mature animals are normally in a state of nitrogen balance, where input and output of nitrogen are equal. These well-known principles are mentioned here only because it is important to realize that it has not yet been possible to study NPN in nematodes under conditions which are physiologically satisfying. All investigations have been based on the collection of material excreted during a 24-hour period or longer, which means, particularly in the smaller species, that the nitrogen metabolism is that of a starving animal and is not necessarily the same as that of animals in nitrogen balance. Earlier studies were also complicated in some cases by the presence of a bacterial flora, but this problem can now be solved by judicious use of antibiotics or bactericides. However, it is hardly necessary to stress that a certain amount of caution is desirable in assessing the results which will now be presented.

Total nitrogen excreted by several animal parasitic species, expressed as mg per 100 gm body weight per 24 hours is as follows:

Ascaris lumbricoides	39	(Savel, 1955a)
Ascaridia galli	29 41	(Rogers, 1952)
Nematodirus spp.	123—159	(Rogers, 1952)
Trichinella spiralis (larvae)	280	(Haskins & Weinstein, 1957a

Clearly, the rate of nitrogen excretion by the small nematodes *Trichinella* and *Nematodirus* is several times as great as in the much larger *Ascaridia* and *Ascaris*. The need for caution in accepting such results can be illustrated by a simple calculation based on the nitrogen excreted in 24 hours by *Trichinella*. It this nitrogen originates in tissue proteins, and the factor for converting nitrogen to protein is 6.25, then protein metabolized is 1.75 gm. Assuming that total solids in *Trichinella* larvae amount to 20 per cent of the fresh weight, and that these solids are 70 per cent protein, it turns out that surviving *Trichinella* larvae catabolize 12.5 per cent of their body proteins evevery 24 hours. This is a very vigorous metabolism, which may acctually be much more extensive than that occurring under natural conditions in the host muscle.

Analysis of the excreted nitrogen in the species listed above has confirmed Weinland's (1904) conclusion that ammonia is a major constituent. As has often been pointed out, this is not surprising, since most nematodes are aquatic invertebrates in which the danger of auto-intoxication by ammonia is probably not great. Its effect on the host in heavy parasitic infections is less certain, although Myuge (1957b) has suggested that it is a major cause of damage in infections of plants by *Ditylenchus destructor*. However, the data summarized in Table shows clearly that the excretions also contain significant amounts of urea, amino acids and peptides. Rogers (1955a) and Haskins and Weinstein (1957b) have analyzed the last two fractions in some detail, and have considered the possibility that these are not normally waste products of metabolism, but appear in large quantities in the excreta as a result of the somewhat abnormal experimental conditions which have already been discussed. It is of interest in this connection that by simply confining individuals of *Ascaris* in such a way that the ingestion and ejection of water from the intestinal tract was not restricted, although the body itself was not immersed in saline, Cavier and Savel (1954c) were able to show that this species can regulate the relative proportions of ammonia and urea which are excreted. Under such conditions urea accounted for 52 per cent of the total nitrogen excreted, ammonia for only 2 per cent. As many nematodes live from time to time, or at certain stages of their life cycle, under conditions of wate restriction, a similar regulation of the excretion products may occur.

One of the more interesting aspects of nitrogen excretion was recently discovered by Weinstein and Haskins (1955) who obtained evidence for the excretion of small amounts of several aliphatic amines from the excretory pore of *Nippostrongylus muris*. *Trichinella* excretes similar amines. Most of these, such as ethyl amine, propyl amine and cadaverine, can be regarded as decarboxylation products of the corresponding amino acids, and some are known to exert a pharmacological action upon the tissues of higher animals. They comparise 7 per cent of the total nitrogen excreted by *Trichinella*.

Tissue minces of *Ascaris Ascaridia* and *Nematodirus* all liberate ammonia from added 1-amino acids presumably as a result of 1-amino acid oxidase action. The most active enzyme is found in the intestine, which, as stated earlier, also contains active proteinases and peptidases. This then, could be one source of excreted ammonia. However, these species also contain urease, an enzyme which is not common in animals, and which converts urea to ammonia and carbon dioxide. As urea is a common excretion product it might also be an immediate precursor of some of the ammonia formed. In higher animals, of course, ammonia formed by deamination reactions is efficiently converted to the less toxic urea by means of the Krebs-Henseleit ornithine cycle. All the species mentioned above seem to possess some form of this cycle, with the enzymes concerned in it being most active in the intestine. Thus, there appear to be alternative routes for the disposal of the primary product of deamination, which is ammonia. It can be excreted directly or it can be converted to urea; the urea can be excreted, or reconverted to ammonia by the action of urease. The balance between these alternatives is possibly determined by the conditions

Table 15.7 Nitrogenous Excretion Products of Animal-Parasitic Nematodes

Spcies	NH_2	*Urea-N*	α-NH_2-N	*Peptide N*	*Volatile amine-N*
			% of total	N	
Ascaris lumbricoides (Savel, 1955a; Rogers, 1955)	69	7	21		
Ascaridia galli (Rogers, 1952)	55	14	present	15	
Nematodirus spp. (Rogers, 1952)	41	14	present	35	
Trichinella spiralis (larvae). (Haskins and Weinstein, 1957a)	33	nil	29	29	7

to which the parasite is exposed from time to time, conditions of water restriction favouring the excretion of urea. On the basis of enzyme distribution in *Ascaris,* the intestine is the probable site of the reactions involved, and may also be the tissue primarily concerned with nitrogen excretion. Cavier and Savel (1954c) provided direct experimental support for this hypothesis in their studies on *Ascaris*. Certainly the curious and unusual juxtaposition of the Krebs-Henseleit urea-synthesizing cycle and the enzyme urease are deserving of more extended study.

Toxicity of Excretion Products

The long-standing curiosity of parasitologists concerning the toxic properties of the secretions and excretions has never been adequately satisfied. Toxicity is only one of the factors which determines the aggressiveness and pathogenicity of the parasite, and it is impossible here to attempt an assessment of its general role in the establishment of parasitic infections. The following comments, therefore, are made only because it has become clear during the preceding discussion that nematodes do in fact secrete or excrete substances which are apt to be toxic to the host.

There seems as yet to be no sound evidence that nematodes secrete toxins comparable to the several well-defined bacterial toxins *i.e.,* substances toxic in very small concentrations which are excreted by the microorganism and which when injected separately from the microorganism produce symptoms that mimic the disease. According to this definition, it will not even be possible to put the criteria for toxin production to an experimental test until at least one pathogenic nematode has been cultured successfully, and the culture fluids examined critically for toxic reactions. An indirect approach to this problem has been made by Nolte and Kohler (1952) who prepared extracts from leaves of plants infected with *Heterodera schachtii, Pratylenchus pratensis* or *Ditylenchus dipsaci* and then exposed leaves of healthy, uninfected plants to the extracts. Curling and wilting similar to that observed in the leaves of infected plants occurred, but, as the authors were aware, such experiments cannot distinguish between nematode toxins and plant toxins produced in response to the infection. Similarly, the grass *Festuca* which has been infected with *Anguina agrostis* becomes toxic to cattle and sheep, which are not harmed by direct injection of *Anguina* into the blood stream. Such observations are interesting, and inconclusive. Once again it is apparent that the solution to an important problem awaits the development of improved methods for studying the nematodes themselves.

The toxicity of excretion products generally, or of substances arising in the host as the result of infection, can be likened to the decomposition products produced by bacteria, and known generically as ptomaines. The doctrine of ptomaines as developed long ago by Brieger has not, in bacteriological science, survived the demonstration that such ptomaines are as a rule only feebly toxic. One should not, however, forget that this kind of restrained toxicity may provide a more satisfactory explanation for the subacute symptoms of many nematode diseases than any other. In *Ascaris* infections, for example, the adults living free in the intestine excrete considerable amounts of five-and six-carbon fermentation acids whose toxicity to the host does not seem to have been studied. Similarly, nematodes commonly undergo several molts which may result in the introduction of a good deal of protein and protein decomposition products into the host tissues, and it has been suggested that the guanidine liberated from muscle during severe *Trichinella* infections may be responsible for many of the symptoms of trichinosis. Where circulation of host fluids is not highly efficient, much of the local damage due to infection may be the result of the normal living habits of the parasite. Myuge (1957b) has suggested, as the result of his experiments with *Ditylenchus destructor* that infection of potatoes can be described in physiological terms as follows. The parasite secretes amylas and protease which bring about extracorporeal digestion of starch and protein, and at the same time it excretes ammonia, which causes the pH to rise and initiates necrosis of the host tissue. The plant responds, as most plants do to infection of any kind, by increasing the local respiratory rate. Meanwhile, changes in osmotic pressure caused by hydrolysis of proteins and starch lead to increased turgor pressure within the cells which results eventually in disruption and rotting of the infected portion of the tuber. On the other hand, similar changes in Qo_2 and pH in cucumber roots infected with *Meloidogyne incognita* lead not to necrosis but to gall formation. Clearly, the host response in *Meloidogyne* infections is more dramatic than in *D. destructor* infections, although the initial stimulus provided by the parasite may be similar.

It is not necessary to accept the details of such hypotheses as these in order to profit from the concept that toxicity to the host may be the result of the parasite's normal physiological functions of secretion and excretion. On the other hand it is essential to remember that the pathogenicity of nematodes may be due to other factors which are not suitable for inclusion in the present discussion, but which are described elsewhere in this book.

WATER BALANCE AND PERMEABILITY

All the space between tissues in nematodes is filled with an aqueous solution, the hemolymph. There are no circular muscles in the body wall, so that the longitudinal muscles contract against the incompressible hemolymph, and, as these muscles are divided into dorsal and ventral groups by the lateral chords, effective motion can only occur in the dorsoventral direction. Considerable muscular tone is normally exerted, with the result that the hemolymph always exerts a hydrostatic, or turgor, presure which is easily demonstrated by cutting the body wall and observing the sudden gush of hemolymph. Fielding (1951) reported the persistence of turgor pressure in individuals of *Anguina tritici* and *Ditylenchus dipsaci* which had been dormant for more than twenty years. The magnitude of the turgor pressure of *Ascaris* surviving in isotonic solution has been measured by Harris and Crofton (1957) who found that it varied in a rhythmical way, occasionally approaching one-half atmosphere, and averaging about 70 mm. of mercury. Harris and Crofton suggested that the peculiarities of musculature and turgor pressure, coupled with the limited distribution of nerve cells which are confined entirely to the head and tail regions, must create problems in neuro-muscular physiology quite different from the reflex mechanism occurring in other animals. Not much has yet been learned about neuro-muscular physiology in nematodes, although the subject is one of great interest from the point of view of drug action. It has been established that acetylcholine stimulates isolated muscle strips of *Ascaris,* and that its action is prolonged, probably owing to the extremely low concentrations of cholinesterase in the muscle. The lips and tails are extremely sensitive and the head region, including the nerve ring, contains about sixteen times more acetylcholine than the somatic muscles. Thus, although acetylcholine appears to play a part in sensory and nerve-muscle physiology in *Ascaris* its precise role, and the possible role of other humoral mediators, has not been accurately defined.

The constant turgor pressure exerted by the hemolymph means that the nematode intestine, which consists of a single layer of cells devoid of muscle fibres, is completely collapsed when empty, and can be filled only by exerting a back pressure of its own. This is made possible by the muscles and valves of the pharynx or esophagus and the anus, which constitute a closed system until a sufficient supply of fluids have been ingested and the hydrostatic pressure has increased to a critical point. The anus then opens and, as anyone who has worked with *Ascaris* knows, the intestinal contents are discharged with considerable violence. Filling of the intestine entails an increase in the total volume of the worm which is possible because of the niceties of the cuticular structure. The fibre layers are so arranged that relaxation of the somatic musculature permits the extension of the fibre "cage" and the elongation of the nematode. The volume also increase provided that the original value of the angle Θ is greater than 55° As in *Ascaris* this angle is usually about 70°

the worm is theoretically capable of very considerable length and volume increase at constant turgor pressure. Presumably other nematodes also make use of this favourable geometry in the fibre layers, for Stephenson (1942) observed that in hypertonic solutions loss of water entailed a 30 per cent *decrease* in the length of *Rhabditis terrestris,* while Vargas (1952) reported that ramoval of the microfilaria of *Onchocerca vovulus* from an isotonic glucose solution into water *increased* the length by 55 per cent.

Thses observations on *Rhabditis* and *Onchoccrca* quite clearly indicate a lack of osmoregulation, and raise the question whether nematodes in general can control the water content of their bodies. Consideration of the results of several investigations show that many species can, and others cannot. *Ascaris* and *Parascaris,* for example, rapidly absorb water from hypotonic colutions, and lose it to hypertonic solutions. Water is apparently lost or gained freely through the cuticle, for ligatering the worm at both ends does not prevent the changes from taking place. *Ascaris* larvae are also sensitive to osmotic changes in the medium. Many other species, however are able to survive aparently unharmed in solutions which are definitely hypotonic, although they are apt to be more sensitive to hypertonic, solutions. *Ostertagia, Haemonthus* larvae, and *Eustrongylides* larvae all survive well in hypotonic saline while adults of *Meloidogyne javanica* live for many days in water alone. Osche (1952) has shown that many saprophytic species readily adapt themselves to the osmotic changes occurring during the progressive decomposition of organic matter, and that a number of free-living species which inhabit environments of low osmotic pressure adapt themselves without difficulty to the higher osmotic pressures normally encountered by parasitic species. In short, nematodes as a class appear to be well-endowed with the ability to regulate their water content in widely differing environments, although a number of parasitic species can no longr do so.

In *Ascaris,* as already mentined, water can be absorbed through the cuticle, which is similar morphologically to the cuticles of other nematodes. It possible, therefore, that the cuticle is generally used for this purpose. In the many species which control their water content, however, it is less certain how water is discharged from the body. A useful study was made by Weinstein (1952), who found that the rate of pulsation of the excretory ampulla in larvae of *Nippostrongylur muris* and *Ancylostoma caninum* is inversely proportional to the degree of hypotonicity of the medium, and that water is actually discharged from the excretory pore of these two species. Additional observations on other species are now needed to test the general validity of the hypothesis that the excretory system plays a major role in water regulation.

Granting that many nematodes are capable of regulating their water content in solutions hypotonic to their own tissues, there remains the problem of salt regulation, concerning which there is still very little reliable information. Despite iitts inability to regulate water, *Ascaris* maintains fairly close control over the concentration of magnesium, calcium and potassium in its hemolymph when placed in solutions of varying ionic composition, and althogh the chloride concentration canbe made to fluctuate widely, it tends to be significantly lower than the concentration in the medium. In this parasite, then, there is a fairly efficient regulation of ions, even though there is essentially no regulation of water. No similar experiment seems to have been made using other species. Variations in the ionic composition of the medium are known to affect survival and respiration, as mentioned earlier, but it is not known whether such effect can be correlated with alternations in the internal ionic environment. Investigation of the ionic composition of soil and plant parasitic nematodes, the degree of control exercised by particular species, and comparison with the composition of the normal environment, might be expected to provide basic bio ogical knowledge of potential practical usefulness.

The permeability of *Ascaris* cuticle to respiratory gases, water, chloride and possible to small ions such as the ammonium ion is perhaps to be construed as a general property of the nematode cuticle. On the other hand, there is no evidence for permeability to nonvolatile organic compounds and the cuticle is rightly regarded as providing an efficient barrier between the nematode and damaging substances in its environment. Digestive enzymes, whether those of the host in parasitic species or of microorganisms in saprophytic species, do not normally attack the cuticle. the resistant properties are primarily those of the external cortex, for it has been relized for many years that the other layers are readity digested by enzymes such a pepsin or trypsin. As previously mentioned there is no general agreement concerning the chemical nature of the external cortex, which is perhaps a somewhat unusual collagen subjected to some degree of tanning. Such proteins are commonly resistant to enzyme or chemical action, and when tanning is heavy, tend also to be impremeable even to water. This change was observed by Ellenby (1946) in cysts of *Heterodera rostochiensis,* which when young are white arer freely permeable to water, but when old and yellow (tanned) are highly resistant to water loss. However, in addition to the external cortex there is present in *Ascaris* an outermost layer which is very thin, osmiophilic and presumably a lipid. This membrane, about which very litle is known, may provide the initial environmental barrier set up the cuticle. Trim (1944) found that antihelminthics such as hexylresorcinol enter *Ascaris* by penetrating the cuticular barrier. Penetration is facilitated by surface active agents which are not themselves absorbed, and worms pre-treated with detergent lose phosphate from their bodies. These observations suggest that a lipid barrier is of great importance in imaintaining the normal cuticular function. Whether the barrier consists solely of a coating over the external cortex, such as Chitwood (1938) claimed to have isolated from *Ditylenchus dipsaci,* and Bird and Deutsch (1957) found in *Ascaris* by electron microscopy, or whether lipids distributed more generally in the cuticle also play their part remains to be seen.

If there is still much uncertainty regarding the intimate cuticular distribution of lipids, it is nevertheless clear that the action of a great many antihelminthics, and most of the useful soil nematocides, is closely relted to their ability to penetrate lipid barriers. The soil nematocides, for example, are relatively non-polar organic molecules with only slight water solubility but pronounced lipid solubility, and which have an appreciable vapour pressure at physiological temperatures. Many organic compounds possessing these properties apparently, have access to nematode tissue by permeation through the cuticle. Their success as nematocides, however, may depend upon a more specific efect on the tissue. This field, which is that of drug action, is still largely unexplord in nematodes. It is worth considering however, in the light of Trim's (1949) observations on the leakage of phosphate through *Ascaris* cuticle, that nematocides of the type discussed above may act in part by altering the cuticular structure insuch as way that unfavourable disturbances in water and salt regulation take place.

During periods in which nematode larvae or adults are more or less dormant, it might be predicted that their permeability to water andother substances would be much reduced. Infective filariform strongyle larvae, for example, are sheathed in the remains of the second stage cuticle, and are generally considered to be much more resistant to adverse environmental conditions than younger larvae, or those which have entered a host animal. However, the physiological basis of this sesistance does not seem to have been systematically explored. similarly, the prolonged survival of some plant parasitic nematodes in soil suggests that important adaptations, among which might be greatly reduced permeability, are made. If this is so, then susceptibility to nematocides may also be altered. Thus, there is obviously a need for much broader physiological studies than have yet been made.

Permeability of Eggs, and Hatching

Nematode eggs present special problems to the investigator for in these it is increasingly apparent that during development and subsequently up to the time a suitable stimulus is received, the eggs have only a very limited permeability, which suddenly increases markedly in response to the stimulus, and as part of the hatching process.

The egg-shell of a great many nematodes consists, as has been pointed out already, of an other layer which may contain chitin or quinone-tanned protein, and an inner vitelline membrane containing a preponderance of lipid-soluble substances. In *Ascaris* and *Parascaris* this material consists of ascarosides, and reasons for believing that similar glycosides may be present in the vitelline membranes of other eggs have already been presented. Recent work suggests that all shells do not conform to this general structure, but since those which do not, apart from the three species of lungworms studied by Monne (*Dictyocaulus viviparous, D. filaria,* and *Metastrongylus elongantus*) are still not well-known, they will not be considered here.

There can be little double that the semi-permeability of *Ascaris* eggs is primarily a function of the vitelline membrane, for if the chitonous shell is dissolved in hypochlorite solution the eggs retain their semi-permeable properties. Moreover, heating the eggs to about 75°C melts this membrane, and it is then found that solutes within the eggs diffuse freely through the hard shell. Both the hard shell and the vitelline membrane are also extremely resistant to the action of many corrosive chemicals. The vitelline membrane, apart from its remarkable chemical intertness, seems also to be truly semi-permeable, *i.e.,* only gases are able to pass back and forth through it, and water leaks from it verys lowly. Permeability to respiratory gases is of course self-evident, as the egg metabolism is aerobic. A striking demonstration of permeability to gases but not to related ions is afforded by the high sensitivity to poisoning by hydrogen cyanide or hydrogen azide (10^{-5}M) and the low sensitivity to cyanide or azide ions (10^{-2}M) Passey and Fair-brain, (1955). The severely restricted permeability to water can be shown by the failure of the embryo to plasmolyze even in concentrated salt solutions.

Although other nematode eggs have not been examined in similar detail, the consensus seems to the that they too are chemically resistant and highly impermeable. Wilson (1958) has found, for example, that the relatively thin-shelled eggs of *Trichostrongylus retortaeformis* are, like those of *Ascaris,* incapable of being plasmolyzed in strong salt solutions. As the development of most eggs occurs outside the body of the host, the protection offered by the shell provides an obvious advantages. At the same time, it presents practical problems of great importance to those who are concerned in public health or agriculture. Fortunately, the lipoid nature of the vitelline membrane tends to produce a sensitivity to organic pesticides similar to that conferred by the cuticle on larvae and adults, with the result that a single agent is sometimes effective against all three stages of development.

Most nematode eggs, including those of a great many parasitic species, hatch soon after the embryo is completely developed. This appears to be spontaneous process requiring no alteration of the external environment, and must therefore be due to a physiological stimulus provided by the embryo. Wilson (1958), for example, found that the best hatching of *Trichostrongylus retortaeformis* eggs occurred in distilled water, and that hatching was delayed, rather than accelerated, by salts or sucrose. Just before hatching, the vitelline membrane suddenly became permeable to water, while changes in the hard outer shell were also noted but not detailed.

In contrast to those eggs which hatch spontaneously, there are a number which are not self-stimulating, but require a more or less specific alteration in their external environment which is normally provided by the host. Until such time as these suitable conditions for hatching are provided, these eggs survive and retain their infectivity. Two such eggs are those of *Ascaris* and *Heterodera,* about which useful information has recently been acquired.

The first change which occurs in *Ascaris* eggs as the result of a hatching stimulus is, an *Trichostrongylus* eggs, the development of permeability in the vitelline membrane. This is followed, as shown by Rogers (1958), by the secretion or activation of a chitinase which permeates the vitelline membrane and dissolves the hard shell. The digestion is usually more extensive in one area than in any other, so that the embryo actually leaves the egg through a hole in the shell. These are chemical events which may also take place in *Trichostrongylus* eggs. *Ascaris* differs chiefly in that hatching normally occurs only in response to a stimulus supplied by the host. The stimulus appears to have at least three components which are (1) a pH near 7.2, (2) a partial pressure of carbondioxide approximating 40 mm of mercury, (3) a low redox potential. As these are conditions to be expected in the small intestine of many species during feeding, they explain adequately the well-known fact that host specificity for *Ascaris* is not determined by conditions for hatching, but by conditions for subsequent development of the larvae. Eggs of the cat and fowl ascarids hatch in response to the same stimulus as *Ascaris* eggs (Rogers, private communication).

At the present time it is not possible to state that hatching in *Heterodera* follows the pattern of stimulus and response characteristic of *Trichostrongylus* and *Ascaris* eggs. It is clear, as the result of many investigations, that *Heterodera rostochiensis* eggs hatch when exposed to a hatching factor secreted from the roots of the potato or tomato, and that this factor has chemical and physiological properties similar to those of the cardiac glycosides. Although recent work strongly supports the view that cardiac glycosides have a profound effect on the permeability of animal membranes there seems to be no clear evidence that hatching factor causes of corresponding increase in the permeability of *Heterodera* eggs. It it does, then it action on the vitelline membrane is direct, and dies not necessarily involve the embryo at all. This does not mean that the embryo can make to contribution to the hatching process, such as digestion of the chitinous shell, but only that the development of permeability in the vitelline membrane is a direct result of the presence of hatching factor in the external environment, rather than an indirect response of the embryo itself to a suitable stimulus.

The stimulus-response phenomenon is also responsible for the exsheathment of infective trichostrongyle larvae in the gastro-intestinal tract of the sheep, in which larvae of *Haemonchus contortus, Trichostrongylus axei* and other species produce an exsheathing fluid in response to a host stimulus which has not been completely identified, but which is similar to the stimulus affecting ascarid eggs. However, the sheath differs from the eggs in being permeable to small molecules and in having no structure comparable to the vitelline membrane. As a result, exsheathing fluid secreted by the stimulated larva has immediate access to the inner surface of the sheath, which is digested directly without any necessity for the preliminary development of permeability.

CONCLUSION

While the foregoing account of nematode physiology and biochemistry makes no pretence at completeness, a dispassionate view can lead only to the conclusion that present knowledge is centered about a few animal parasites with occasional excursions into the realm of plant-parasitic and free-living species. In discussing "the extraodinary similarity in from and organization between the very large number of species, genera, orders and families which constitute this class of the very diversiform phylum Aschelminthes," Harris and Crofton (1957) comment that "the elementary student may be forgiven at times for thinking that there is only one nematode but that the model comes in different size and with a great variety of life histories." It is unfortunately not yet possible to formulate criteria for judging the applicability of this view to nematode physiology, although its validity is certainly questionable. In considering respiration, for example, a reasonable amount of evidence assembled in the preceding pages supports the view that animal parasites, including those dwelling in essentially anaerobic habitats such as the mammlian intestine, are facultative aerobes which continue to ferment actively even while respiring. On the other hand, anaerobic and aerobic fermentations in free-living or plant-parasitic species have not been investigated, although an active respiration has been demonstrated. The questions then arise, whether oxidative metabolism in free-living nematodes is of the cytochrome-catalyzed type, with little ability to ferment and a well-defined Pasteur effect, and are the observed deviations from this pattern in parasites a reflection of adaptations to parasitism ? These are obvious questions, representative of many others which might be asked, and which are at the moment unanswerable. In so far as the physiology and biochemistry of parasitism by nematodes is concerned, it is clear that the solution to many such problems depends upon the acquisition of broad comparative knowledge, much of which is now within technical reach of the patient investigator.

5